GCSE MATHEMATICS

Course Notes, Examples and Exercises

R. C. SOLOMON M.A., Ph.D.

Master at University College School

DP PUBLICATIONS LTD
12 Romsey Road,
Eastleigh, Hants. SO5 4AL
1987

ACKNOWLEDGEMENTS

My thanks are due to the following:-

To the examining authorities for their help, and for permission to reproduce formula sheets:

London and East Anglian Group
East Anglian Examination Board
University of London School Examination Board
London Regional Examining Board.

Midland Examining Group
East Midland Regional Examinations Board
Oxford and Cambridge Schools Examination Board
Southern Universities Joint Board for School Examinations
West Midlands Examinations Board
University of Cambridge Local Examinations Syndicate

Northern Examining Association
Associated Lancashire Schools Examining Board
Joint Matriculation Board
North Regional Examinations Board
North West Regional Examinations Board
Yorkshire and Humberside Regional Examinations Board

Southern Examining Group
South East Regional Examinations Board
Associated Examining Board
Southern Regional Examinations Board
Oxford Delegacy of Local Examinations
South Western Examinations Board

to Hoverspeed for permission to reproduce their fare schedules.

to British Rail Board at Euston House, 24 Eversholt Street London NW1 for permission to reproduce timetables.

to Rupert Sydenham for his assistance with the solutions for the test papers.

ISBN 0905435 80 X

Printed by The Guernsey Press Company Ltd
Braye Road, Vale,
Guernsey, Channel Island

CONTENTS

Preface

AIM

The aim of this book is to provide all the support needed for a course in GCSE Mathematics. It is assumed that the principles of the subject have been taught, and the book contains all the extra material needed by students, such as course notes, examples, exercises, examination questions and answers.

NEED

The need was seen for a book which didn't **duplicate** the teaching of the lecturer. In the author's experience students **understand** the principles as explained by the lecturer but they have **problems** in remembering key points and in **applying** their knowledge.

This book supports **any** and all levels of the GCSE by providing the lecturer and student with only what they want - thus making it **easier to follow** whilst **minimising** the cost.

APPROACH

The book sets out to achieve its aim in a number of ways:

1. **Graded examples and exercises.** The many worked examples and exercises (with answers) are graded in line with the three levels of the examination and provide all the practical self assessment needed. This helps the student **identify** the level reached and enables **progression** to the next.

2. **GCSE style examination questions.** There are nineteen test papers modelled on the sample papers of the various GCSE boards. The answers are provided with **marking guides** for student self assessment.

3. **Common errors.** In every topic of Mathematics there are mistakes which occur frequently. Each chapter of the text contains a **list** of the common errors relevant to that chapter. This enables students to spot their mistakes and to correct them.

4. **Summaries of principles.** All the principles, as taught by the lecturer, are summarised at each stage of the text (as appropriate) for student reference.

5. **Topic analysis.** A table is included showing the topics covered by the various examining boards.

TERMINOLOGY

Each group will set papers for three levels. These levels are:

> London Group: Levels X, Y, Z.
> Midland Group: Levels F, I, H.
> Northern Group: Levels P, Q, R.
> Southern Group: Levels 1, 2, 3.

Throughout the book the levels are referred to as 1, 2, 3.

Study and Examination Hints

1. STUDY HINTS

This book has been written to enable you to pass the GCSE examination in Mathematics. Of course it must be used in an efficient way. The following hints will help you to make good use of your study time.

A good rule to follow is "Little but Often". An hour a day for 7 days is more useful than 7 hours of work on the same day.

Make sure you get a full 60 minutes out of each hour. Do your study in as quiet a room as possible, with all your equipment to hand, and with as few distractions as possible. Do not try to combine study with watching television or eating.

Know what your syllabus contains. There is no point in studying a topic which your board or level does not require you to know. Remember though that the syllabus for level 2 includes that for level 1, and the syllabus for level 3 includes that for level 2.

Write as well as read. There is little to be gained from simply reading Mathematics. Read the topic outlines and the worked examples, and then try the exercises on the paper.

The first few questions in each block of exercises are straightforward, and the last few are often more testing. Make sure that you can do the simple ones before you go on to the harder ones.

Do not give in too soon. There are answers at the back of the book, but use them for checking that you are right, rather than for finding out how to do a problem.

The test papers at the back of the book should be done under test conditions, with a time limit and no reference books. In all the GCSE examinations you are allowed to use a calculator; make sure that you are familiar with it.

2. EXAMINATION HINTS

How well you do in an examination depends mainly on how well you have prepared for it. Many candidates do poorly because of bad examination technique. The following hints will help you make the best out of your examination.

Read the question carefully. Very many marks are lost by candidates who have misunderstood a question. Not only do they lose the marks for that question, but they also lose time because the question is often made more complicated by their mistake.

Do your best questions first. Try to get as many marks as possible, before you tackle the harder questions.

Do not spend too much time on one question. There is no point in spending a third of the examination on a question which carries a twentieth of the marks. Leave that question, and come back to it later if you have time at the end of the examination.

Show your working. If you obtain the wrong answer just because of a slight slip, then you will not lose many marks provided that you show your working. If you put down the wrong answer without any explanation then you get no marks at all.

Except in multiple choice examinations, examiners are not trying to catch you out. If a question has several parts, then each part leads on to the next. Part (a) helps you with part (b), part (b) with part (c), and so on.

Make sure that your answer is sensible. Examination questions are set so that the solutions are reasonable. If your answer gives eggs weighing 5 kilograms each or mothers who are younger than their children, then almost certainly it is you who have made a mistake rather than that the question is wrongly set. Go through your working again, but do not spend too much time checking a question which is only worth a couple of marks.

Try to make your solutions neat. Use enough paper to ensure that your answer can be read easily, but do not spend too much time on this. There is no point in spending 5 minutes on a beautiful diagram for a question which is only worth 2 marks.

All these hints can be summarized by: *make good use of your time!*

During the months that you are preparing yourself for your examination ensure that each hour of learning or revision is well spent. During the actual examination make sure that you do not waste the precious minutes!

1 Numbers

1.1 THE FOUR OPERATIONS

Add two numbers to get the *sum*. The sum of 3 and 5 is 8.

Subtract two numbers to get the *difference*. The difference of 9 and 4 is 5.

Multiply two numbers to get the *product*. The *product* of 2 and 3 is 6.

Divide one number by another to get the *quotient*. The amount left over is the *remainder*. When 13 is divided by 5, the quotient is 2 and the remainder is 3.

When a number is the product of two other numbers, those numbers are *factors*. 2 and 3 are factors of 6.

Numbers which are divisible by 2 are called *even*: numbers not divisible by 2 are *odd*.

A *prime* number has no factors except 1 and itself.

1.1.1 Examples for level 1

(1)   The postage for an overseas letter is 36p. How can I make this amount up out of 10p and 16p stamps?

Solution   If I use one 16p stamp, that leaves 20p. This can be paid for by two 10p stamps.

One 16p and two 10p stamps.

(2)   Jason needs £235 to buy a bike. He has saved up £57. How much more does he need?

Solution   He must get the difference between £57 and £235.

He needs £235 - £57 = £178.

(3)   I have 20 eggs to be shared fairly among 6 children. How many does each get, and how many eggs are left over?

Solution   Divide 20 by 6. 6 goes into 20 3 times, with remainder 2.
    Hence:

Each child gets 3 eggs, and 2 are left over.

(4)   Write down the factors of (a) 35 (b) 13.

Solution   (a) 35 can be divided by 5, with quotient 7.

$$35 = 5 \times 7$$

(b) 13 can only be divided by 1 and 13.

13 is a prime number.

1.1.2 Exercises for level 1

(1)   You have 12p stamps and 17p stamps. How can you use them to make up the following amounts:

(a) 29p (b) 36p (c) 41p (d) 70p?

(2)   Your bus fare costs 39p, and the bus driver does not give change. How can you pay your fare out of three 10p, three 5p, and four 2p coins? Is there more than one way of paying?

(3)   You have two jugs which you can fill from the tap. Jug A holds 5 pints, Jug B holds 3 pints. How can you use them to pour out a quantity of:

(a) 9 pints (b) 11 pints (c) 17 pints.

(4)   A dress costs £23. How much change will there be from three £10 notes?

(5)   A 25 year mortgage will be paid off in 1990. When was it taken out?

(6)   A dealer buys a table for £230 and sells it for £355. How much was her profit?

(7)   A woman was born in 1934. How old will she be in 1992?

(8)   A cricket team scores 153. Their opponents have scored 34. How many more runs do they need (a) to draw (b) to win?

(9)   Find the quotient and remainder when:

(a) 40 is divided by 3 (b) 72 is divided by 11 (c) 100 is divided by 7.

(10)   30 marbles are shared fairly between 4 boys. How many does each boy get, and how many are left over?

(11)   Jane has 83 p to spend on chocolate bars. If each bar costs 12p, how many can she buy and how much will be left over?

(12)   Suppose that the calendar is to be decimalised so that a week is 10 days long. How many weeks will there be in a year of 365 days, and how many days will be left over at the end?

(13)   The hero of a fairy story wins a pair of 7 league boots, in which each stride takes him 7 leagues. (1 league = 3 miles). If he is to travel 300 miles, how many strides should he take? How many miles will be left to travel after he has taken the boots off?

(14) Write down the following numbers in figures:

(a) Four thousand three hundred and sixty three.

(b) Two hundred and twenty seven thousand, four hundred and twelve.

(c) Five hundred thousand and sixty two.

(15) What is the next whole number after:

(a) 2901 (b) 3980 (c) 4849 (d) 1999?

(16) What is the last whole number before:

(a) 213 (b) 4210 (c) 2000 (d) 9,000,000?

(17) Fill in the boxes so that the following equations are correct:

(a) $\square + 4 = 13$ (b) $\square \times 5 = 35$

(c) $28 \div \square = 7$ (d) $16 - \square = 4$

(18) Paul wrote out some addition and subtraction problems. Ink was spilt on his paper, covering some of the numbers. Complete the sums:

```
(a)    241       (b)     52-5      (c)       24-1
    +  -62           +  173-            +  -38-
       40-              -969              -1820


(d)    746       (e)     5-93      (f)     -378-
    -  -24           -  353-            -  9457
       42-              -756              -3-6
```

(19) Which of the following numbers are prime? Write down factors of those numbers which are *not* prime.

(a) 11 (b) 6 (c) 23 (d) 25 (e) 29 (f) 35 (g) 41.

(20) Anne has 12 square tiles, and wishes to glue them to the wall in the shape of a rectangle. One of the ways is shown below. What are the other possible rectangles she could make?

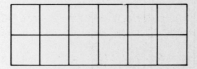

Fig 1.1

3

For Level 2

The greatest number which is a factor of two or more numbers is the *highest common factor (HCF)* or the *greatest common divisor (GCD)* of the numbers.

The smallest number which is a multiple of two or more numbers is the *least common multiple (LCM)* of the numbers.

1.1.3 Examples for level 2

(1)   Find the HCF and the LCM for 12 and 15.

Solution   The factors of 12 are 2, 2 and 3, and the factors of 15 are 3 and 5. The greatest common factor is 3.

The HCF of 12 and 15 is 3

If both 12 and 15 are factors of a number, that number must be divisible by 2, 2, 3 and 5.

The LCM of 12 and 15 is 2 x 2 x 3 x 5 = 60

(2)   On the left side of a road the lamp-posts are 5 metres apart, and on the right side they are 6 metres apart. How far is it before two lamp-posts are directly opposite each other?

Solution   There will be a lamp-post on the left side after every multiple of 5 metres, and a lamp-post on the right after every multiple of 6 metres. There will be a lamp-post on both sides after the LCM of 6 and 5 metres.

Two lamp-posts are opposite after 30 metres.

1.1.4 Exercises for level 2

(1)   Find the HCF for the following pairs of numbers:

(a) 6 and 10 (b) 4 and 14 (c) 9 and 15 (d) 12 and 30
(e) 14 and 15.

(2)   Find the LCM for the pairs of numbers in question 1.

(3)   For each pair of numbers in question 1 multiply together the HCF and the LCM. What do you notice?

(4)   You have several 2 p coins and several 5 p coins. What is the least sum of money that can be made up entirely from 2 p coins or entirely from 5 p coins?

(5)   You have a 3 pint jug and a 5 pint jug. What is the least volume which can be filled using either the 3 pint jug only or the 5 pint jug only?

(6)   Two hands move round a dial: the faster moves round in 6 seconds, and the slower in 8 seconds. If the hands start together at the top, when are they next together at the top?

(7)   Ben and Fred are playing with a pair of kitchen scales. Ben puts marbles which weigh 6 grams each into the left side, and Fred puts ball-bearings which weigh 15 grams into the right side. What is the smallest weight for which the sides balance?

(8)   A rectangular area of 150 cm by 120 cm is to be covered with square tiles. We want to use the largest tiles we can, and we don't want to break any. How big should the tiles be?

(9)   You have a source of water and a 3 pint jug and a 5 pint jug. You are allowed to pour from one jug to the other. How can you obtain volumes of:

(a) 2 pints (b) 7 pints (c) 1 pint ?

1.2 NUMBER PATTERNS

A number pattern is a sequence of numbers following a rule.

1.2.1 Examples for level 1

(1)   Find the next 3 terms of the sequence: 1, 4, 7, 10, ...

Solution   Notice that the terms are going up in steps of 3. Hence the next terms are:

13, 16, 19

(2)   Fig 1.2 shows a pattern of triangles. Write down the number of dots for each triangle. Continue the pattern of numbers for two more triangles.

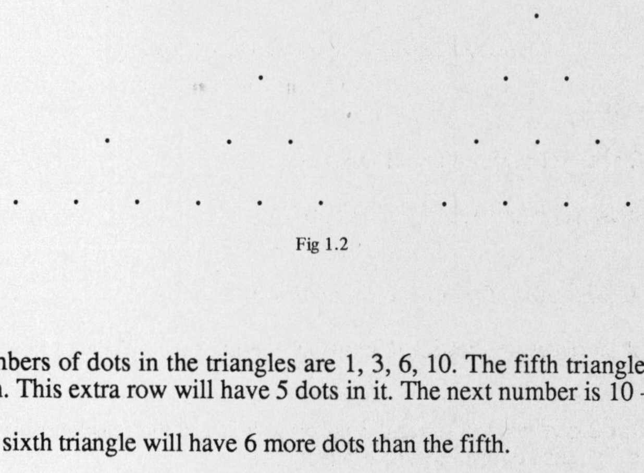

Fig 1.2

Solution   The numbers of dots in the triangles are 1, 3, 6, 10. The fifth triangle will have one more row than the fourth. This extra row will have 5 dots in it. The next number is $10 + 5 = 15$.

The sixth triangle will have 6 more dots than the fifth.

The next two numbers in the pattern are 15, 21.

1.2.2 Exercises for level 1

(1)   Find the next 2 numbers in the following patterns:

(a) 1, 3, 5, 7, 9, ...

(b) 1, 6, 11, 16, 21, ...

(c) 2, 6, 10, 14, 18, ...

(d) 1, 2, 4, 8, 16, ...

(e) 1, 3, 9, 27, ...

(f) 3, 6, 12, 24, ...

(g) 128, 64, 32, 16, ...

(h) 162, 54, 18, ...

(i) 1, 4, 9, 16, 25, ...

(j) 1, 8, 27, 64, ...

(k) 2, 8, 18, 32, ...

(l) 2, 5, 10, 17, 26, ...

(2)   For the numbers in question 1 part (e) write down an odd number and a prime number.

(3)   For the numbers in question 1 part (l) write down an even number and a prime number.

(4)   Add together three of the numbers in question 1 part (b) to obtain a total of 28.

(5)   Add together three of the numbers in question 1 part (l) to obtain a total of 41.

(6)   Write down the next two lines in the following sets of equations:

(a) $2 \times 3 + 1 = 7$
    $3 \times 4 + 1 = 13$
    $4 \times 5 + 1 = 21$

(b) $1 \times 3 - 1 = 2$
    $2 \times 4 - 1 = 8$
    $3 \times 5 - 1 = 14$

(c) $2 \times 2 = 4$
    $2 \times 2 \times 2 = 8$
    $2 \times 2 \times 2 \times 2 = 16$

(d) $1 \times 2 = 2$
    $1 \times 2 \times 3 = 6$
    $1 \times 2 \times 3 \times 4 = 24$

(7)   In each of the following sequences of figures, write down the numbers of dots in each figure. Find the next two terms in each of the number patterns.

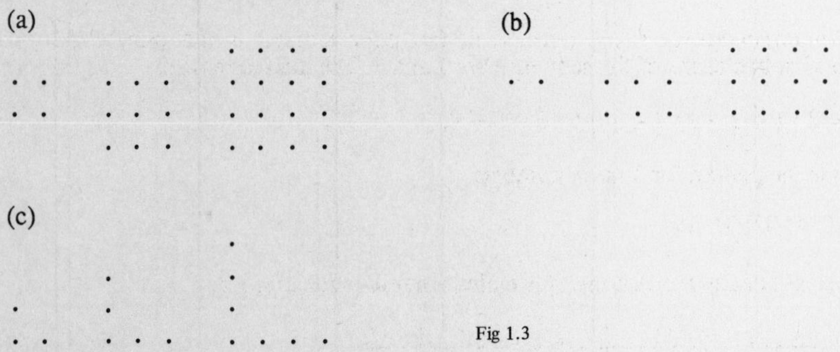

(a)

(b)

(c)

Fig 1.3

6

(8) Each of the following figures is made out of matches. Write down the number of matches for each figure. Find the next two terms in the number pattern.

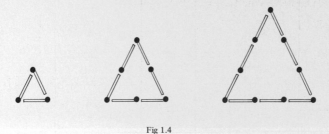

Fig 1.4

(9) Repeat question 8 for the following sequences of figures:

(a)

(b)

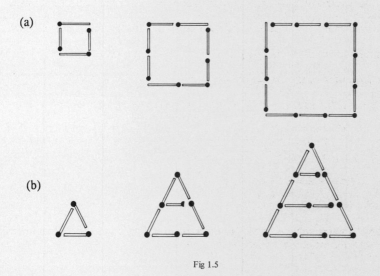

Fig 1.5

(10) The *Fibonacci numbers* are defined so that each number is the sum of the previous two numbers. The first two terms of the sequence are 1 and 1. The next terms are:

$1 + 1 = 2 : 1 + 2 = 3 : 2 + 3 = 5 : 3 + 5 = 8.$

Continue the pattern for 3 more numbers.

1.3 DIRECTED NUMBERS

A line with numbers marked on it is called a number line. Fig 1.6.

-3 -2 -1 0 1 2 3 4 5 6

The numbers to the right of 0 are *positive*, or *plus* numbers. The numbers to the left of 0 are *negative*, or *minus* numbers.

7

1.3.1 Examples for level 1

(1)   Fig 1.7 shows a thermometer marked in Centigrade degrees. What are the temperatures labelled A and B? What is the change in temperature between these values? Mark on the thermometer the point corresponding to -15°.

Fig 1.7

Solution   A is 16 degrees above 0. B is 11 degrees below 0.

A is 16°, B is -11°

The difference in temperature is the sum of 11 and 16.

The change in temperature is 27°

The point corresponding to -15° is marked by X on fig 1.7.

(2)   The table below gives the highest and lowest temperatures at several towns over the year.

Town	York	Paris	Cairo	Moscow	Sydney
Highest temp.	26	31	37	29	34
Lowest temp.	-8	-5	5	-23	4

(a) Which town recorded the lowest temperature?

(b) What was the change in temperature over the year for York?

Solution   (a) The lowest temperature is the most negative number, -23 °.

Moscow recorded the lowest temperature

(b) The change in temperature at York is the difference between its highest and its lowest value.

The change for York was 34 °

1.3.2 Exercises for level 1

(1)   Make a copy of the number line at the beginning of this section. Mark on it the following values:

(a) 3   (b) 2   (c) 0   (d) -2

(e) $1\frac{1}{2}$ (f) $-1\frac{1}{2}$ (g) $-\frac{1}{4}$ (h) $-2\frac{3}{4}$.

(2) John keeps a record of the temperature over a week in winter. His results are given in the following table:

```
Day             |Mon|Tue|Wed|Thu|Fri|Sat|Sun|

Highest Temp.  | 5 | 4 | 4 | 3 | 6 | 5 | 5 |

Lowest  Temp.  | 0 | -2| -1| -5| -3| -1| 1 |
```

(a) What was the lowest temperature recorded?

(b) Which day had the greatest change in temperature?

(3) A ship sails from 20 miles South of the equator to 130 miles North of the equator. How far North has it sailed?

(4) I have £150 in my bank account, but last week I had -£55. How much has gone into my account over the week?

(5) In Greece the clock is 2 hours ahead of London time, and in California it is 8 hours behind. If you fly from California to Greece, by how much do you alter your watch?

(6) Jane stands on the edge of a cliff and throws a stone up in the air. After 1 second it is 3 m. high. After 5 seconds it has fallen to the bottom of the cliff 50 m. below. What distance has it travelled through between these times?

(7) The altitudes in metres above sea-level of various places are given in the following table:

```
Place    | London | Nairobi | Amsterdam | Dead Sea

Altitude|   30   |  1800   |   - 2     |  - 400
```

(a) Which is the lowest place?

(b) How far do you climb when travelling from London to Nairobi?

(c) How far do you climb when travelling from the Dead Sea to Amsterdam?

(d) How far do you descend when travelling from Nairobi to Amsterdam?

(8) The mileometer of a car has 5 digits, so that it only goes up to 99,999 miles. It now registers 98,567. What will it register after 3,945 more miles?

(9) The level of a river is measured in inches above its average level. During a drought the level of the river is measured each week as follows:

```
Week  |  1  |  2  |  3  |  4  |  5  |  6  |

Level |  -3 |  -7 | -12 | -15 |  -3 |  1  |
```

(a) What was the highest level? What was the lowest level?

(b) What was the greatest weekly change?

(c) By how much did the river rise from its lowest point to its highest?

(10)   How many years are there between 150 BC and 370 AD? What was the date 1250 years before the Battle of Hastings? (1066 AD)

(11)   A man was born in 27 BC, and lived for 62 years. What was the date when he died?

(12)   A woman was born in 158 BC, and lived for 84 years. What was the date when she died?

(13)   A man died aged 53 in 21 AD. What was his birth date?

(14)   The French revolutionaries reformed the calendar by measuring dates after the fall of the Bastille. (1789 AD).

(a) What date corresponds to year 8? What corresponds to year -15?

(b) What date would they give to 1800 AD? What to 1746 AD? What to 213 BC?

1.4 FRACTIONS AND DECIMALS

Fractions express quantities which are not whole numbers.

If a quantity is divided into 8 equal parts, then each of those parts is an *eighth*, or $1/8$, of the whole.

Decimals are another way of writing fractions. The digit one place to the right of the decimal point is the number of tenths, the digit two places to the right is the number of hundredths, and so on.

A fraction is unchanged if the top and the bottom are multiplied or divided by the same number.

1.4.1 Examples for level 1

(1)   (a) Convert $1/4$ to a decimal.

(b) Express 0.3 as a fraction.

(c) Simplify $5/15$.

Solution   (a) Divide 4 into 1.

$$^1/_4 = 0.25$$

(b) 0.3 represents three tenths.

$$0.3 = {}^3/_{10}$$

(c) Divide top and bottom of this fraction by 5.

$$^5/_{15} = {}^1/_3$$

(2)   A man can walk at 6 km. per hour. How far does he walk in 5 minutes? Express your answer both as a fraction and as a decimal.

Solution   5 minutes is $^1/_{12}$ of an hour. The distance travelled in this time is therefore:

$$^6/_{12} = {}^1/_2 \text{ km.} = 0.5 \text{ km.}$$

1.4.2 Exercises for level 1

(1)   Express as fractions: a half, a third, a quarter, three quarters, five eighths.

(2)   Express in words: $^1/_5, {}^1/_9, {}^5/_6, {}^7/_8$.

(3)   Convert to decimal form: $^1/_2, {}^1/_4, {}^3/_8, {}^7/_{10}$.

(4)   Express as fractions: 0.5, 0.25, 0.7, 0.35.

(5)   Simplify as far as possible the following fractions:

$$^2/_4, {}^3/_9, {}^4/_{12}, {}^7/_{14}, {}^5/_{20}.$$

(6)   Jasmine was attempting to simplify fractions. She had to leave some of them unfinished. Complete them for her:

(a) $^2/_- = {}^1/_2$ (b) $^-/_6 = {}^2/_3$ (c) $^{15}/_{20} = {}^-/_4$ (d) $^4/_{28} = {}^1/_-$

(7)   2 cakes are divided fairly between 10 children. What fraction of a cake does each child get?

(8)   Which is the greatest of $^1/_3, 0.3, {}^3/_8$? Which is the least?

(9)   Fig 1.8 shows several shapes in which part has been shaded. In each case find the fraction which has been shaded.

(a)   (b)   (c)   (d)

   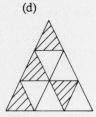

Fig 1.8

(10)   A supermarket sells rice in 500 g bags for 50 p, or in 200 g bags for 25 p. How much does a gram of rice cost in each bag?

(11) The petrol tank of my car holds 20 gallons. How much petrol is there left when the needle is in the position shown?

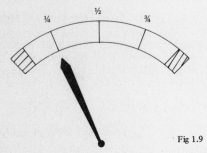

Fig 1.9

Mark on the diagram the position of the needle when there are 15 gallons left.

(12)   (a) How many minutes are there in half an hour? in three quarters of an hour?

(b) What fraction of an hour is 10 minutes? 5 minutes?

(13)   Out of 1,000 voters questioned, 450 said they voted Conservative. What fraction voted Conservative?

(14)   A school has 1,500 pupils, of which one fifth are in the sixth form. How many are not in the sixth form?

For level 2

The top part of a fraction is the *numerator*. The bottom part is the *denominator*.

A fraction such as $^3/5$, in which the denominator is greater than the numerator, is called *proper*. If the denominator is less than the numerator, as in $^7/5$, it is called *improper*.

A number such as $5^3/8$, containing both fractions and whole numbers, is a *mixed number*.

Positive whole numbers are called *natural numbers.*

Whole numbers, whether positive or negative, are called *integers.*

Fractions are sometimes called *rational* numbers.

Several numbers, in particular $\sqrt{2}$ and π, cannot be written as fractions. These numbers are called *irrational..*

1.4.3 Examples for level 2

(1) (a) Express $^{15}/8$ as a mixed number.

(b) Express $5\,^3/4$ as an improper fraction.

Solution (a) Divide 15 by 8. The result is 1 with remainder 7.

$$^{15}/8 = 1\,^7/8$$

(b) Multiply 5 by 4. 5 is equal to $^{20}/4$

$$5\,^3/4 = {}^{20}/4 + {}^3/4 = {}^{23}/4$$

(2) Three quarters of a school are under 16. 200 are 16 or over. How many are there in the school?

Solution One quarter of the school are 16 or over. There are 200 in this quarter.

The school has 4 x 200 = 800 pupils

1.4.4 Exercises for level 2

(1) Express the following as mixed numbers:

(a) $^5/4$ (b) $^7/3$ (c) $^{20}/3$ (d) $^{12}/8$ (e) $^{27}/6$.

(2) Express the following as improper fractions:

(a) $1\,^1/2$ (b) $2\,^3/4$ (c) $3\,^3/8$ (d) $3\,^4/5$ (e) $2\,^2/7$ (f) $5\,^3/10$

(3) Write the following fractions so that their denominators are 100. Hence convert them to decimals.

(a) $^1/50$ (b) $^2/25$ (c) $^7/10$ (d) $^3/4$

(4) Convert the following decimals into improper fractions and into mixed numbers:

(a) 1.5 (b) 2.3 (c) 3.25 (d) 1.125.

(5) Two thirds of a class pass a French exam. 8 fail the exam. How many are there in the class?

(6)   Five eighths of a tennis club are women. There are 33 men in the club; how many women are there?

(7)   A man leaves a quarter of his money to his wife, and the remainder is divided equally between his two sons. What fraction does each son get? If each son gets £3,000, how much does the wife get?

COMMON ERRORS

(1)   **The four operations.**

(a) Be sure that you do the correct operation.

If 5 people each have £300, then the total is found by *multiplying* £300 by 5.

If £600 is shared between 4 people, then each share is found by *dividing* £600 by 4.

(b) The LCM of two numbers is not necessarily the product of the two numbers. Certainly the product is a common multiple, but it may not be the *least* common multiple.

Similarly, when finding the HCF of two numbers, be sure that you get the *greatest* common factor.

(2)   **Patterns**
Do not be too hasty when continuing a sequence. Look at all the terms you are given before deciding how they should go on. There are many possible patterns which begin 1, 2, 4, ... You must know the fourth term at least, before you can continue the pattern.

(3)   **Directed numbers**
(a) The lowest of a set of negative numbers is the one which is most negative. So for example -17 is less than -3. Do not get this the wrong way round.

(b) Be careful when finding the difference between directed numbers. The difference between -4 and 9 is 13, not 5.

(4)   **Fractions and decimals**
(a) Notice carefully where the decimal point is. 0.05 is not the same as 0.5.
(b) Do not confuse, for example, $\frac{1}{3}$ and 0.3.

2 Calculation and Arithmetic

2.1 SIMPLE CALCULATIONS

Modern calculators follow the natural order of operations. To find out what - *Five plus six equals* - press the buttons in that order.

$$\boxed{5} \quad \boxed{+} \quad \boxed{6} \quad \boxed{=}$$

The answer 11 appears.

Multiplication and division are done before addition and subtraction. If you want to do the addition or subtraction first then you must use *brackets*.

$$2 \times 3 + 4 = 6 + 4 = 10 \qquad\qquad 2 \times (3 + 4) = 2 \times 7 = 14$$

2.1.1 Examples for level 1

(1)   Evaluate 4.3 + 5.2 + 11.9, and carry out a simple check of accuracy

Solution   Press the buttons in the order of the sum.

$$\boxed{4}\ \boxed{.}\ \boxed{3}\ \boxed{+}\ \boxed{5}\ \boxed{.}\ \boxed{2}\ \boxed{+}\ \boxed{1}\ \boxed{1}\ \boxed{.}\ \boxed{9}\ \boxed{=}$$

4.3 + 5.2 + 11.9 = 21.4

A simple check consists of taking each number to the nearest whole number.

$$4 + 5 + 12 = 21.$$

So the answer is reasonably accurate.

(2)   Nancy and Peggy go on holiday to France. Nancy takes £25, Peggy takes £34. How much do they have in French money, if there are 10.6 French Francs to the £?

Solution   Their total sum of money is £(25 + 34). In FF this becomes

$$(25 + 34) \times 10.6$$

Because the addition sum is in brackets, it must be done before the multiplication.

$$25 + 34 = 59$$

Now multiply by 10.6:

They have 625.4 FF.

2.1.2 Exercises for level 1

(1)   Use your calculator to evaluate the following. In each case make a simple check that your answer is correct.

  (a) 2.7 + 5.6 + 1.1   (b) 6.3 + 6.6 + 9.8   (c) 12.43 + 43.65 + 23.84

  (d) 9.5 - 5.6   (e) 12.43 - 8.65   (f) 2.6 + 4.8 - 5.9

  (g) 2.65 x 4.65   (h) 54.91 x 35.20   (i) 2.5 x 5.4 x 8.2

  (j) 10.34 ÷ 2.2   (k) 10.85 ÷ 3.1   (l) 109.48 ÷ 9.52   (m) 4.6 ÷ 0.23

(2)   Find the total price of a meal costing £5.64 and drinks costing £3.17.

(4)   David is 12.3 km North of his home, and then walks 7.8 km further North. How far is he from home now?

(5)   I pay for a £7.89 pair of shoes with a £10 note. How much change should I get?

(6)   Patricia buys 10 metres of flex, and cuts off 3.65 metres. How much is there left?

(7)   The total weight of a full bottle is 1,019 grams. If the liquid weighs 923 grams, how much does the glass weigh?

(8)   The total cost of a shirt and a tie was £11.34. If the tie cost £3.72 find the cost of the shirt.

(9)   Find the cost of 3 kg of meat at £3.88 per kg.

(10)   Find the cost of 6 gallons of petrol at £1.67 per gallon.

(11)   Find the cost of 3.42 metres of flex at 50 p per metre.

(12)   Cinema tickets cost £2.50 per adult and £1.60 per child. How much does it cost for a family of two adults and three children to go to the cinema?

(13)   5 pounds of apples cost £1.85. How much does each pound cost?

(14)   Seven yards of curtain material cost £10.01. How much does each yard cost?

(15)   £2,001 is shared equally between three sisters. How much does each sister get?

(16)   How many cakes will Joe be able to buy for £1.68, if each cake costs 24 p?

(17) Complete the bill of fig 2.1.

4 galls. petrol @ £1.75 =		a
2 litre oil	@ £1.31 =	£2.62
Total		b

Fig 2.1

How much change will there be from a £20 note?

(18) The Jackson family stayed for a weekend at a hotel. Fig 2.2 shows their bill. Fill in all the figures.

6 nights at £4.50 each = £27

6 meals at £3.10 each = [a]
_____
Total = [b]

(19) Mr Singh receives his electricity bill as shown in fig 2.3. Fill in all the numbers.

Units used	Unit Price	Amount
580	4.75	a
	Standing charge	£9.73
	Total	b

(20)   A certain book has 432 pages, with an average of 392 words per page. Find the total number of words in the book.

When the book comes out in paperback there will be an average of 576 words per page. How many pages will there be?

(21) Evaluate the following:

(a) 5 x 2 + 3

(b) 5 x (2 + 3)

(c) 3 x 7 - 5

(d) 3 x (7 - 5)

(e) 5 + 8÷4

(f) (5 + 8)÷4

(g) 15 - 9÷3

(h) (15 - 9)÷3

(22)   Insert brackets into the following equations to make them correct.

(a) 3 x 4 + 6 = 30

(b) 12 - 5 x 2 = 14

(c) 7 - 1 + 2 = 3

(d) 6 + 17 - 14 = 2

2.1.3 Examples for level 2

(1)   Evaluate 3 x 2.4 + 5 x 4.7 + 7 x 9.3.

Solution   If you have a scientific calculator, then you can press the keys in the same order as the problem.

$$\boxed{3}\ \boxed{\times}\ \boxed{2}\ \boxed{\cdot}\ \boxed{4}\ \boxed{+}\ \boxed{5}\ \boxed{\times}\ \boxed{4}\ \boxed{\cdot}\ \boxed{7}\ \boxed{+}\ \boxed{7}\ \boxed{\times}\ \boxed{9}\ \boxed{\cdot}\ \boxed{3}\ \boxed{=}$$

If you have an elementary calculator, it may not be able to work out the problem all in one go. In this case you must perform all the multiplications and then add the results.

$$3 \text{ x } 2.4 = 7.2 : 5 \text{ x } 4.7 = 23.5 : 7 \text{ x } 9.3 = 65.1$$

$$7.2 + 23.5 + 65.1 = 95.8$$

(2)   Evaluate $\dfrac{13.3 \times 4.17}{2.4 - 2.21}$

Perform a simple check that you have the correct answer.

Solution   It is best to work out the denominator separately.

$$2.4 - 2.21 = 0.19.$$

Now do the multiplication and the division.

$$\textbf{13.3} \times \textbf{4.17} \div \textbf{0.19} = \textbf{291.9}$$

As a check, simplify the problem to:

$$13 \times 4 \div 0.2 = 52 \div 0.2 = 260.$$

This is approximately correct.

2.1.4 Exercises for level 2

(1)   Evaluate the following, making simple checks on your answers:

(a) $2 \times 9.3 + 3 \times 3.4 + 7 \times 2.9$

(b) $11 \times 3.3 + 21 \times 6.8 - 42 \times 1.9$

(c) $1.43 \times 5.73 + 4.43 \times 9.21 + 8.67 \times 2.21$

(d) $34.2 \times 0.3 + 45.7 \div 0.2$

(2)   Evaluate the following:

(a) $3.7 \times (3.8 + 2.8)$

(b) $98.34 \times (5.32 - 5.29)$

(c) $(34.83 + 57.39) \times (45.43 - 39.87)$

(d) $(4.95 - 2.99)/(6.39 + 2.36)$

(3)   A second-hand book dealer is buying books: he buys 43 at 25p, 87 at 35p and 85 at 55p. How much does he pay in all?

(4)   Muhammed is thinking of decorating his flat. He reckons that he will have to buy:

6 litres of emulsion paint at £1.75 per litre.
2 litres of gloss paint at £3.77 per litre.
2 brushes at £2.55 each.
1 litre of white spirit at £1.05.

How much will he have to spend in all?

(5)   A full bottle of whisky weighs 823 grams, of which 748 grams is the weight of the whisky. How many bottles can be made from 1,200 grams of glass?

2.2 FRACTIONS AND NEGATIVE NUMBERS

When fractions are added or subtracted, they must first be put over the same denominator.

2.2.1 Example for level 1

Evaluate $^2/_5 + {}^1/_4$, leaving your answer as a fraction.

Solution   Write both the fractions with a denominator of 20.

$$^2/_5 + {}^1/_4 = {}^8/_{20} + {}^5/_{20} = {}^{13}/_{20}$$

2.2.2 Exercises for level 1

(1)   Evaluate the following, leaving your answers as fractions.

(a) $^2/_3 + {}^1/_4$ (b) $^3/_5 + {}^1/_6$ (c) $^1/_7 + {}^2/_3$ (d) $^3/_4 - {}^2/_5$

(e) $^1/_4 + {}^1/_2$ (f) $^1/_{10} + {}^3/_5$ (g) $^5/_7 - {}^1/_3$ (h) $^7/_8 - {}^1/_5$

(2)   In a horse race there was a very close finish. Amanda was $^1/_3$ of a length ahead of Brenda, and Brenda was $^1/_2$ of a length ahead of Celine. By what fraction of a length did Amanda beat Celine?

(3)   George buys $1^3/_4$ yards of wire, and cuts off $^1/_3$ of a yard. How much does he have left?

(4) $2\,{}^1/_4$ pints are poured from a 15 pint container. How much is left?

(5)   A man has three children: in his will he leaves half his estate to his wife, a quarter to his eldest child and a fifth to the middle child. What fraction does the youngest child get?

For level 2

When fractions are multiplied, the tops are multiplied together and the bottoms are multiplied together.

To *divide* by a fraction $^a/_b$, *multiply* by $^b/_a$

The product of two negative numbers is positive.

$$-3 \times -2 = 6.$$

The product of a negative number and a positive number is negative.

$$3x - 2 = -6.$$

2.2.3 Examples for level 2

(1)   Evaluate (a) $^3/_4 \times {}^2/_5$ (b) $^3/_4 \div {}^2/_5$.

Solution   (a) Multiply the tops together and the bottoms together.

$$^3/_4 \times {}^2/_5 = {}^6/_{20} = {}^3/_{10}.$$

(b) To divide by $^2/_5$ is to multiply by $^5/_2$.

$$3/_4 \div {}^2/_5 = {}^3/_4 \times {}^5/_2 = {}^{15}/_8 = 1{}^7/_8$$

(2)   Evaluate (3.2 - 5.8) x (2.7 - 4.1)

Solution   The subtractions in brackets must be evaluated first.

(3.2 - 5.8) x (2.7 - 4.1) = (-2.6) x (-1.4)

This is now the product of two negative numbers.

-2.6 x -1.4 = 3.64

2.2.4 Exercises for level 2

(1)   Evaluate the following, leaving your answers as fractions.

(a) $^2/_3$ x $^4/_7$ (b) $^5/_6$ x $^2/_{15}$ (c) $^3/_2$ x $^3/_5$ (d) $^2/_7$ x $^5/_3$ (e) $^1/_3$ x 9

(f) $1^1/_3 + 2^1/_4$ (g) $4\,^1/_5 - 3^2/_3$ (h) $2^7/_8 - 3^1/_4$ (i) $^1/_{10}$ - 4.25

(j) $^2/_3 \div {}^4/_7$ (k) $^1/_2 \div {}^3/_5$ (l) $^5/_9 \div 3$ (m) $^{11}/_{20} \div {}^1/_{40}$

(2)   Evaluate the following:

(a) -3.2 x -4   (b) -12 ÷ -6   (c) 7 - (3 - 5)   (d) (-1.71)/(3.5 - 7.3)

(e) -2 x -3 x 4   (f) -4 x -0.5 x -6   (g) $(-3)^2$ (h) $(-2)^5$.

(3)   How many $^1/_2$ pint jugs can be filled from a 20 pint container?

(4)   I buy a batch of 32 apples, of which a quarter are bad. How many are good?

(5)   A ship sails at $12^1/_2$ km.p.h. How long will it take to cover 75 km.?

(6)   Michael walks one quarter of the distance to school, and then takes the bus. If the bus ride is 1 $^1/_2$ miles, how far does he walk?

(7)   Ailsa and Beatrice share a sum of money so that Ailsa gets two thirds of it. If Beatrice receives £70, how much was the sum?

(8)   A swimming bath is 50 metres long. Sharon can swim $2^1/_2$ lengths. How far can she swim?

(9)   Gabriel takes $1^1/_2$ minutes to peel an apple. How many apples can he peel in 9 minutes?

(10)   The profit of a firm is divided equally between its three partners. Each partner then has to pay three-tenths of his share in tax. What fraction of the original profit does each partner keep?

(11)   A cook uses eleven twelfths of a pound of flour to make 20 cakes. How much flour does each cake contain?

COMMON ERRORS

(1)   Use of calculator
(a) Don't forget to press the = button after a calculation. And don't press it *twice* - if you do so the results are unpredictable.

(b) Be sure that you use brackets correctly. Do not confuse 2 x 3 + 4 and 2 x (3 + 4).

2 x 3 + 4 means: *Multiply 2 and 3 and then add 4*. The result is 10.

2 x (3 + 4) means: *Add 3 and 4 and then multiply by 2*. The result is 14.

(c) Checks of accuracy are very useful, but they are only rough. Do not expect too much of them. They ensure that, for example, your answer is not 10 times too big or 10 times too small.

(2)   Fractions
(a) When adding fractions, do not add the numerators together and the denominators together.

$$^2/_3 + ^4/_5 \neq ^6/_8.$$

(b) Do not invert fractions.

$$^1/_3 + ^1/_6 = ^1/_2. \text{ But } 3 + 6 \neq 2.$$

(3)   Negative numbers
Be careful when multiplying or dividing negative numbers. It is easy to forget the rules:

plus times plus is plus
plus times minus is minus
minus times minus is plus.

3 Approximation

3.1 ROUNDING

When a number is *rounded* to a whole number, it is written as the nearest whole number.

When a number is rounded to the *nearest 100*, it is written as the nearest multiple of 100.

3.1.1 Examples for level 1

(1)   Round to a whole number:

   (a) 1.35   (b) $7^5/8$ (c) -2.8°.

Solution   (a) 1.35 is nearer 1 than 2.

Round to 1

(b) $7^5/8$ is nearer 8 than 7.

Round to 8

(c) -2.8° is nearer -3° than -2°.

Round to -3°

(2)   Round 234,635 (a) to the nearest 100 (b) to the nearest 1,000 (c) to the nearest 10.

Solution   (a) 35 is nearer 0 than 100.

Round to 234,600

(b) 635 is nearer 1,000 than 0.

Round to 235,000

(c) 5 is halfway between 0 and 10. It could be rounded up or down. By convention it is usually rounded up.

Round to 234,640

(3)   A train travels for 7 hours at 46 m.p.h. How far has it travelled, to the nearest 10 miles?

Solution   The distance is the product of the speed and the time.

$$7 \times 46 = 322 \text{ miles.}$$

320 miles to the nearest 10 miles.

3.1.2 Exercises for level 1

(1) Round the following numbers to the nearest whole number:

(a) 3.56   (b) 4.267   (c) 129.67   (d) 987.46   (e) 5.5   (f) -8.45   (g) -76.76   (h) 0.67   (i) 0.43
 (j) -0.87   (k) -0.34   (l) $9^1/4$ (m) $76^1/3$ (n) $^6/7$

(2)   Round 1,097,382 to the nearest (a) 1,000 (b) 100 (c) 10.

(3)   Round 463 to the nearest (a) 10 (b) 100 (c) 1,000.

(4)   Round -67.4 to the nearest (a) whole number (b) 10 (c) 100.

(5)   Work out the following, giving your answer to the nearest whole number:

(a) 34.7 + 54.86 (b) 2.9 - 1.7 (c) 3.5 + 2.9 + 6.4 (d) 10.6 + 2.3 - 8

(6)   Work out the following, giving your answer to the nearest whole number.

(a) 3.6 x 4.7   (b) 2.8 x 3.2 x 9.6

(c) 93 ÷ 7   (d) $^{213}/_{31}$

(e) 0.012 ÷ 0.0023   (f) (2.43 - 2.39)/0.047

(7)   What is the cost of 17 litres of petrol at 47.8 p per litre? Give your answer to the nearest penny.

(8)   What is the cost of 144 kg of tomatoes, if 1 kg costs 74 p? Give your answer to the nearest £.

(9)   I buy 46 yards of material for £632. What is the cost per yard, given to the nearest penny?

(10)   To hire a tennis court costs £1.30 per hour. What is the cost per minute, given to the nearest penny?

(11)   I travel at 55 m.p.h. for $3^1/2$ hours. How far have I travelled? Give your answer to the nearest 10 miles.

(12)   I drove for a distance of 200 miles at a speed of 70 m.p.h. How long did it take me, expressed to the nearest hour?

(13)   £10,000 is shared out equally between 17 people. How much does each get, expressed to the nearest £10?

3.2 SIGNIFICANT FIGURES AND DECIMAL PLACES. Level 2

The left-most non-zero digit of a number is the *first significant figure*. The next digit, even if it is zero, is the second significant figure.

When a number is rounded to *2 significant figures* the digits to the right of the second significant figure are rounded up or down.

The third digit after the decimal point of a number is in the *3rd decimal place*.

When a number is rounded to *3 decimal places* the digits to the right of the third decimal place are rounded up or down.

3.2.1 Examples for level 2

(1)   Express the following to 3 significant figures:

(a) 234,576 (b) 1,003 (c) 0.00000012743.

Solution   (a) The third significant figure is the 4. The figures to the right of it must be rounded up, giving:

235,000

(b) The third significant figure is the 0 in front of the 3. The 3 is rounded down, to give:

1,000

(c) The third significant figure is the 7. Round down the remaining figures:

0.000000127

(2)   Express the following to 2 decimal places:

(a) 12.358 (b) 5.1029

Solution   (a) The digit in the second decimal place is the 5. The next figure is 8, so it is rounded up:

12.36

(b) The digit in the second decimal place is the 0. The next figure is 2, so round down:

5.10

(3)   When rounded to the nearest £10, the amount of money in my bank account is £120. What are the greatest and the least possible amounts that there could be?

Solution   An amount over £125 would be rounded up to £130. An amount below £115 would be rounded down to £110.

The greatest possible amount is £125, and the least £115

3.2.2 Exercises for level 2

(1)   Round the following to 2 significant figures:

(a) 2,347 (b) 656,265 (c) 0.004654 (d) 0.0249 (e) -3.47 (f) 1.03 (g) 2.09 (h) 497 (i) 602 (j) 5.096 (k) 4.000098

(2)   Round 4,738 to (a) 3 sig. figs. (b) 2 sig. figs. (c) 1 sig. fig.

(3)   Round the following to 3 decimal places:

(a) 3.5764 (b) 5.2349 (c) 23.03823 (d) 65.47583 (e) 0.3457 (f) 0.1234 (g) 0.00476 (h) 0.00012 (i) 0.12034 (j) 0.26976

(4)   Round 0.13582 to (a) 4 dec. places (b) 3 dec. places (c) 2 dec. places.

(5)   Evaluate the following, giving your answer to 2 decimal places.

(a) 4.564 + 2.372 (b) 2.4638 - 0.9564 (c) 2.432 x 3.276 (d) $^5/_7$ (e) 2.574 ÷ 0.945 (f) 1.35^2 (g) (1.032 - 1.029)/7

(6)   17 cm^3 of a chemical compound weigh 25 grams. What is the density of the compound? Give your answer to 2 decimal places.

(7)   What is the weight of 288 ball-bearings, each of which weighs 2.3 grams? Give your answer to 2 significant figures.

(8)   What is the cost of 27 litres of petrol at 48.3 p per litre? Give your answer to the nearest £.

(9) The figure shows the display of a calculator after 25 has been divided by 7. Express the result to (a) 4 significant figures (b) 2 decimal places.

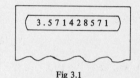

Fig 3.1

(10)   A temperature is given as 34°, rounded to the nearest whole number. What is the greatest possible value of the temperature? What is the least?

(11)   The weight of a car is given as 910 kg, rounded to the nearest 10 kg. What is the greatest possible weight of the car?

(12)   The population of a town is given as 80,000, to the nearest 10,000. What are the greatest and the least possible values of the population?

(13)   The weight of a book is given as 0.54 kg, rounded to 2 decimal places. What are the greatest and the least possible values for the weight?

(14)   The distance between two cities is given as 320 miles, to 2 significant figures. What are the greatest and the least possible values for the distance?

(15)   A boy can throw a ball for 57 metres. How large an area is within his range? Give your answer to (a) 3 significant figures (b) 2 significant figures (c) 1 significant figure
    (The area of a circle with radius r is πr^2)

3.3 APPROPRIATE LIMITS OF ACCURACY. Level 2

A measurement should be given to an *appropriate* number of decimal places or an *appropriate* number of significant figures

3.3.1 Examples for level 2

(1)   The radius of the earth is 6,400,000 metres, to 2 significant figures. Find the circumference of the earth, giving your answer to an appropriate degree of accuracy.

Solution   The formula for the circumference of a circle of radius r is $2\pi r$. Using this formula, the circumference of the earth is:

$$2 \times \pi \times 6{,}400{,}000 = 40{,}212{,}386 \text{ to the nearest metre.}$$

This answer gives a misleading impression of accuracy. It is given to 8 significant figures. The radius of the earth is only given to 2 significant figures, so the answer can only be given to 2 significant figures.

The circumference of the earth is 40,000,000 metres.

(2)   A rectangle was measured as 13.5 cm long and 26.7 cm wide. What is the area of the rectangle? Give your answer to an appropriate degree of accuracy.

Solution   The area is $13.5 \times 26.7 = 360.45$ cm^2. This gives an answer to 2 decimal places, while the original lengths were only given to 1 decimal place.

The length of the rectangle could be as high as 13.55. The width could be as high as 26.75. Its area could then be as high as 362 cm^2.

The answer can only be given to 2 significant figures:

The area is 360 cm^2

3.3.2 Exercises for level 2

(1)   9 inches of rain fell in one week. What was the average rainfall per day? Give your answer to an appropriate degree of accuracy.

(2)   A ball-bearing weighs 3.2 grams, to 1 decimal place. How much do 111 similar ball-bearings weigh?

(3)   6 cakes weigh 400 grams. How much does each cake weigh?

(4)   The distance of the earth from the sun is 90,000,000 miles. Assuming the orbit of the earth to be a circle, how long is the earth's orbit? (A circle with radius r has length $2\pi r$.)

(5)   Evaluate the following, giving your answer to an appropriate degree of accuracy.

(a) $3.5 + 4.6 + 7.3 + 5.5$ (b) $8.3 + 9.4 - 10.345$

(c) 7.32×4.17 (d) 34.3×83.4

(e) $2/.07$ (f) $23.6 \div 8.45$

(6)   What would be appropriate levels of accuracy for the following measurements:

      (a) The distance from London to Paris
      (b) The population of Birmingham
      (c) Your weight
      (d) Your height
      (e) Your mother's height
      (f) The top speed of Concorde
      (g) The price in June 1986 of a gallon of four-star petrol.

COMMON ERRORS

(1)   Rounding

(a) When you round a number to the nearest 100 do not discard the 0's. That would mean that you have *divided* the number by 100.

3,468 to the nearest 100 is 3,500, not 35.

(b) When you are asked to give the answer to a calculation in a rounded form, do the rounding right at the end of your calculation. For example:

3.4 + 4.3 = 7.7 = 8 to the nearest whole number.

Do not do the rounding before. This would give the wrong answer:

3.4 + 4.3 = 3 + 4 = 7.

(c) If a quantity is given as 240 to the nearest 10, then the greatest possible value it could have is 245, not 250. Similarly the least value it could have is 235, not 230.

(2)   Significant figures and decimal places

(a) The first significant figure is the first *non-zero* figure. So the first significant figure of 0.0023 is 2, not 0.

(b) After the first significant figure, zeros count as well as non-zero digits. The first 2 significant figures of 103 are 1 and 0, not 1 and 3.

(c) If you are rounding to 2 decimal places, and the digit in the second place is 0, then leave it in. This shows the degree of accuracy of the calculation or measurement.

To 2 decimal places, 2.497 = 2.50, not 2.5.

(d) Save the rounding until the very end of your calculation. A useful rule is to work to one more decimal place than you are asked to give. So if your final answer is to be given to 2 decimal places, use 3 decimal places until the end of your calculation.

(3)   Appropriate accuracy

Do not give your answer to too many decimal places or too many significant figures.

If a cake weighing 1 kg is divided into 3 equal portions, then your calculator will tell you that each portion is 0.33333333 kg. It would be misleading to give this as the answer. Round it to 0.3 or 0.33 kg.

4 Indices

4.1 SQUARES, CUBES, SQUARE ROOTS

When a number is multiplied by itself, the result is the *square* of the number.

$$3 \times 3 = 3^2$$

If a number is multiplied by itself 3 times, the result is the *cube* of the number.

$$4 \times 4 \times 4 = 4^3$$

49 is the square of 7. So 7 is the *square root* of 49.

$$49 = 7^2, \text{ so } 7 = \sqrt{49}.$$

To use a calculator to find a square root, press the number and then the square root button. (Do not press the = button). To find $\sqrt{1024}$, press:

$$\boxed{1} \; \boxed{0} \; \boxed{2} \; \boxed{4} \; \boxed{\sqrt{}}$$

The answer 32 will appear.

4.1.1 Examples for level 1

(1)　Evaluate the following:

(a) 5^2　(b) 3^3　(c) $\sqrt{64}$.

Solution　Use the definitions.

$$\text{(a) } 5^2 = 5 \times 5 = 25$$

$$\text{(b) } 3^3 = 3 \times 3 \times 3 = 27$$

(c) Note that $8^2 = 64$. Hence:

$$\sqrt{64} = 8$$

(2)　(a) What is the area of a square whose side is 4 inches?

(b) A square has area 36 cm^2. What is the side of the square?

Solution   (a) The area of the square is found by squaring the side. This gives:

$$\text{Area} = 4^2 = 16 \text{ square inches.}$$

(b) When the side is squared, the result is 36 cm^2. This gives:

$$\text{The side is } \sqrt{36} = 6 \text{ cm.}$$

4.1.2 Exercises for level 1

(1)   Evaluate the following:

(a) 3^2 (b) 8^2 (c) 2^2 (d) 9^2 (e) 10^2 (f) 12^2 (g) 100^2

(2)   Give the values of the following:

(a) $\sqrt{4}$ (b) $\sqrt{9}$ (c) $\sqrt{16}$ (d) $\sqrt{64}$ (e) $\sqrt{100}$ (f) $\sqrt{121}$

(3)   Evaluate the following:

(a) 2^3 (b) 4^3 (c) 10^3 (d) 1.5^2 (e) $\frac{1}{2}^2$ (f) $(\frac{1}{3})^2$ (g) 0.5^3

(4)   Work out the following:

(a) $\sqrt{10,000}$ (b) $\sqrt{81}$ (c) $\sqrt{144}$ (d) $\sqrt{1}$ (e) $\sqrt{1/4}$ (f) $\sqrt{9/4}$

(5)   Complete the following by filling in the boxes:

(a) $7^2 + 6^2 = 49 + \square = \square$ (b) $5^2 - 4^2 = 5^2 - \square = \square$

(c) $\sqrt{9} + \sqrt{36} = 3 + \sqrt{36} = \square$ (d) $\sqrt{81} - \sqrt{49} = \sqrt{81} - 7 = \square$

(6)   A square lawn has side 12 feet. What is the area?

(7)   A card-table has a square top, of side 40 cm. What is the area of the top?

(8)   A square of carpet is 4 m^2 in area. What is the side of the carpet?

(9)   A square of material is 25 cm^2. What is the side of the material?

(10)   A cube has side 3 cm. What is the volume of the cube?

(11)   The volume of a cube is 64 cm^3. What is the side of the cube?

4.2 INTEGER POWERS. Level 2

When a number is multiplied by itself n times, the result is the *n'th power* of the number.

$$3 \times 3 \times 3 \times 3 \times 3 = 3^5 = 243.$$

The *negative* power of a number is 1 over the positive power.

$$3^{-2} = (\tfrac{1}{3})^2 = \tfrac{1}{9}$$

When powers are multiplied, the indices are *added*.

$$3^4 \times 3^5 = 3^9$$

When powers are divided, the indices are *subtracted*.

$$5^6 \div 5^3 = 5^3.$$

4.2.1 Examples for level 2

(1)   Evaluate the following:

(a) 2^5 (b) 4^{-2}

Solution   Use the definition:

$$(a)\ \ 2^5 = 2\times2\times2\times2\times2 = 32$$

$$(b)\ \ 4^{-2} = \frac{1}{4}^2 = \frac{1}{16}$$

(2)   Express as a single power of 5:

$$\frac{5^3 \times 5^5}{5^6}$$

Solution   The indices are added along the top line of the fraction, and the bottom index is subtracted. This gives:

$$5^{3+5-6} = 5^2$$

4.2.2 Exercises for level 2

(1)   Evaluate the following:

(a) 3^4 (b) 4^4 (c) 1^6 (d) 2^{10} (e) $\frac{1}{2}^3$ (f) 0.3^3 (g) 1.5^3 (h) $(-2)^3$

(2)   Evaluate the following:

(a) 2^{-1} (b) 3^{-3} (c) 5^{-3} (d) 4^{-3} (e) 1^{-7} (f) $\frac{1}{2}^{-1}$ (g) $(1/3)^{-2}$

(3)   Simplify the following:

(a) $3^2 \times 3^4$ (b) $2^4 \times 2^3$ (c) $4^3 \div 4^2$ (d) $3^5 \times 3^{-4}$ (e) $7^4 \times 7^{-3}$

(4)   Simplify the following:

(a) $4^3 \div 4^{-2}$ (b) $3^5 \div 3^{-1}$ (c) $\frac{1}{2}^3 \times \frac{1}{2}^{-3}$ (d) $(^3/_2)^5 \div (^2/_3)^4$

(5)   Express as a single power of 2:

$$\frac{2^5 \times 2^5}{2^3 \times 2^6}$$

(6)   Express as a single power of 3:

$$\frac{3^8}{3^7 \times 3^4}$$

(7)   Use your calculator to find the following, giving your answer to 3 decimal places:

(a) $\sqrt{2}$ (b) $\sqrt{3}$ (c) $\sqrt{65}$ (d) $\sqrt{0.2}$ (e) $\sqrt{0.00005}$.

(8)   Find $\sqrt{1,000}$, giving your answer to:

(a) 1 significant figure (b) 2 significant figures (c) 5 significant figures.

(9)   Evaluate the following, giving your answer to 2 decimal places:

(a) $2 + \sqrt{3}$ (b) $35 - \sqrt{7}$ (c) $(\sqrt{2} + \sqrt{5})/3$ (d) $5/(\sqrt{6} - \sqrt{3})$

(10)   Insert $\sqrt{}$ signs to make the following true:

(a) $9 = 3$ (b) $4 = 16$ (c) $100 = 10,000$

(11)   Insert indices to make the following true:

(a) $4 = 2^*$ (b) $8 = 2^*$ (c) $27 = 3^*$ (d) $3^* = 81$ (e) $100,000 = 10^*$

4.3 FRACTIONAL INDICES AND LAWS OF INDICES. Level 3

Any non-zero number raised to the power 0 is 1.

$$a^0 = 1 \quad \text{(Provided that } a \neq 0\text{)}$$

A fraction index $^1/_n$ gives the nth root.

$$a^{1/n} = \sqrt[n]{a}$$

Indices obey the following rules:

$$a^n \times a^m = a^{n+m}$$

$$a^n \div a^m = a^{n-m}$$

$$(a^n)^m = a^{nm}$$

$$a^{-n} = {}^1/_{a^n}$$

$$(ab)^n = a^n b^n.$$

4.3.1 Examples for level 3

(1)   Evaluate the following:

(a) 3^0 (b) $4^{1/2}$

Solution   Use the definitions:

$$\text{(a)} \ \ 3^0 = 1$$

$$\text{(b)} \ \ 4^{1/2} = \sqrt{4} = 2$$

(2)   Simplify the following:

(a) $5^3 \times 25^2 \div 125^2$ (b) $16^{3/4}$

Solution   (a) The numbers concerned here are all powers of 5. Re-write as:

$$5^3 \times (5^2)^2 \div (5^3)^2$$

$$5^3 \times 5^4 \div 5^6$$

Add and subtract the indices:

$$5^{3+4-6} = 5^1 = 5$$

(b) Re-write $16^{3/4}$ as $(16^{1/4})^3$

The fourth root of 16 is 2. This gives:

$$2^3 = 8$$

(3)   Solve the equation $3^x = 9^{x+1}$.

Solution   Write 9 as a power of 3.

$$3^x = (3^2)^{x+1}$$

$$3^x = 3^{2x+2}$$

Equate the powers of 3, to obtain $x = 2x + 2$.

$$x = -2$$

4.3.2 Exercises for level 3

(1)   Evaluate the following:

(a) $9^{1/2}$ (b) $100^{1/2}$ (c) $8^{1/3}$ (d) $49^{-1/2}$ (e) $64^{-1/3}$ (f) $1000^{1/3}$

(2)   Evaluate the following:

(a) $8^{2/3}$ (b) $9^{3/2}$ (c) $1000^{2/3}$ (d) $(1/4)^{-2}$ (e) $(1/4)^{-1/2}$ (f) $0.01^{-1/2}$

(3)   Simplify the following:

(a) $4^3 \times 2^6 \div 8^3$ (b) $27^2 \times 9^2 \times 3^{-7}$ (c) $10^{1/2} \times 100^{1/4}$

(4)   Simplify the following:

(a) $x^{1/2} \times x^{1/2}$ (b) $y^{1/4} \times y^{3/4}$ (c) $3^x \times 9^x \div 27^x$

(5)   Solve the following equations:

(a) $2^x = 4^{x-1}$ (b) $25^{2x+1} = 125^{x+2}$ (c) $8^x = 32$

(6)   Express the following as products of prime numbers, using the index notation.

(a) 8   (b) 12   (c) 75   (d) 36   (e) 1024   (f) 180

(7)   What is the largest power of 2 which divides 56?

(8)   Which is the largest odd number which divides 360?

4.4 STANDARD FORM. Level 2

Standard form is used for very large or very small numbers. A number is in standard form when there is only one digit to the left of the decimal point. For example:

$$256,700 = 2.567 \times 10^5$$

4.4.1 Examples for level 2

(1)   Write the following in standard form:

(a) 34,560,000   (b) 0.000000045

Solution   (a) Move the decimal point 7 places to the left. This is balanced by increasing the power of 10 by 7.

$$3.456 \times 10^7$$

(b) Move the decimal point 8 places to the right. This is balanced by decreasing the power of 10 by 8.

$$4.5 \times 10^{-8}$$

(2)   Evaluate the following, leaving your answers in standard form.

(a) $3 \times 10^6 \times 4 \times 10^5$ (b) $4 \times 10^9 + 5 \times 10^8$

Solution   (a) Multiply the two numbers together:

$$3 \times 4 \times 10^6 \times 10^5$$

$$12 \times 10^{11}$$

This is numerically correct, but it is not in standard form.

$$\mathbf{1.2 \times 10^{12}}$$

(b) Before these numbers can be added, they must have the same power of 10.

$$40 \times 10^8 + 5 \times 10^8$$

$$45 \times 10^8$$

This is numerically correct, but it is not in standard form.

$$\mathbf{4.5 \times 10^9}$$

4.4.2 Exercises for level 2

(1)   Express the following in standard form:

(a) 23,000 (b) 876,000,000 (c) 0.000123 (d) $^1/_{1000}$ (e) 20^5

(2)   Evaluate the following, leaving your answer in standard form:

(a) 1000×100 (b) $3 \times 10^4 \times 7 \times 10^4$ (c) $^{900}/_{10,000}$

(d) $1.6 \times 10^8 \div 2$ (e) $5.2 \times 10^{-5} \times 2 \times 10^{-8}$ (f) $1.2 \times 10^5 \div 4 \times 10^2$

(g) $5 \times 10^5 \times 4 \times 10^{-3}$ (h) $(5 \times 10^3)^2$ (i) $(2 \times 10^4)^4$ (j) $(4.2 \times 10^{-3})^4$

(k) $(1.25 \times 10^{-4})^{-2}$.

(3)   Evaluate the following, leaving your answer in standard form:

(a) $4 \times 10^5 + 8 \times 10^6$ (b) $3.3 \times 10^6 + 2.7 \times 10^5$ (c) $9.5 \times 10^5 + 8 \times 10^4$

(d) $7.2 \times 10^7 - 8 \times 10^6$ (e) $4 \times 10^8 - 3 \times 10^9$ (f) $9 \times 10^{-3} + 8 \times 10^{-4}$

(4)   Light travels at 1.86×10^5 miles per second. How far does it travel in an hour? How far does it travel in a year?

(5)   The weight of a carbon atom is 2×10^{-27} grams. How many carbon atoms are there in 10 grams of pure carbon?

(6)   On the number line below indicate the numbers (a) 2 (b) 1.1×10^1 (c) 2×10^{-1} (d) -2×10^{-1}

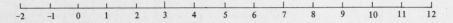

COMMON ERRORS

(1) **Squares and square roots.**
Be careful not to confuse the square of a number with twice that number.

$$a^2 \neq 2a.$$

$$\text{Similarly, } \sqrt{a} \neq {}^1/_2\, a.$$

(2) **Arithmetic of indices**
There are very many common errors when calculating with indices. Usually these occur because *taking powers* has been confused with *multiplying or dividing*. Here are some common mistakes to watch out for:

$$5^3 \times 5^2 \neq 5^6$$

$$5^3 + 5^2 \neq 5^5$$

$$5^3 \times 4^2 \neq 20^5$$

$$(3^5)^2 \neq 9^{10}$$

(3) **Standard form**
(a) Be sure that there is only one digit to the left of the decimal point. If there is more or less than one then the number is not in standard form.

(b) The errors of indices mentioned above also occur when dealing with standard form. Do not make the following mistakes:

$$2 \times 10^5 \times 3 \times 10^6 \neq 6 \times 10^{30}$$

$$4 \times 10^6 + 5 \times 10^6 \neq 9 \times 10^{12}$$

(c) You can only add or subtract numbers in standard form when they have the same power of 10. (Just as we cannot directly add centimetres and kilometres.) The numbers must be changed so that they do have the same power of 10, and then they can be added or subtracted.

5 Ratio and Proportion

Numbers can be compared by *ratios, scales,* and *proportions*. These are all ways of expressing fractions.

5.1 RATIOS AND SCALES

If two numbers A and B are in the *ratio* 3:4, that means that A is $^3/_4$ the size of B.

The *scale* of a map compares the lengths on the map to the lengths on the ground. A scale of 1 cm. to 2 km means that 1 cm. on the map represents 2 km. on the ground.

5.1.1 Examples for level 1

(1)   A school contains 350 boys and 250 girls. Find the ratio of the number of boys to the number of girls, giving your answer in as simple a form as possible.

Solution   The ratio is 350:250. Ratios obey the same rule as fractions. Both terms can be divided by 50. The ratio is then:

$$7:5$$

(2)   Two partners Smith and Jones invest in a business in the ratio 4:5. If Smith put in £20,000, how much did Jones put in?

Solution   Jones invests 5/4 of the amount that Smith invested. Hence Jones invested;

$$5/4 \text{ x } £20,000 = £25,000$$

(3)   The scale of a map is 1 cm for 2 km. A distance on the map is measured as 5 cm. How far is it on the ground?

Solution   For this scale, 1 cm represents 2 km. So 5 cm represents 5 times as much. The distance on the ground is:

$$5\text{x}2 = 10 \text{ km.}$$

5.1.2 Exercises for level 1

(1)   Joan has £20 and John has £15. What is the ratio between these amounts?

(2)   A girl is 5 ft high and her father is 6 ft high. What is the ratio of their heights?

(3)   A man is 32 and his daughter is 8. What is the ratio of their ages?

(4)   A recipe for pastry requires 8 oz of flour to 4 oz of butter. What is the ratio of flour to butter?

(5)   There are twice as many boys as girls in a certain school. Express this as a ratio of the number of boys to the number of girls.

(6)   Three quarters of the population of a nation live in towns. Express this as a ratio of the town-dwellers to the country-dwellers.

(7)   For a two-stroke motorcycle the fuel consists of petrol and oil in the ratio 20:1. How much petrol should be bought with $^1/_2$ litre of oil?

(8)   Joan and Jackie have weights in the ratio 5:4. If Joan weighs 50 kg, how much does Jackie weigh?

(9)   An alloy consists of copper and tin in the ratio 5:2. How much copper goes with 12 kg of tin?

(10)   Two friends Jane and Sandra share a flat, and agree to pay the rent in the ratio 8:5. If Sandra contributes £20 per week, how much does Jane pay?

(11)   The scale of a map is 1 cm per 1 km. A distance on the map is measured as 3 cm. How far is it on the ground?

(12)   The scale of a map is 1 cm for 2 km. The distance between two towns is 10 km. How far apart are the towns on the map?

(13)   The model of a aircraft is in the scale 1:80. If the real aircraft is 40 metres long, how long is the model?

(14)   An architect's model of a block of flats is in the scale 1:50. If the model is 0.5 metres wide, how wide is the block of flats?

(15)   A model ship is $^1/_2$ metre long, and the real boat is 50 metres long. What is the scale of the model?

(16)   A photograph is enlarged in the ratio 2:3. If the photo was 6 cm high before, how high is it after the enlargement?

5.1.3 Examples for level 2

(1)   A legacy of £8,000 is divided between two people in the ratio 3:5. How much does each get?

Solution   The sum is divided up into 3 + 5 = 8 equal shares.

Each share is worth £8,000 ÷ 8 = £1,000.

The first person gets 3 of these shares, and the second person gets 5 shares.

The shares are £3,000 and £5,000

(2)   Gunpowder contains salt-petre, charcoal and sulphur in the ratio 6:1:1. How much salt-petre does 40 pounds of gunpowder contain?

Solution   This example shows that ratios may be used to compare the sizes of three or more things. In this case there are 6+1+1 = 8 shares, each of which is $^{40}/_8 = 5$ pounds. The salt-petre has 6 of these shares, and hence it weighs:

6 x 5 = 30 pounds.

5.1.4 Exercises for level 2

(1)   In a will money is left to Alfred and Bonnie in the ratio 3:4. If the total sum of money was £21,000, how much did Alfred get?

(2)   An amalgam mixes mercury and silver in the ratio 3:2. How much silver is needed to make 10 kg of amalgam?

(3)   A line is divided in the ratio 2:13. If the line is 30 cm long, how long is the larger part?

(4)   A tennis club has men and women members in the ratio 4:5. If there are 270 members, how many of these are women?

(5)   Mr Robinson and Mr Gillespie buy a race horse; Mr Robinson contributes £8,000 and Mr Gillespie contributes £4,000.

      (a) What is the ratio of their contributions?

      (b) The horse wins a race. If they divide the prize money of £6,000 in the same ratio, how much does Mr Robinson get?

(6)   Building works for a school are to be funded by the local council and by the government in the ratio 4:1. If the building costs £50,000, how much does the government pay?

(7)   An exam consists of two papers. The first lasts 1 hour and the second $1^1/2$ hours. If marks are awarded in the same ratio as the time, how many marks out of 100 are accounted for by the first paper?

(8)   Money is left to three sons in the ratio 3:4:5. If the total sum of money left is £24,000, how much does each son get?

(9)   A mixed fruit drink contains orange juice, pineapple juice and lemon juice in the ratio 6:3:1. How much pineapple juice is there in 800 cc of the mixture?

(10)   A man finds that he divides his day between work, play and sleep in the ratio 5:3:4. For how many hours per day is he asleep?

(11)   The three movements of a concerto are in the ratio 5:2:3. If the concerto lasts for 20 minutes, how long are the movements?

For level 3

If the scale of a map is written as 1:100,000, this is called the *representative fraction* of the map.

If two lengths are in the ratio a:b, then the corresponding areas are in the ratio $a^2:b^2$, and the corresponding volumes are in the ratio $a^3:b^3$.

5.1.5 Examples for level 3

(1)   The representative fraction of a map is 1:50,000. A lake on the map covers 5 cm^2. How big is the real lake?

Solution   The lake on the map can be regarded as being 1 cm by 5 cm. The scale of the map means that 1 cm represents $^1/_2$ km, so the map lake represents $^1/_2$ km by $2^1/_2$ km on the ground. Hence the area of the lake is:

$$^1/_2 \text{ x } 2^1/_2 = 1^1/_4 \text{ km}^2$$

(2)   A model airplane is in the scale 1:80. If the model is 200 cm^3 in volume, what is the volume of the real airplane?

Solution   The height, the length and the width of the real plane are each 80 times as big as in the model. Hence the volume of the real plane is 80x80x80 times as big as the model's volume.

$$\textbf{Real volume} = 80^3 \text{ x } 200 = 1.024\text{x}10^8 \text{ cm}^3 = 102.4 \text{ m}^3$$

5.1.6 Exercises for level 3

(1)   The scale of a map is 1 cm to $^1/_2$ km. A field on the map is 1 cm^2. How large is the real field?

(2)   The scale of a map is 1 cm to 1 km. How large on the map is a forest of 20 km^2?

(3)   The representative fraction of a map is 1:20,000. An area on the map is 3 cm^2. What real area does it correspond to?

(4)   A map of England is in the scale 1:1,000,000. What area on the map corresponds to a county of 500 km^2?

(5)   An architect's model of a house is in the scale 1:10. If the living-room is 600 m^3, what is the volume of the model of the living-room?

(6)   Find the representative fractions for maps with the following scales:

(a) 1 cm to 1 km. (b) 1 cm to 50 km. (c) 2 cm to 5 km.

(7)   How much does 1 cm correspond to on maps with the following representative fractions:

(a) 1:100,000   (b) 1:50,000   (c) 2:50,000?

(8)   On a map a road of length $^1/_2$ km is represented by a distance of 2 cm. What is the scale of the map? What is the representative fraction?

(9)   Find the scale of a map, if 7 km is represented by $3^1/_2$ cm. What is the representative fraction?

(10)   A model of a ship is in the scale 1:200. If the model has volume 500 cm^3, what is the volume of the ship?

(11)   A model plane is in the scale 1:60.

(a) The model is 8 cm long. How long is the plane?

(b) The area of the wings is 10 m^2. What is the area of the wings of the model?

(c) The volume of the fuselage of the model is 70 cm^3. What is the volume of the real fuselage?

(d) The model has 2 wings. How many wings are there on the real plane?

(12)   A model locomotive is in the scale 1:90.

(a) The locomotive is 15 metres long. How long is the model?

(b) The front of the model is 20 cm^2 in area. What is the area of the front of the real locomotive?

(c) The locomotive weighs 80 tonnes. Assuming that the model is made out of exactly the same materials as the real locomotive, what is the weight of the model?

(13)   A child has a model of Superman. The model is 10 cm high, while Superman is 2 metres high.

(a) What is the scale of the model?

(b) Superman's vest is 0.8 m^2 in area. What is the area of the vest on the model?

(c) The model has ten fingers. How many fingers does Superman have?

(14)   A photograph is enlarged in the ratio 2:3. If sky occupies 10 cm^2 of the original photo, how much area does it occupy of the enlargement?

(15)   A map shows a park of area 3 km^2 as a region of 12 cm^2. What is the scale of the map? What is the representative fraction?

5.2 PROPORTION

If two quantities increase at the same rate, so that if one is doubled then the other is doubled, the quantities are *proportional*.

If two quantities increase at opposite rates, so that if one is doubled then the other is halved, the quantities are *inversely proportional*.

5.2.1 Examples for level 1

(1)   A pastry recipe calls for 8 oz of flour for 4 oz of butter. How much butter should be used to go with 16 oz of flour?

Solution   Here the weight of flour is proportional to the weight of butter. The amount of flour has been doubled. Hence the amount of butter should also be doubled.

$$2 \times 4 = 8 \text{ oz of butter}$$

(2)   A man normally drives to work at 30 m.p.h., and it takes him 10 minutes. On a fine day he decides to cycle instead, and he can cycle at 10 m.p.h. How long does it take him?

Solution   Here the time taken is inversely proportional to the speed. The speed has been divided by 3, and so the time taken will be multiplied by 3.

Time taken = 3 x 10 = 30 minutes.

5.2.2 Exercises for level 1

(1)   A recipe for mayonnaise tells the cook to use one egg yolk for each 100 ml of oil. How much oil will mix with 3 egg yolks?

(2)   A journey of 100 miles costs £6. Assuming that the cost per mile is constant, how much will a journey of 200 miles cost?

(3)   A car travels 34 miles on one gallon of petrol. How far will it get on a tankful of 8 gallons?

(4)   40 c.c. of a metal weigh 120 grams. What is the weight of 30 c.c.?

(5)   10 metres of curtain material cost £32. How much do 15 metres cost?

(6) The recipe of fig 5.1 for kofte is for 6 people. Write down the ingredients for 9 people.

(7) Fig 5.2 shows the ingredients for Welsh Rarebit. Write down the quantities if 12 ounces of cheese are to be used.

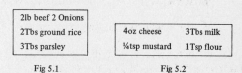

| 2lb beef 2 Onions |
| 2Tbs ground rice |
| 3Tbs parsley |

Fig 5.1

| 4oz cheese | 3Tbs milk |
| ¼tsp mustard | 1Tsp flour |

Fig 5.2

(8)   A certain journey takes 3 hours at 40 m.p.h. How long would it take at 80 m.p.h.?

(9)   When a sum of money is divided between 8 children, each gets £3. If there were only four children, how much would each get?

(10)   It takes 3 men 4 days to complete the painting of a house. How long would it take 6 men?

(11)   After a motorway is completed, the average speed increases from 40 m.p.h. to 60 m.p.h. A certain journey took 3 hours before. How long does it take now?

5.2.3 Examples for level 2

(1)   The extension E of a spring is proportional to the tension T. A tension of 5 N. gives an extension of 4 cm. Find an equation giving E in terms of T, and use it to find the extension for a load of 7 N.

Solution   E is proportional to T. So E is a multiple of T. Write:

$$E = kT$$

Here k is a constant to be found from the figures given. E = 4 when T = 5, so:

$$4 = kx5$$

$$k = {}^4/_5$$

The equation is E = ^{4}/₅xT

41

Put T = 7 into this equation:

$$E = {}^4/_5 x 7 = 5.6 \text{ cm}$$

(2)   For a fixed voltage, the current I along a wire is inversely proportional to the resistance R. I = 4 amps when R = 60 ohms. Find an equation linking I and R, and find the resistance necessary for a current of 6 amps.

Solution   Here the equation is of the form:

$$I = k/R.$$

Put in the values given;

$$4 = k/60$$

$$k = 240.$$

The equation is I = ^{240}/R.

Put I = 6 into this equation:

$$6 = {}^{240}/R$$

R = ^{240}/$_6$ = 40 ohms.

5.2.4 Exercises for level 2

(1)   y is proportional to x. y = 5 when x = 10. Find an equation giving y in terms of x, and find y when x = 5.

(2)   R is proportional to S. When S = 6, R = 18. Find (a) R when S = 24, (b) S when R = 90.

(3)   For a wire of fixed resistance, the current I is proportional to the voltage V. I = 7 amps for V = 12 volts. (a) Find the current if V = 6 volts. (b) Find the voltage necessary to transmit a current of 70 amps.

(4)   Distances on a map are proportional to distances on the ground. If 5 cm. corresponds to 20 km, what does 25 cm correspond to?

(5)   p is inversely proportional to q. For p = 7, q = 10. Find an equation linking p and q. Use this equation to find q when p = 4.

(6)   F and G are inversely proportional, and F is 6 when G = 8. (a) Find F when G = 12. (b) Find G when F = 3.

(7)   The pressure P of a fixed mass of gas is inversely proportional to the volume V. P = 50 kg/m^2 when V = 4 m^3. Find the equation which gives V in terms of P. Find the volume of the gas when the pressure is 40 kg/m^2.

(8) The wavelength 1 of certain waves is inversely proportional to the frequency f. When f = 250,000 cycles per second the wave length is 1000 metres.

(a) Find the frequency of waves with length 5,000 m.
(b) Find the length of waves with frequency 1,000,000 cycles per second.

For level 3

The statement "y is proportional to x" is often written as:

$$y \propto x$$

5.2.5 Example for level 3

The mass M of a metal sphere is proportional to the cube of the radius r. If the mass of a sphere radius 2 cm is 100 grams, find an equation giving M in terms of r. Find the radius of a sphere whose mass is 800 grams.

Solution   The formula is:

$$M \propto r^3.$$

This is equivalent to the equation:

$$M = kr^3.$$

Put M = 100 and r = 2, to obtain the equation:

$$100 = k \times 8$$

$$k = 12^1/2$$

$$\mathbf{M = 12^1/_2\, r^3}$$

Put M = 800 into the equation:

$$800 = 12^1/2\, r^3$$

$$r^3 = 64$$

$$\mathbf{r = 4\ cm.}$$

5.2.6 Exercises for level 3

(1) y is proportional to the square of x. y = 16 when x = 2.

(a) Find y in terms of x.
(b) Find y when x = 3.
(c) Find x when y = 4.

(2)   T is proportional to the square of S. When S is 3 then T is 7.

(a) Find the equation linking T and S.
(b) Find T when S is 6.
(c) Find S when T is 63.

(3)   z is proportional to the cube of w. z = 4 when w = 2. Find z when w = 5.

(4)   P is proportional to the cube of Q. P =^{1}/2 when Q = 6. Find the equation between P and Q. Find Q when P is ^{1}/54.

(5)   k is proportional to the square root of j. When j = 9 then k = 4. Find the equation linking k and j, and use it to find k when j is 49.

(6)   m is proportional to the cube root of n. For n = 8, m = 5. Find m in terms of n. Find n when m = 8.

(7)   The mass M of a cube is proportional to the cube of the length l of the side. A cube of side 4 cm^3 has mass 256 grams. Find M in terms of l. Find the mass of a cube side 3 cm, and the side of a cube which has mass 13.5 grams.

(8)   The energy of a moving body is proportional to the square of the speed. A body moving at 5 m/sec has 50 Nm of energy. Express the energy in terms of the speed. If the body has energy 200 Nm, how fast is it moving?

(9)   The power generated by an electrical circuit is proportional to the square of the current. When the current is 3 amps the power is 24 Watts. Find an expression for the power P in terms of the current I. Use your expression to find P when I = 5 amps.

(10)   The energy E stored in a spring is proportional to the square of the extension e. Express this fact using the symbol $\propto$. If the extension is 3 for an energy of 7, find an equation linking e and E. Find E when e = 1.

(11)   The period P of a simple pendulum is proportional to the square root of the length l. A pendulum of length 3.24 m. has period 6.3 seconds. Find P in terms of l. Find the period of a pendulum of length 25 cm. Find how long a pendulum should be if its period is to be 4.2 secs.

(12)   When a ship is at sea, the distance that can be seen from the top of the mast is proportional to the square root of the height of the mast. From a mast of height 16 m a look-out can see for 10 km. How far could he see from a mast of height 25 m.?

(13)   T is inversely proportional to the square of R. T = 5 when R = 7.

(a) Find the equation giving T in terms of R.
(b) Find T when R = 6.
(c) Find R when T = 125.

(14)   B is inversely proportional to the square of C. B = ^{1}/2 for C = 2. Find the expression linking B and C. Find B when C = ^{1}/2, and find C when B = 8.

(15)   The illumination from a light source is inversely proportional to the square of the distance from that light source. If a light gives 4 candle-power at a distance of 4 m, find the illumination at a distance of 6 metres.

(16)   The resistance R of a fixed length of wire is inversely proportional to the square of the radius r. When the radius is $1/2$ cm, the resistance is 3 ohms. Find R in terms of r. Find the resistance if the radius is $1/4$ cm. What radius should be chosen to give a resistance of $3/4$ ohms?

(17)   Cylindrical cans are to be made containing a fixed volume. It can be shown that the height is inversely proportional to the square of the radius. A can of height 20 cm has a base radius of 4 cm. What is the base radius of a can of height 45 cm?

(18)   The square of the time for a planet to revolve around the sun is proportional to the cube of its distance from the sun. The Earth is 93 million miles from the sun. Mars is 142 million miles from the sun. Find the length of the Martian year.

(19)   y is proportional to a power of x. The following table gives various values of y and x. Find the power, and find y when x = 6.

x	2	3	4
y	12	27	48

(20)   It is thought that T is inversely proportional to a power of S. When S is doubled T decreases by a factor of 8. What is the power? If S is divided by 3, what is T multiplied by?

COMMON ERRORS

(1)   **Ratios**   Be careful that you express the ratio in the correct order. If it is said that the ratio of A to B is 6:5, then that means that:

$$A/B = {}^6/_5.$$

Be sure not to get this the wrong way round.

(2)   **Proportional Division**
When a quantity is divided in the ratio 3:9, then the first part is $3/12$ of the whole, not $3/9$ of the whole.

(3)   **Area and Volume ratios**
When one object is a model of another, the scale of the model is the ratio of the lengths. It is not the ratio of the areas or of the volumes. The area ratio is the *square* of the length ratio, and the volume ratio is the *cube* of the length ratio.

(4)   **Proportion**

The $\propto$ symbol is often misused. The statement y $\propto$ x is not an equation. It can be converted by putting in a constant k, to give the equation y = kx.

Similarly, it is unnecessary to write y $\propto$ kx. Either write y $\propto$ x or y = kx.

(5)   **Arithmetic**
Arithmetic errors are very common. Be sure that you do the correct operation when finding the constant k. For example:

From 4 = k2 we get k = 2. (Not k = $1/2$ or k = 8)

From 4 = $k/2$ we get k = 8. (Not k = 2 or k = $1/2$)

6 Percentages

A *percentage* is a form of fraction, in which the denominator is always 100. For example, 15 percent (written 15%) is the same as the fraction 15/100.

6.1 Examples for level 1

(1)   (a) Convert 25% to a fraction.

(b) Convert $^4/_5$ to a percentage.

(c) Express 0.55 as a percentage.

(d) Express 15% as a decimal.

Solution   (a) 25% represents 25 over 100. The fraction is:

$$^{25}/_{100} = ^1/_4$$

(b) Multiply the fraction by 100:

$$^4/_5 \text{ x } 100 = ^{400}/_5 = 80\%$$

(c) Percentages are fractions of 100. To convert decimals to percentages move the decimal point two places to the right.

$$0.55 = 55\%$$

(d) Going the other way, to convert percentages to decimals move the decimal point two places to the left.

$$15\% = 0.15$$

(2)   A car window costs £50 before VAT at 15%. How much is the VAT?

Solution   15% represents $^{15}/_{100}$. The VAT is therefore:

$$^{15}/_{100} \text{ x } £50 = £7.50$$

(3)   A woman earning £8,000 is given a 6% pay rise. What is her new salary?

Solution   Her rise is $^6/_{100}$ x £8,000 = £480. Her new salary is:

$$£8,000 + 480 = £8,480$$

6.2 Exercises for level 1

(1)   Convert the following percentages to fractions:

(a) 10% (b) 20% (c) 60% (d) 5% (e) $12^1/2$% (f) 150% (g) 100%

(2)   Express the following fractions as percentages:

(a) $^1/2$ (b) $^1/4$ (c) $^3/4$ (d) $^3/5$ (e) $^1/20$ (f) $^{17}/20$ (g) $1^3/4$

(3)   Express the following decimals as percentages:

(a) 0.12 (b) 0.74 (c) 0.03 (d) 0.6 (e) 1.54 (f) 2.7

(4)   Express the following percentages as decimals:

(a) 23% (b) 54% (c) 5% (d) 20% (e) 150% (f) 104%

(5)   A meal costs £15 plus 10% service charge. How much is the service charge?

(6)   The cost of some building work is £800. VAT at 15% is added. How much is the VAT?

(7)   A house agent asks for 2% of the price of a house. If the house sells for £50,000, how much does the agent get?

(8)   During a sale a clothing shop takes 20% off the price of its suits. How much is taken off a suit costing £55?

(9)   An antique dealer reckons that she can make 25% profit on her sales. If she buys a desk for £160, how much profit will she make on it?

(10)   A brand of whisky contains 40% alcohol. How much alcohol is there in 20 litres of the whisky?

(11)   An alloy contains 30% of silver. How much silver is there in 40 kg of the alloy?

(12)   The pass mark for an exam is 45%. If the maximum number of marks is 200, how many must a candidate get to pass?

(13)   The height of a tree increases by 20%. If it was 10 feet high before, how high is it now?

(14)   A woman finds that her salary has increased by 15%. If she used to earn £14,000, how much does she earn after the increase?

(15)   After going on a diet, George finds that his weight has decreased by 25%. If he weighed 16 stone before, how much does he weigh now?

(16)   A union is asking for wage increases of 7% for all its members. If the total wage bill for a company is now £200,000, how much will it be if it agrees to the increases?

(17)   A shirt is originally priced at £12, but during a sale all prices are reduced by 20%. How much does the shirt cost now?

(18)   A pensioner invests her savings of £20,000 at 8% interest. How much income will she receive each year?

(19) A man borrows £5,000 at 12% interest. How much interest does he have to pay back each year?

(20) The Johnson household takes a daily paper costing 18p for six days of the week, and a Sunday paper costing 32p. There is a delivery charge of 5%. What is the total bill for the week?

(21) Kieron wants to pave part of his garden. He buys 32 paving stones costing £2 each and concrete costing £4.60. VAT at 15% is added to the bill. How much does he have to pay in all?

(22) Brenda's telephone bill consists of a rental charge of £16.45 together with 219 units at 5p each. VAT at 15% is added to the bill. How much in all does she have to pay?

(23) A surveyor measures a distance as 125 metres. He knows that his instruments may give an error of up to 2%.

        (a) What is the greatest possible error?

        (b) What is the greatest possible value of the distance?

        (c) What is the least possible value of the distance?

6.3 Examples for level 2

(1) In a school of 800 pupils, 450 are boys. What percentage are girls?

Solution 350 out of the 800 are girls. The fraction of girls is $350/800 = 7/16$. Convert this to a percentage:

$$7/16 \times 100 = 43.75\% \text{ are girls.}$$

(2) A dealer buys a car for £1,200 and sells it for £1,500. What is the percentage profit?

Solution The profit is £300. Express this as a percentage of the buying price:

$$300/1200 \times 100 = 25\%$$

6.4 Exercises for level 2

(1) At a concert, 650 out of an audience of 1,000 were women. What percentage were women?

(2) A man works out that of his salary of £12,000 he pays £3,000 in tax. What percentage of his salary goes in tax?

(3) A kilogram of beef contains 150 grams of fat. What percentage is lean meat?

(4) I calculate that I sleep for 8 hours each day. For what percentage of the day am I awake?

(5) Out of 10,000 candidates who take a certain exam, 5,500 will pass. What is the percentage pass rate?

(6) An exam is marked out of 250, and the pass mark is 100. What percentage mark is needed to pass?

(7) The manager of a bookshop buys books at £1.50 each and sells them at £2.50. What is the percentage profit?

(8) After going on a diet, Jane decreased her weight from 10 stone to 9 stone. What was her percentage loss in weight?

(9) The price of a car is cut from £3,600 to £2,700. What is the percentage cut?

(10) A salary increase raises my wages from £120 to £132. What is the percentage increase?

(11) The budget of a council is £130 million, of which £26 million is spent on housing. What is the percentage spent on housing?

(12) Tanya puts an apple on her kitchen scales, and finds that it weighs 50g. She knows that the scales may be inaccurate by up to 2 grams. What is the percentage error in her measurement?

6.5 Examples for level 3

(1) £400 is left in a building society at 8% compound interest. The interest is allowed to accumulate. How much is there after (a) 3 years (b) 10 years.

Solution   (a) After 1 year the interest is £400×^{8}/$_{100}$ = £32, and the total is £432.

After 2 years the interest is £432×^{8}/$_{100}$ = £34.56, and the total is £466.56

After 3 years, the interest is £466.56×^{8}/$_{100}$ = £37.32.

The amount after 3 years is £37.32 + £466.56 = £503.88

(b) The method of part (a) would require 10 lines, calculating the interest in 10 successive years. A shorter method is as follows:

When a sum of money is increased by 8%, every £100 is increased to £108. Hence the sum of money is multiplied by a factor of ^{108}/$_{100}$ = 1.08.

After 2 years the sum of money is multiplied by 1.08^2.

After 3 years the sum of money is multiplied by 1.08^3.

After 10 years the money will be multiplied by 1.08^{10}. The amount will be:

£400 x 1.08^{10} = £863.57

(2)   After a salary increase of 8%, a man's salary is £8,640. What was it before the increase?

Solution   In the previous example it was shown that adding 8% was equivalent to multiplying by 1.08.

So the salary before the rise is £8,640 *divided* by 1.08

The salary before is £8,640 ÷ 1.08 = £8,000

6.6 Exercises for level 3

(1)   £200 is invested at 10% compound interest. How much is there after 3 years?

(2)   £1,000 is invested at 5% compound interest for 3 years. How much will it amount to at the end of this period?

(3)   £400 is put into a government loan, which gives 7% compound interest. How much will there be after 4 years?

(4)   A car depreciates in value at 15% each year. If it was originally worth £5,000, how much is it worth after 3 years?

(5)   The population of a country is increasing at 3% each year. If it is 20 million now, how large will it be in 3 years time?

(6)   £600 is invested at 10% compound interest. How much will there be after (a) 2 years (b) 10 years?

(7)   £300 is invested at 12% compound interest. How much will there be after (a) 4 years (b) 18 years?

(8)   £1,200 is invested at 9% compound interest. How much will there be after (a) 2 years (b) 12 years?

(9)   The population of a country is increasing at 2%. If it is 15 million in size now, how large will it be in 20 years time?

(10)   The value of a car depreciates at 12% each year. If it cost £10,000 when new, how much will it be worth in 8 years time?

(11)   A radioactive material decays at the rate of 10% each year. If there is 3 kg now, how much will be left after (a) 2 years (b) 10 years?

(12)   I have invested my money at 8% compound interest. If I have £324 now, how much did I have one year ago?

(13)   A man can invest money at 12%. How much should he invest now, to ensure that he has £1,680 after one year?

(14)   A dealer sells a car for £2,000, making 25% profit. How much was the car bought for?

(15)   A population is increasing at 2%; if it is 20,400,000 now, how large was it a year ago?

(16)   After a 20% wage increase, a man's salary is £13,200. How much did he earn before the increase?

(17)   An antique dealer buys a chair, but has to sell at a loss of 10%. If she sold it for £450, how much did she pay for it?

(18)   After a 20% reduction in the sales the price of a pair of trousers is £20. How much did they cost before the reduction?

(19)   The price of a meal including **VAT** at 15% was £18.40. How much was the VAT?

COMMON ERRORS

(1)   **Conversion**
When converting percentages to decimals or fractions, be careful to divide by 100, not by 10. Do not write 5% as 0.5 instead of 0.05.

The same applies when converting fractions or decimals to percentages. 0.3 is 30%, not 3%.

(2)   **Percentage change**
The percentage change of a quantity is always the percentage of the *original* value, not the final value.

If an item is bought for £20 and sold for £25, the percentage profit is the percentage of the buying price not the selling price. i.e.:

$$\text{Percentage profit} = {}^{5}/_{20} \times 100 = 25\%.$$

You must be especially careful when doing reverse percentages. Suppose a dealer sells an item for £50, at a profit of 25%.

The *buying* price is £50 ÷ 1.25 = £40.

If your answer is £37.50 then you have taken 25% off the *selling* price.

7 Measures

7.1 UNITS AND CHANGE OF UNITS

There are two systems of units in common use. They are the *metric* system and the *imperial* system.

The most important units are as below. Abbreviations are in brackets.

Length

Metric
The basic unit is the *metre* (m).

1 metre = 100 centimetres (cm) = 1000 millimetres (mm).

1000 m. = 1 kilometre (km).

Imperial
12 inches (in) = 1 foot (ft). 3 feet = 1 yard (yd). 1760 yds = 1 mile.

1 inch is approximately 2.54 cm.

Weight

Metric
The basic unit is the *gram* (g).

1 gram = 1000 milligrams (mg). 1000 grams = 1 kilogram (kg).

Imperial
16 ounces (oz) = 1 pound (lb). 14 lb = 1 stone. 160 stone = 1 ton.

1 pound is approximately 454 grams.

Area

Metric
There are 100^2 = 10,000 cm^2 in 1 m^2.

10,000 m^2 = 1 hectare.

Imperial
There are 4,840 square yards in an acre.

1 acre is approximately 4,047 square metres.

Volume

Metric

The basic unit is the *litre* (l).

1 litre = 100 centilitres (cl) = 1,000 millilitres (ml).

1 millilitre = 1 cubic centimetre. (c.c. or cm^3).

Imperial

The basic units are the *pint* and the *gallon*. 8 pints = 1 gallon.

1 gallon is approximately 4.54 litres.

7.1.1 Examples for level 1

(1)   Angie measures the length of her bedroom, and finds that it is 15 feet. How many inches is that? Express the length in centimetres and metres.

Solution   There are 12 inches in 1 foot.

The room is 12 x 15 = 180 in.

Using the rate of 2.54 cm per inch:

The room is 180 x 2.54 = 457.2 cm

There are 100 cm in a metre.

The room is 457.2 ÷ 100 = 4.572 m.

(2)   At a certain time 1 pound (£1) is worth 1.5 dollars ($1.50). The chart of fig 7.1 gives the conversion of £ to $ and from $ to £.

(a) How much is £14 in dollars?

(b) How much is $30 in pounds?

(c) After a month the exchange rate changes so that £1 is worth $1.40. Draw a new line on the chart to show the new rate.

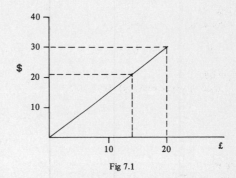

Fig 7.1

Solution   (a) Start at 14 on the £ axis, and go up until the line is reached. Now go across to the $ axis, and read off the amount.

£14 is worth $21

(b) Start at 30 on the $ axis, go across to the line and down to the £ axis. Read off the result.

$30 is worth £20

(c) The new line must start from the bottom left corner as before. £10 is now worth $14, so mark that point in and join up. The result is shown in fig 7.2.

(3)   A farm is 350 hectares. How many acres is that?

Solution   There are 10,000 m² in a hectare.

The farm contains 10,000 x 350 = 3,500,000 m².

There are 4,047 m² in an acre.

The farm contains 3,500,000 ÷ 4,047 = 865 acres.

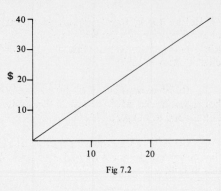

Fig 7.2

7.1.2 Exercises for level 1

(1)   Convert the following to centimetres:

(a) 3 metres (b) 3.6 m. (c) 36 mm. (d) 0.065 km. (e) 5 inches (f) 3.4 ft.

(2)   Convert the following to metres:

(a) 40 cm. (b) 213 mm. (c) 1.39 km. (d) 27 in. (e) 5 yards.

(3)   Convert the following to grams:

(a) 300 mg. (b) 15.34 kg. (c) 12.3 lb. (d) 0.23 lb. (e) 17 oz.

(4)   Convert the following to kilograms:

(a) 867 g. (b) 65 g. (c) 3 lb. (d) 4.28 lb (e) 34 oz.

(5)   Convert the following to litres:

(a) 4,000 cm³ (b) 277 cm³ (c) 3 gallons (d) 5 pints.

(6)   Convert 18 acres to square metres and to hectares.

(7) The chart of fig 7.3 converts miles to kilometres. Use the chart to find:

(a) What is 14 miles in km?

(b) How far is 26 km. in miles?

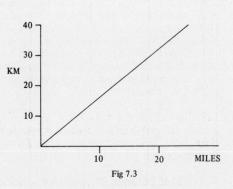

Fig 7.3

(8) The chart of fig 7.4 gives the exchange rate of the French Franc (FF) to the pound. Use the chart to find out:

(a) How many francs do I get for £5?

(b) How much is 85 FF worth in pounds?

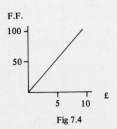

Fig 7.4

(9) Before taking her car abroad, Mavis prepares a chart to convert pounds per square inch (lb/in^2) to kilograms per square centimetre. (kg/cm^2). The chart is shown in fig 7.5.

(a) She keeps her front tyres at 28 lb/in^2 and her back tyres at 30 lb/in^2. What do these correspond to in kg/cm^2?

(b) After a day's drive she finds at a garage that one tyre has a pressure of 1.8 kg/cm^2. What is this in lb/in^2?

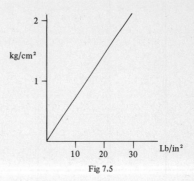

Fig 7.5

(10)   Convert to inches:

(a) 5 ft. (b) 16 yards (c) 10 cm. (d) 24.5 cm. (e) 0.34 metres.

(11)   Convert to pounds:

(a) 3 stone (b) 32 ounces (c) 4,565 g. (d) 23 kg.

(12)   Convert to acres:

(a) 44,000 m^2 (b) 500 hectares.

(13) If there are 3.75 German Marks (DM) to the pound, draw a line on the chart of fig 7.6 to convert from £ to DM. Use it to find the number of DM equal to £16, and the value in pounds of 10 DM.

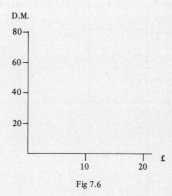

Fig 7.6

(14) There are approximately 2.2 lb in 1 kg. Draw a line on the chart of fig 7.7 to convert from kg to lb. Use it to find the weight in kg of 20 lb, and the weight in pounds corresponding to 5 kg.

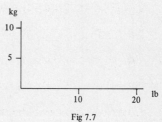

Fig 7.7

(15) Fig 7.8 shows the buying and selling rates for pounds and dollars. To change from £ to $ use the filled-in line, and to change from $ to £ use the dotted line.

(a) If I change £50, how many dollars do I get?

(b) If I now change these back to pounds, how many do I get? How much have I lost?

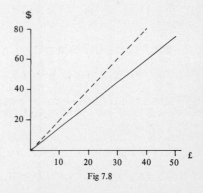

Fig 7.8

7.2 RATES AND AVERAGES

A *rate* is obtained when one quantity is divided by another. Examples of rates are:

Miles per hour: Price per kilogram: Wages per week.

The *average* speed during a journey is the total speed divided by the total time.

7.2.1 Examples for level 1

(1)   Petrol is sold at £1.64 per gallon. How much will 4 gallons cost? How much petrol can be bought for £10?

Solution   Multiply £1.64 by 4.

4 gallons cost 4 x £1.64 = £6.56

Divide £10 by 1.64.

£10 will buy 10 ÷ 1.64 = 6.10 gallons

(2)   In fig 7.9 two bottles of olive oil are shown. Which brand offers cheaper oil?

Solution Puccini brand offers 1,000 c.c. for £2.40. The price per c.c. is:

£2.40 ÷ 1,000 = 0.24 p.

The price per c.c. for Verdi oil is £1 ÷ 400 = 0.25 p.

Puccini oil is cheaper

Fig 7.9

(3)   I drive for 86 miles at 43 m.p.h., then for $1^{1}/2$ hours at 64 m.p.h. Find the total distance travelled and the total time taken. What has been my average speed?

Solution   For the second part of the journey I travelled $1^{1}/2$ x 64 = 96 miles.

The total distance was 96 + 86 = 182 miles.

The first part of the journey took 86 ÷ 43 = 2 hours.

The total time was 2 + $1^{1}/2$ = 3 $^{1}/2$ hours.

Average speed is total distance divided by total time.

The average speed was 182 ÷ $3^{1}/2$ = 52 m.p.h.

7.2.2 Exercises for level 1

(1)   Work out how much the following cost:

(a) 5 kilograms of flour at 40 pence per kg.

(b) 7 metres of rope at 15 p. per metre.

(c) 15 litres of petrol at 47.5 p. per litre.

(d) 3 m^2 of cloth at £1.20 per m^2.

(2)   I have £20 to spend. How much could I buy of the following:

(a) Petrol at £1.25 per gallon.

(b) Beef at £2.00 per pound.

(c) Cloth at £2.50 per m^2.

(d) Flex at 50 p. per metre.

(3)   (a) A 5 lb chicken cost £4.30. What is the cost per pound?

(b) 12 gallons of oil cost £72. What is the cost per gallon?

(c) To hire a tennis court for an hour costs £1.80. What is the cost per minute?

(d) 7 metres of electrical flex cost £1.33. What is the cost per metre?

(4)   A candle is 20 cm. high. It burns at a rate of 4 cm per hour.

(a) How long will it burn?

(b) After 3 hours, how much of the candle is left?

(c) How long will it be before the candle is half finished?

(5)   Roger makes a water-clock by punching a hole in the bottom of a 600 c.c. can and filling it with water. The water drips out at a steady rate, and the can is empty in 30 minutes.

(a) What is the rate of water loss in c.c. per minute?

(b) What is the rate of water loss in c.c. per second?

(c) How much water is there left after 20 minutes?

(6)   A long-playing record rotates at 33 revolutions per minute. If it rotates 726 times, for how long does it last?

(7)   James drives for 100 miles, and it takes him $2^{1}/_{2}$ hours. What has been his average speed?

(8)   I walk for two hours at 4 km. per hour, then for 1 hour at 5 km. per hour. How far have I gone? What has been my average speed?

(9) George buys 10 pencils at 12 p. each and 5 pencils at 15 p. each. What is the average price of the pencils?

(10) I buy 6 cans of orange juice containing 1 litre each, and 24 cans containing 400 c.c. each. What is the average volume of the cans?

(11) A car travels at a steady speed of 45 m.p.h. How far does it go in 3 hours? How long does it take to travel 180 miles?

(12) Dolores is driving down the M1 towards London. She sees the sign (i) at 11 a.m. and the sign (ii) at 1 p.m.

(a) What is her average speed?

(I) | LONDON 135 MILES |

(b) If she keeps up this speed, when will she be in London?

(II) | LONDON 27 MILES |

Fig 7.10

(13) A car does 12 km per litre of petrol. How many litres will it take to cover 156 km?

(14) A car travels 160 miles, and uses 5 gallons of petrol. What is the rate of petrol consumption in miles per gallon? At this rate how far will 8 gallons take the car?

(15) The charge for a telephone consists of a standing charge of £16 and 5 p. per unit used. If a subscriber uses 55 units, how much is the total bill?

(16) The standing charge for an elecricity bill is £9, and each unit of electricity cost 6 p. If a household uses 800 units, how much is the bill?

(17) The instructions for cooking a joint of beef are that it should be roasted for 30 minutes plus 20 minutes per pound. How long should a 6 lb joint stay in the oven?

(18) The entrance fee for a squash club is £25, and the annual subscription is £36. Naomi joins in May, when there is only 5 months of the squash year left to run. If the annual subscription is reduced proportionally, how much does she pay?

7.3 MONEY MATTERS

Wages are usually expressed as a rate per hour or per week. *Salaries* are usually expressed as a rate per month or per year.

Overtime consists of extra hours of work. Overtime is often paid at "time and a half", which is $1\frac{1}{2}$ times the usual rate.

Income tax is levied by the government on incomes. *Rates* are levied by local governments on the *rateable value* of property.

Value added tax (VAT) is a tax taking a fixed percentage of the price of goods.

Goods are bought on *hire-purchase* by a down payment and several monthly or weekly payments.

7.3.1 Examples for level 1

(1)   Jill sees several jobs advertised. One is for a filing clerk at £6,000 per year, one for a shop assistant at £110 per week, one for a packer at £2.50 per hour for a 40 hour week. Which job will give her the most money?

Solution   The second job will pay 52 x £110 = £5,720 per year.

The third job will pay 52 x 40 x £2.50 = £5,200 per year.

The filing-clerk job will give the most money.

(2)   A borough sets a rate of 156 pence in the pound.

(a) Mr Smith's house has a rateable value of £560. How much does he pay?

(b) Mrs Patel pays £702 in rates. What is the rateable value of her house?

(c) In the next borough Mr Jones pays £936 on a house with rateable value £650. What are the rates in this borough?

Solution   (a) Multiply £560 by 1.56.

He pays £873.60

(b) Divide £702 by 1.56.

The rateable value of her house is £450.

(c) Divide £936 by £650.

The rate is $^{936}/_{650}$ = 1.44 = 144 pence in the pound.

7.3.2 Exercises for level 1

(1)   How much per year are the following equivalent to?

(a) £65 per week   (b) £185 per week

(c) £900 per month   (d) £2.10 per hour for a 40 hour week

(e) £1.80 per hour for a 35 hour week.

(2)   How much per week are the following equal to?

(a) £5,824 per year   (b) £11,960 per year.

(3)   Rupert earns £4 per hour for proof-reading. Would he be better paid working in an office at £147.60 for a 36 hour week?

(4)   Stan earns £2.70 per hour for a 36 hour week. Overtime is paid at time and a half. How much is he paid if he works 43 hours?

(5)   Pearl earns £2.30 per hour for a 30 hour week. Overtime is at time and a half. How much is she paid for 36 hours?

      If she is paid £82.80 for a week's work, how much overtime has she done?

(6)   Alf is paid at £4.15 per hour for a 38 hour week. Overtime is at time and a half, except at weekends when it is double time. How much is he paid if he works 46 hours, of which 4 hours are at the weekend?

(7)   The rates in a certain borough are 175 pence in the pound. Find the rates for houses with the following rateable values:

(a) £660   (b) £1,020

(c) £466   (d) £553

(8)   In another borough the rates are 180 pence in the pound. Calculate the rateable value of houses which pay the following in rates:

(a) £1,656   (b) £766.8

(c) £405   (d) £943.20

(9)   Mr Robinson pays £1,333 in rates for a house with rateable value £860. What is the rate for his borough?

(10)   The total rateable value for a borough is £23,000,000. The total cost of the services the borough provides is £28,750,000. What rate should be set?

(11)   If the borough of question 10 sets a rate of 110 p. in the pound what will be the revenue?

(12)   For the purposes of income tax citizens are given a tax-free allowance, and tax is then charged at 30% on the income above that allowance. How much tax is paid by the following:

      (a) A single woman with an allowance of £3,000 and an income of £9,500.

      (b) A married man with an allowance of £5,500 and an income of £12,000.

(13)   A man with an allowance of £2,500 pays £2,400 tax at 30%. What is his income before tax?

(14)   If the rate of VAT is 15% find the tax on the following:

      (a) A radio costing £35.

      (b) Paint costing £20.

      (c) Building work costing £420.

(15)   Find the total cost for each of the following items bought on hire purchase:

(a) A television, for £100 down payment and 12 monthly payments of £24.

(b) A car for £550 down payment and 36 monthly payments of £60.

(16)   Mike sees a computer for sale. He can buy it for £65 cash, or on hire purchase for £10 down payment and 12 monthly instalments of £5. How much more will he pay if he buys it by hire purchase?

7.4 CHARTS AND TABLES

Information about times, distances, prices etc. is often presented in the form of *charts* and *tables*.

Timetables often give the times of flights or trains in the *24 hour clock*. Afternoon times with the 24 hour clock are 12 hours greater than with the 12 hour clock. For example:

17 35 (24 hour clock) = 35 minutes past 5 p.m. (12 hour clock).

7.4.1 Examples for level 1

(1)   Fig 7.11 shows the dials of the gas meter in a house. Write down the reading.

After two months an extra 3,547 units have been used. Mark on the dials the new position of the hands.

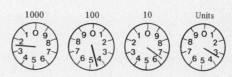

Fig 7.11

Solution   The hand of the 1,000 dial is between 2 and 3. Hence the reading is above 2,000 but below 3,000. The first digit of the reading is therefore 2. The other readings follow similarly.

The reading is 2,463

After two months the reading should be 2,463 + 3,547 = 6,010. The hands of the dials will be in the position shown in fig 7.12.

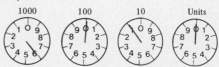

Fig 7.12

(2)   Fig 7.13 gives part of the timetable for trains from Bedford to Bletchley.

(a) If I take the first train of the day, how long will the whole journey take?

(b) I want to arrive in Fenny Stratford before 4 p.m. What is the last train I could take from Stewartby?

Miles																
0	Bedford Midland . . . . d	0554	0655	0705	0744	....	0824	0846	0957	1057	1205	....	1300	1330	1400	1513	1615
¼	Beford St. Johns . . . . d	0557	0658	0708	0747	..	0827	0849	1000	1100	1208	..	1303	1333	1403	1516	1618
3¾	Kempston Hardwick . . . . . d	0600	0707	0717	0756	..	0836	0858	1009	1109	1217	....	1312	1342	1412	1525	1627
5½	Stewartby . . . . d	0610	0711	0721	0800	..	0840	0902	1013	1113	1221	..	1316	1346	1416	1529	1631
6¾	Millbrook (Bedfordshire). . . d	0614	0715	0725	0804	....	0844	0906	1017	1117	1225	..	1320	1350	1420	1533	1635
8¼	Lidlington . . . . d	0618	0719	0729	0808	..	0848	0910	1021	1121	1229	....	1324	1354	1424	1537	1639
10	Ridgmont . . . . d	0622	0723	0733	0812	....	0852	0914	1025	1125	1233	....	1328	1358	1428	1541	1643
11¾	Aspley Guise. . . . . d	0626	0727	0737	0816	..	0856	0918	1029	1129	1237	..	1332	1402	1432	1545	1647
12¾	Woburn Sands. . . . . d	0629	0730	0740	0819	....	0859	0921	1032	1132	1240	....	1335	1405	1435	1548	1650
14¾	Bow Brickhill . . . . d	0634	0735	0745	0824	..	0905	0926	1037	1137	1245	..	1340	1410	1440	1553	1655
15¾	Fenny Stratford. . . . . . . d	0637	0738	0748	0827	....	0908	0929	1040	1140	1248	....	1343	1413	1443	1556	1658
16¾	Bletchley. . . . . d	0640	0741	0751	0830	..	0912	0932	1043	1143	1251	..	1346	1416	1446	1559	1701

Fig 7.13

Solution (a) The time taken is the difference between the departure time for Bedford Midland and the arrival time for Bletchley. The train leaves at 6 minutes to 6 and arrives at 40 minutes past 6.

Time taken = 46 minutes.

(b) 4 p.m. corresponds to 16 00 in the 24 hour clock. The last train which gets to Fenny Stratford before this arrives at 15 56.

I must catch the 15 29 at Stewartby.

(3)   Fig 7.14 show the prices of the Hovercraft service between Dover and Calais. The Newman family consists of two adults and two children aged 8 and 3: their car is 4.3 metres long. How much will it cost them to cross the Channel at Tariff C?

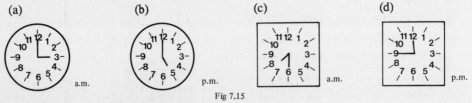

Fig 7.14

Solution The car will cost £53. Each adult costs £12, the elder child costs £6 and the younger goes free. The total cost is:

£53 + 2 x £12 + £6 = £83

7.4.2 Exercises for level 1

(1)   Write the following times in terms of the 24 hour clock:

(a) 11 a.m. (b) half past 7 in the morning (c) 4 p.m. (d) Noon. (e) Quarter to 8 in the evening (f) 5 minutes past 11 at night.

(2)   Write the following times in terms of the 12 hour clock:

(a) 08 00 (b) 10 15 (c) 12 15 (d) 16 45 (e) 20 30.

(3)   Write down in terms of the 24 hour clock the times shown:

(a)   (b)   (c)   (d)

Fig 7.15

(4)   How much time has passed between the following times:

(a) 6.30 a.m. and 11 a.m. (b) 9.45 a.m. and 1 p.m. (c) 04 45 and 16 23.

(5) Fig 7.16 shows the calendar for 1986. In that year Rachel started a holiday job on Monday 21 July and ended it on Friday 5 September.

    (a) How many weeks did she work?

    (b) In each month she was paid on the last Thursday she worked. On which days was she paid?

Calendar   **1986**

```
January              February             March
M T W T F S S        M T W T F S S        M T W T F S S
    1 2 3 4 5                    1 2                    1 2
6 7 8 9 10 11 12      3 4 5 6 7 8 9        3 4 5 6 7 8 9
13 14 15 16 17 18 19  10 11 12 13 14 15 16 10 11 12 13 14 15 16
20 21 22 23 24 25 26  17 18 19 20 21 22 23 17 18 19 20 21 22 23
27 28 29 30 31        24 25 26 27 28       24 25 26 27 28 29 30
                                           31

April                May                  June
M T W T F S S        M T W T F S S        M T W T F S S
  1 2 3 4 5 6              1 2 3 4                        1
7 8 9 10 11 12 13     5 6 7 8 9 10 11      2 3 4 5 6 7 8
14 15 16 17 18 19 20  12 13 14 15 16 17 18 9 10 11 12 13 14 15
21 22 23 24 25 26 27  19 20 21 22 23 24 25 16 17 18 19 20 21 22
28 29 30              26 27 28 29 30 31    23 24 25 26 27 28 29
                                           30

July                 August               September
M T W T F S S        M T W T F S S        M T W T F S S
  1 2 3 4 5 6                  1 2 3       1 2 3 4 5 6 7
7 8 9 10 11 12 13     4 5 6 7 8 9 10       8 9 10 11 12 13 14
14 15 16 17 18 19 20  11 12 13 14 15 16 17 15 16 17 18 19 20 21
21 22 23 24 25 26 27  18 19 20 21 22 23 24 22 23 24 25 26 27 28
28 29 30 31           25 26 27 28 29 30 31 29 30

October              November             December
M T W T F S S        M T W T F S S        M T W T F S S
      1 2 3 4 5                    1 2     1 2 3 4 5 6 7
6 7 8 9 10 11 12      3 4 5 6 7 8 9        8 9 10 11 12 13 14
13 14 15 16 17 18 19  10 11 12 13 14 15 16 15 16 17 18 19 20 21
20 21 22 23 24 25 26  17 18 19 20 21 22 23 22 23 24 25 26 27 28
27 28 29 30 31        24 25 26 27 28 29 30 29 30 31
```

(6)   In 1986, how many Sundays were there between the last Friday in January and the first Friday in May?

(7)   What is the electricity reading from the dials below?

    (a)   (b)

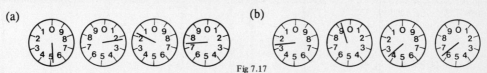

Fig 7.17

(8)   Fill in the dials below to show a reading of (a) 7354 (b) 1009.

    (a)   (b)

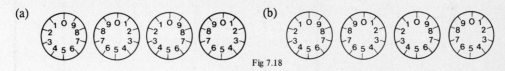

Fig 7.18

(9)   The table shows the distances between several towns in England. Use the table to find the following:

    (a) The distance between York and Leeds

    (b) The distance between Bristol and London.

    (c) The direct distance from Birmingham to York, and the distance if one goes through Leeds.

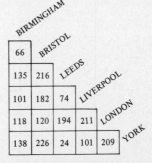

BIRMINGHAM	BRISTOL	LEEDS	LIVERPOOL	LONDON	YORK
66					
135	216				
101	182	74			
118	120	194	211		
138	226	24	101	209	

Fig 7.19

(10)   Fig 7.20 shows part of the timetable for the service from Reading to Redhill.

A

```
Reading ............. 148  d  16 16 16 55  ...  ....  17 08 17 43 18 15  ..  18 36 18 45 19 10  ...  ...  19 16 20 10 20 14  ...  21 16  ...  22 16 23 25  ...  ...
Earley .............. 148  d  16 21 ...   ....  ...  17 13 ...   ...   ..  18 50 ....  ...  ...  19 21 ...  20 21  ...  21 21  ...  22 21 23 31  ....  ..
Winnersh ............ 148  d  16 25 ...   ....  ...  17 17 ...   ...   ..  18 54 ...   ...  ...  19 25 ...  20 25  ...  21 25  ...  22 25  ...
Wokingham ........... 148  d  16 31 17 06  ...  ...  17 27 17 53 18 25  ..  19 00 ...   ...  ...  19 31 ...  20 31  ...  21 31  ...  22 31 23 38  ...
Crowthorne ..........      d  16 38 17 12  ...  ...  17 29 17 59 18 31  ..  19 08 ...   ...  ...  19 38 ...  20 38  ...  21 38  ...  22 38 23 44  ...
Sandhurst (Berks).....     d  16 42 17 15  ...  ...  17 31 18 03 18 34  ..  19 12 ...   ...  ...  19 42 ...  20 42  ...  21 42  ...  22 42 23 47  ...
Blackwater...........      d  16 46 17 19  ...  ...  17 37 18 07 18 38  ..  19 16 ...   ...  ...  19 46 ...  20 46  ...  21 46  ...  22 46 23 51  ...
Farnborough North....      d  16 51 17 24  ...  ...  17 42 18 12 18 43  ..  19 21 ...   ...  ...  19 51 ...  20 51  ...  21 51  ...  22 51 23 56  ...
North Camp .........       d  16 55 17 29  ...  ...  17 47 18 16 18 48  ..  19 25 19 31 ...  ...  19 55 20 31 20 55  ...  21 55  ...  22 55 23 59  ...
Ash ................. 148  d  17 00 ...   ...   ..  17 51 18 21 ...    ..  19 30 ...   ...  ...  20 00 ...  21 00  ...  22 00  ...  23 00 00 05  ...
Wanborough.......... 148   d  17 04 ...   ...   ..  17 56 18 25 ...    ..  19 34 ...   ...  ...  20 04 ...  21 04  ...  22 04  ...  23 04 00 09  ...
Guildford ........... 148  d  17 12 17 42  ...  ..  18 03 18 33 19 02  ..  19 07 19 42 19 45 19 42  ..  20 12 20 43 21 12  ...  22 12  ...  23 13 00 17  ...

156 London Waterloo ......a  17 57 18 25  ...  ..  18 57 19 25 19 55  ..  20 25  ...  ..   ...  20 55  ...  21 55  ...  22 55  ...  23 59 ...

156 London Waterloo ......d  16 20 ...    ...  ..  17 32 17 50 ...    ..  18 52 ...   ...   ..  19 20 ...  20 20  ...  21 20  ...  22 20 ...

Guildford ...........      d  17 17 17 43  ...  ..  18 15 18 35 19 11  ..  19 09 19 53 19 46 19 53  ..  20 18 20 50 21 17  ...  22 17  ...  23 17 ...
Shalford ............      d  17 22 ...   ...   ..  18 21 18 40 ...    ..  ...  19 58 ...  20 23 ...  21 22  ...  22 22  ...  23 22 ...
Chilworth ...........      d  17 26 ...   ...   ..  18 25 18 44 ...    ..  ..  20 02 ...   20 27 ...  21 26  ...  22 26  ...  23 26 ...
Gomshall............       d  17 34 ...   ...   ..  18 33 18 52 ...    ..  ..  20 10 ...   20 35 ...  21 34  ...  22 34  ...  23 34 ...
Dorking Town.........      d  17 41 ...   ...   ..  18 41 18 59 ...    ..  ..  20 17 ...   20 42 ...  21 41  ...  22 41  ...  23 41 ...
Deepdene ............      d  17 44 ...   ...   ..  18 44 19 02 ...    ..  ..  20 20 ...   20 45 ...  21 44  ...  22 44  ...  23 44 ...
Betchworth ..........      d  17 49 ...   ...   ..  18 49 19 07 ...    ..  ..  20 25 ...   20 50 ...  21 49  ...  22 49  ...  23 49 ...
Reigate..............      d  17 55 18 07 18 39  .. 18 55 19 13 ...  19 42  ..  20 31 20 42 20 56 ...  21 56  ...  22 55  ...  23 55 ...
Redhill..............      d  18 00 18 12 18 43  .. 19 00 19 18 19 38 19 47  ..  20 14 20 36 20 47 21 02 21 17 22 00 ...  23 00  ...  23 59 ...
```

Fig 7.20

(a) If I catch the 16 16 at Reading, how long will it take me to get to Redhill? How long does the train wait at Guildford?

(b) If I arrive at Reading station at 5 p.m., when do I arrive at Dorking?

(c) When must I arrive at Sandhurst station to ensure that I reach Betchworth by 10 in the evening?

(11)   The Cohen family, consisting of two adults, Daniel (14), Jonathan (10) and Ruth (8) are going on holiday to Greece. How much would it cost them to stay in each of the following hotels for 2 weeks?

(a)

HOTEL DIMITRA			
Cost per person	per week	per 14 days	Reductions per day
	£250	£320	2-5 yrs £1 6-11 yrs £0.60 12-16 yrs £0.40

(b)

HOTEL IANNIS			
Cost per person	7 days	14 days	Reductions per week
	£220	£295	2-5 yrs £10 6-16 yrs £8

Fig. 7.21

(12)   The table shows the cost for hiring a car. How much will it cost Kevin to hire a car for one day, in which he covers 180 miles?

DAILY RATE £10.00
FREE MILEAGE UP TO 100M
5p PER MILE EXCESS MILEAGE

Fig. 7.22

(13) Fig 7.23 shows some of the programmes on television on a certain evening.

(a) How long does the film last?

(b) Jason watches television from 7.15 for $2\frac{1}{4}$ hours. What programmes has he seen? How much of the film was there left to run?

```
6.45   News
7.30   Comedy Theatre
8.30   Zoo news
8.55   Film: "The Blob"
10.35  News
```

Fig 7.23

COMMON ERRORS

(1)   **Units**
(a) When converting from one unit to another, be careful that you go the right way.

To go from inches to centimetres *multiply* by 2.54.

To go from centimetres to inches *divide* by 2.54.

Check your answer by considering whether it is sensible. An inch is larger than a centimetre, so a distance will contain less inches than centimetres.

(b) Be very careful with area and volume units. There are 100 cm in a metre, but there are 10,000 cm^2 in a m^2. There are 1,000,000 cm^3 in a m^3.

(2)   **Averages**
The average cost of several items is the total cost divided by the number of items.

If I buy some items at 40 p. each and some other items at 60 p. each, then the average cost is *not* necessarily 50 p. To find the average cost divide the total cost by the total number of items.

(3)   **Rates**
Mistakes are often made with the quantities distance, speed and time. Make sure that you divide and multiply in the right situations, as follows:

$$\text{Speed} = \text{Distance} \div \text{Time.}$$
$$\text{Distance} = \text{Speed} \times \text{Time.}$$
$$\text{Time} = \text{Distance} \div \text{Speed.}$$

(4)   **Clocks** (a) Recall that there are 60 minutes in an hour. So 7.55 is 5 minutes to 8, not 45 minutes to 8.

(b) The 24 hour clock adds on 12 hours to afternoon times. A common mistake is to think that 16 00 represents 6 in the afternoon. The correct time is 4 in the afternoon.

8 Algebraic Expressions

Algebraic expressions contain letters as well as numbers. Some examples are:

$$(F - 32) \times {}^5/_9$$

(Which converts temperature in Fahrenheit to Centigrade)

$$\sqrt{a^2 + b^2}$$

(Which is the formula for the hypoteneuse of a right-angled triangle)

8.1 THE USE OF LETTERS FOR NUMBERS

8.1.1. Examples for level 1.

(1) We are given the following sequence of equations:

$$5 \times 1 = 5 : 5 \times 2 = 10 : 5 \times 3 = 15 : 5 \times 4 = 20 : 5 \times x = y.$$

Find the values of x and y.

Solution   The sequence 1, 2, 3, 4, x is completed by putting:

$$x = 5.$$

The value of y is given by:

$$y = 5x5 = 25$$

(2) Anne has 10 more pounds than Bill. Bill has x pounds.
(a) How much does Anne have?
(b) How much do they have in total?

Solution   (a) Add 10 to Bill's money.

Anne has x + 10 pounds

(b) Add x and x + 10.

Together they have 2x + 10 pounds

(3) What is the cost of 5 sweets, which are z pence each?

Solution   Multiply z by 5.

The cost is 5z pence

8.1.2. Exercises for level 1.

(1)   In the sequence 2, 4, 8, x, 32, what is x?

(2)   A multiplication table gives the following:

7x1 = 7 : 7x2 = 14 : 7x3 = 21 : 7xp = q.
Find p and q.

(3)   George weighs 7 kg more than Edward. If Edward weighs x kg, what does George weigh?

(4)   Mr Smith earns £x per hour. Mr Jones earns £2 more than him. How much does Mr Jones earn in an hour?

(5)   A plank 10 metres long has x metres sawn off. How much is left?

(6)   Two angles add up to 180°. One is d°. What is the other?

(7)   A machine can produce x items in one hour. How many items can it produce in 8 hours?

(8)   A room is 10 metres long and b metres broad. What is the area?

(9)   A record lasts for 20 minutes. There are two tracks on it, and the first track lasts for x minutes. How long is the second track?

(10)   A machine can produce 100 components per hour. How many can it produce in h hours?

(11)   Shoes cost £10 per pair. How much do p pairs cost?

(12)   £P is shared equally between 8 people. How much does each person get?

(13)   x and y are two numbers which add up to 10. Write down an equation in x and y.

(14)   N is 5 greater than M. Write down an equation giving N in terms of M.

(15)   Find the values of x which will make the following equations true:

(a) $x + 3 = 7$ (b) $5 + x = 19$ (c) $x \times 2 = 6$ (d) $4 \times x = 24$.

8.1.3 Examples for level 2

(1)   I buy 10 cakes at x pence each and 15 buns at y pence each. How much do I spend in all?

Solution   The cakes cost 10x p. and the buns 15y p. The total is:

$$\textbf{10x + 15y pence}$$

(2)   t bricks weigh 50 kg in all. How much do s bricks weigh?

Solution   Each brick weighs $^{50}/_t$ kg. So s bricks will weigh:

$$\mathbf{s \times {}^{50}/_t \textbf{ kg.}}$$

(3)   A car travels at x m.p.h. for 1 hour, then at y m.p.h. for 2 hours. How far has it travelled? What has been the average speed?

Solution   x×1 miles is travelled in the first stage, then y×2 in the second stage. The total distance is:

$$x + 2y \text{ miles}$$

The average speed is the total distance divided by the total time. This gives:

$$(x + 2y)/3 \text{ m.p.h.}$$

8.1.4 Exercises for level 2

(1)   A room is L metres long and B metres broad. What is the area and the perimeter?

(2)   A motorist drives for t hours at 68 m.p.h. If he covers x miles, find x in terms of t.

(3)   A dealer buys a car for £m and sells it for £n. What is the profit?

(4)   For an n-sided figure, to obtain the sum of the interior angles we multiply the number of sides by 180° and then subtract 360°. Write the sum of the angles in terms of n.

(5)   How many minutes are there between 12 o'clock and x minutes past 2 o'clock?

(6)   A darts game ends after scoring 301. If a player has scored s, how much more must he score?

(7)   The cost of hiring a television is as follows: £30 down-payment plus £20 for each month. How much does it cost to hire it for m months?

(8)   A child thinks of a number, doubles it, and then adds 5. If the original number was N, what is the final result?

(9)   In a rectangle which is s cm by r cm a square of side t cm is cut out. What is the remaining area?

(10)   At a theatre, a cheap seat costs £2.50 and an expensive one costs £4. How much do x expensive and y cheap seats cost?

(11)   I buy 10 shirts at £x each and 8 shirts at £y each. What is the total cost of the shirts? What is the average cost?

(12)   A man is paid £x per hour at standard rates, and double rates for weekends. How much did he earn in a week in which he worked 46 hours, of which 8 were at the weekend?

(13)   The angles of a triangle are x°, y° and z°. The sum of the angles in a triangle is 180°. Write x in terms of y and z.

(14)   In an innings a batsman hit x sixes, y fours and z singles. What was his total score?

8.1.5 Examples for level 3

(1)   A cricketer makes x runs in t completed innings. In the next match he is out for 43. What is his new average?

Solution   He has now made x + 43 runs in t + 1 innings. His average is now:

$$(x + 43)/(t + 1).$$

(2)   A girl runs at x m.p.h for 2 miles, then increases her speed by 2 m.p.h. and runs for a further $^1/_2$ mile. How long did she take?

Solution   The first stretch took her $^2/x$ hours. Then her speed became x + 2, and she took $^1/_2/(x + 2)$ hours. The total time is:

$$^2/x + ^1/_2/(x + 2) \text{ hours.}$$

(3)   An exam is taken by g girls and b boys. The boys score an average of p and the girls an average of q. Find the average for the whole exam.

Solution   The total mark for the girls is gq, and for the boys bp. The average is the total mark divided by the number of pupils:

$$(gq + bp)/(g + b).$$

8.1.6 Exercises for level 3

(1)   A group of 10 people have £x between them. What is their average wealth? An eleventh person with £30 joins them. What is the new average wealth?

(2)   A rectangle is twice as long as it is broad. If it is x cm long find its area and its perimeter.

(3)   A horse was bought for £h, and sold at a profit of 20%. What was the selling price?

(4)   T boys play tennis, C boys play cricket, and B boys play both.
     (a) How many play tennis but not cricket?
     (b) How many play tennis or cricket?

(5)   A line of length G cm is divided in the ratio m:n. How long is each section?

(6)   In a pipe the water flows at v cm per second. The pipe delivers L litres per second. Find the area of cross-section in terms of v and L.

(7)   The manageress of a clothing shop bought x black jackets at an average of £P each, and y blue jackets at an average of £Q each. Find the average cost of the jackets that she bought.

(8)   The area of a square side s is equal to the area of a circle of radius r. Find r in terms of s. (The area of a circle with radius r is πr^2.)

8.2 SUBSTITUTION

An algebraic expression contains letters in place of numbers. When numbers are *substituted* for the letters, the numerical value of the expression is found.

8.2.1 Examples for level 1

(1)   Find the value of $(F - 32) \times ^5/9$ when F = 50.

Solution   Put F = 50 in the expression. The result is:

$$(50 - 32) \times ^5/9 = 18 \times ^5/9 = 10$$

(2)   It is given that p = 3q + 2r.
Find p when q = 5 and r = 7.

Solution   Put q = 5 and r = 7 in the expression.

$$p = 3x5 + 2x7 = 15 + 14 = 29$$

8.2.2 Exercises for level 1

(1)   If M = 3N + 2, find M when N = 5.

(2)   Current, voltage and resistance are related by the formula I = $^V/_R$. Find I when V = 200 and R = 40.

(3)   In the formula v = at + b, find v when a = 2, b = 30, t = 6.

(4)   If A = lb, find A when l = 3 and b = 6.

(5)   If A = $^1/_2$bh, find A when b = 4 and h = 7.

(6)   In the formula R = 2h - 4, find R when h = 3.

(7)   If Q = $^1/_4$(10 - x), find Q when x = 2.

(8)   The area of a circle is given by A = πr^2. Find A when r = 7, taking $\pi = ^{22}/_7$.

(9)   From the formula I = $^{PRT}/_{100}$ find I when P = 100, R = 8 and T = 4.

(10)   If R = $^{5s}/_t$ find R when s = 30 and t = 25.

(11)   The area of a trapezium is given by the formula:

$$A = H(x + y)/2.$$

Find A when H = 7, x = 3 and y = 5.

(12)   Find the value of 3X + 4Y when X = 2 and Y = 5.

8.2.3 Examples for level 2

(1)   In the formula r = 2s - 7, find s when r = 3.

Solution   Put r = 3 into the formula:

$$3 = 2s - 7$$

Add 7 to both sides, and then divide by 2.

$$s = (3 + 7)/2 = 5$$

(2)   It is given that $1/R = 1/V - 1/U$. Find R when V = 3 and U = 4.

Solution   Put V = 3 and U = 4.

$$1/R = 1/3 - 1/4 = 1/12$$

R = 12

(3)   From the formula $c = \sqrt{a^2 + b^2}$ find c when a = 7 and b = 24.

Solution   Put a = 7 and b = 24.

$$c = \sqrt{7^2 + 24^2}$$

$$\mathbf{c = \sqrt{49 + 576} = \sqrt{625} = 25}$$

8.2.4 Exercises for level 2

(1)   In the formula $s = ut + 1/2at^2$, find s when u = 5, t = 3 and a = 7.

(2)   The quantities P, V, T and k are related by PVT = k.

      (a) Find k when P = 10, V = 120, T = 50.
      (b) With this value of k, find V when P = 15 and T = 80.

(3)   Illumination is given by the formula $I = c/d^2$. If c = 72 and d = 2 find I.

(4)   The sum of the first n numbers is $1/2n(n + 1)$. Find the sum of the first 15 numbers.

(5)   The volume of a cone is given by $V = \pi r^2(h/3)$. Taking π to be $22/7$, find the volume of a cone for which r = 3 and h = 14.

(6)   In the formula T = a + (n - 1)d, find T for a = 3, n = 10, d = 8.

    If T = 33, a = 3, d = 5 then find n.

(7)   Use the formula $y = (1 + x)/(1 - x)$ to find y when $x = 1/2$.

(8)   C is given in terms of x and a by the formula $C = a(1 + x)^3$. Find C when a = 5 and x = 2.

(9)   r, u and v are related by $1/r = 1/u - 1/v$. Find r when u = 5 and v = 6.

(10)   If $H = n(n + 1) + 1/4$ find H when n = 3.

(11)   Use the formula $A = 1/2(a + b)h$ to find A when a = 7, b = 3 and h = 7. Find b if A = 9, h = 6 and a = 5.

(12)   If $a = \sqrt{25 - h^2}$, find a when h = 3.

(13)   d is given in terms of t and m by the formula: d = t(m - 1)/m. Find d when t = 6 and m = 3.

(14)   The surface area of a cylinder is given by the formula: $A = 2\pi r(r + 1)$. Use $\pi = 22/7$ to find A when r = 3 and l = 4.

8.3 THE USE OF BRACKETS. Levels 2 and 3

The order in which algebraic operations are done is important. Taking powers (i.e. squaring or cubing etc) is done first, then multiplying or dividing, then adding or subtracting. If the operations are to be done in a different order then they must be put in *brackets*.

The process of eliminating brackets from an expression is called *expansion*.

8.3.1 Examples for level 2

(1)   Expand and simplify $3(x + y)$

Solution   When the brackets are removed both terms are multiplied by 3.

$$3x + 3y$$

(2)   Expand and simplify $2(3x + 5) + 4(2x - 7)$.

Solution   Multiply through to obtain:

$$6x + 10 + 8x - 28$$

The x terms can be collected together, and the number terms can be collected:

$$14x - 18$$

(3)   Expand and simplify $(x - 5)(x + 8)$.

Solution   Hold the first pair of brackets, and multiply out:

$$(x - 5)x + (x - 5)8$$

Multiplying out again,

$$x^2 - 5x + 8x - 40$$

$$x^2 + 3x - 40$$

8.3.2 Exercises for level 2

Expand the following and simplify as far as possible:

(1) $5(x + 7)$ (2) $4(a - b)$

(3) $7(2x + y)$ (4) $2(3x - 5y)$

(5) $3(x - 4) + 5(x + 6)$ (6) $7(y - 3) - 2(3 + y)$

(7) $6(z - 3) - 8(z - 5)$ (8) $5(2w + 3) + 9(3w - 4)$

(9) $3(x + y) + 4(x - y)$ (10) $2(a + b) + 3(2a + 3b)$

(11) $4(x + y) - 2(x + y)$

(12) $3(c - 2d) - 2(c + 4d)$

(13) $3(5r - 2s) - 2(2r - 3s)$

(14) $4(p + 2q) - 2(p - 3q)$

(15) $(x + 1)(x + 5)$

(16) $(y + 3)(y + 7)$

(17) $(p + 3)(p - 4)$

(18) $(z - 4)(z + 6)$

(19) $(w - 7)(w - 11)$

(20) $(q - 2)(q - 5)$

8.3.3 Examples for level 3

(1) Expand out $(a + b)(a - b)$

Solution   Hold the second pair of brackets, and multiply through:

$$a(a - b) + b(a - b)$$

Multiply through again:

$$a^2 - ab + ba - b^2$$

The two middle terms cancel, leaving:

$$\mathbf{a^2 - b^2}$$

(Note: this formula is known as the *difference of two squares*)

(2) Expand and simplify $(2x - 3)(3x - 5)$

Solution   Hold the second pair of brackets and multiply out:

$$2x(3x - 5) - 3(3x - 5)$$

$$6x^2 - 10x - 9x + 15$$

$$\mathbf{6x^2 - 19x + 15}$$

(3) Expand and simplify $(p + q)(r + s)$

Solution   Hold the second bracket and multiply out the first.

$$p(r + s) + q(r + s)$$

Now multiply through to obtain:

$$\mathbf{pr + ps + qr + qs}$$

8.3.4 Exercises for level 3

Multiply out and simplify:

(1) $(x + y)(x - y)$

(2) $(2x + y)(2x - y)$

(3) $(3p - 2q)(3p + 2q)$

(4) $(6r - 4s)(3r + 2s)$

(5) $(x + 8)(3x - 5)$

(6) $(3y - 4)(2y + 7)$

(7) $(3a - 2b)(3a + 2b)$

(8) $(2w + 5z)(3w - 2z)$

(9) $(x + 3)^2$

(10) $(2y - 3)^2$

(11) $(x + 2y)^2$

(12) $(3z - 2w)^2$

(13) $(a + b)(c + d)$

(14) $(x - y)(z + w)$

(15) $(2t + s)(3r + p)$

(16) $(f - 3g)(j - 5k)$

(17) $(2y + 3x)(5z - 7w)$

(18) $(5s + 1)(3r - 2t)$

8.4 CHANGING THE SUBJECT OF A FORMULA. Levels 2 and 3

The *subject* of a formula is the letter which is expressed in terms of the other letters. When the subject is changed the formula is re-arranged so that a different letter is expressed in terms of the others.

8.4.1 Examples for level 2

(1)   If $y = mx + c$, put x in terms of y, m and c.

Solution   Subtract c from both sides, to obtain $y - c = mx$. Then divide both sides by m, to obtain:

$$(y - c)/m = x.$$

(2)   If $C = (F - 32) \times {}^5/_9$, express F in terms of C.

Solution   Multiply both sides by 9, to get $9C = (F - 32) \times 5$.

Divide both sides by 5, to get ${}^{9C}/_5 = F - 32$.

Add 32 to get:

$${}^{9C}/_5 + 32 = F.$$

(3)   Make h the subject of the formula $V = \pi r^2 h/3$.

Solution   Multiply both sides by 3, to obtain $3V = \pi r^2 h$.

Divide both sides by π and by r^2, to get:

$${}^{3V}/_{\pi r^2} = h.$$

8.4.2 Exercises for level 2

In each of these questions, change the subject to the letter in brackets.

(1) $x = 3y + 2$ (y)

(2) $a = b - c + d$ (d)

(3) $F = {}^{9C}/5 + 32$ (C)

(4) $rt = 3sp$ (s)

(5) $y = 3ax^2$ (a)

(6) $j = 59 - mn$ (m)

(7) $b = {}^1/2(c + a)$ (a)

(8) $d + 3 = {}^b/2a$ (b)

(9) $V = 4\pi r^3/3$ (π)

(10) $(X + 2Z)/5 = 4Y$ (X)

(11) $2(x + y) = 4z$ (x)

(12) $5p(q - r) = 7$ (p)

8.4.3 Examples for level 3

(1)   Make r the subject of the formula $V = \pi r^2 h/3$.

Solution   Multiply by 3 and divide by πh, to get ${}^{3V}/\pi h = r^2$. Now square root both sides to get:

$$\sqrt{{}^{3V}/\pi\mathbf{h}} = \mathbf{r}.$$

(2)   Make x the subject in the equation: $ax + b = cx - d$.

Solution   Add d to both sides, and subtract ax from both sides. The equation is now $b + d = cx - ax$. Write this as

$$b + d = (c - a)x.$$

Now divide both sides by (c-a). The final answer is:

$$\mathbf{x = (b + d)/(c - a)}.$$

8.4.4 Exercises for level 3

In each of these equations, change the subject to the letter in brackets.

(1) $y = 4ax^2$ (x)

(2) $V = 4\pi r^3/3$ (r)

(3) $T + R = S^2 - 3$ (S)

(4) $1 = \sqrt{h^2 + 4r^2}$ (h)

(5) $h + 3 = 3m\sqrt{n}$ (n)

(6) $t = 2\pi\sqrt{l/g}$ (l)

(7) $a = (c + 1)/(b - 3)$ (b)

(8) ${}^1/f = {}^1/u - {}^1/v$ (u)

(9) $mg - T = ma$ (m)

(10) $x - 5a = 3x + 2b$ (x)

(11) $ap + bq = cp + dq$ (q)

(12) $a(x + 3) = b(7 - x)$ (x)

(13) $m = R/(R + r)$ (R)

(14) $y = (x + 1)/(x - 3)$ (x)

8.5 FACTORIZATION. Levels 2 and 3

Factorization is the opposite operation to expansion. When an expression is factorized, it is written as the product of two or more simpler expressions.

8.5.1 Examples for level 2

(1)   Factorize 3xt - 6xs.

Solution   The letter x can be taken outside a pair of brackets.

$$x(3t - 6s)$$

Both the terms inside the brackets are divisible by 3. Take 3 outside the brackets:

$$3x(t - 2s)$$

(2)   Factorize $x^2 + 3x$.

Solution   Here both terms contain an x. The x can be taken outside a pair of brackets to obtain:

$$x(x + 3)$$

8.5.2 Exercises for level 2

Factorize the following as far as possible

(1) xy + xz

(2) pq - 3q

(3) 3s + rs

(4) qt + rt

(5) rt - 5rs

(6) 3xy - 4xz

(7) 2x + 4y

(8) 3t - 9r

(9) 2rs + 4rq

(10) 9xy - 3zy

(11) 6fg + 9fh

(12) 7t + 21tq

(13) $a^2 + ab$

(14) $z - z^2$

(15) $5x^2 - 15x$

(16) $21y^2 - 14y$

(17) $8z^2 + 4cz$

(18) $3d + 9cd^2$

For level 3

Recall that factorization is the opposite of expansion. In section 8.3 there were the following expansions:

(A) $(a + b)(a - b) = a^2 - b^2$ (Example 1 of 8.3.3)

(B) $(x - 5)(x + 8) = x^2 + 3x - 40$ (Example 3 of 8.3.1)

(C) $(2x - 3)(3x - 5) = 6x^2 - 19x + 15$ (Example 2 of 8.3.3)

(D) $(p + q)(r + s) = pr + ps + qr + qs$ (Example 3 of 8.3.3)

Each of these expansions corresponds to a factorization.

8.5.3 Examples for level 3

(1) Factorize $4t^2 - s^2$

Solution   This expression consists of one square subtracted from another. It matches the right-hand side of (A) above. Hence it can be factorized to the left-hand side of (A). The a term corresponds to 2t, and the b term corresponds to s.

$$(2t + s)(2t - s)$$

(2) Factorize $y^2 + 5y + 6.$

Solution   This expression is similar to the right-hand side of (B) above. To match the left-hand side of (B) the factorization must be:

$$(y + a)(y + b)$$

Here a and b are numbers whose product ab is 6 and whose sum (a + b) is 5. Try the factors of 6 until you find the pair whose sum is 5. The factors must be 2 and 3:

$$(y + 2)(y + 3)$$

(3) Factorize $6x^2 - x - 2.$

Solution   This is similar to the right-hand side of (C). When it is factorized it must be:

$$(ax + b)(cx - d)$$

Here a, b, c and d are numbers such that ac = 6 and bd = 2. The first pair could be 6x1 or 3x2, and the second pair must be 2x1. Try the possible arrangements until the correct values are found:

$$(2x + 1)(3x - 2)$$

(4) Factorize xa - yb + ya -xb.

Solution   This is similar to the right-hand side of (D). Re-arrange the terms so that the first two have a common factor:

$$xa + ya - yb - xb$$

$$(x + y)a -(y + x)b$$

Now there is a common factor of $(x + y)$. Divide by it:

$$(x + y)(a - b)$$

8.5.4 Exercises for level 3

Factorize the following as far as possible:

(1) $p^2 - q^2$ (2) $t^2 - 1$

(3) $4 - n^2$ (4) $q^2 - 9$

(5) $9 - s^2$ (6) $7 - 7s^2$

(7) $2 - 8x^2$ (8) $9 - 81a^2$

(9) $49t^2 - 9s^2$ (10) $16m^2 - 25n^2$

(11) $y^2 - {}^1/_4$ (12) ${}^1/_9 - z^2$

(13) $x^2 + 3x + 2$ (14) $x^2 + 4x + 4$

(15) $x^2 + 7x + 12$ (16) $x^2 - 11x + 24$

(17) $x^2 + x - 6$ (18) $y^2 - y - 12$

(19) $z^2 + 5z - 6$ (20) $w^2 - 4w - 12$

(21) $x^2 - 5x - 24$ (22) $p^2 + 7p - 30$

(23) $2x^2 + 5x - 3$ (24) $3x^2 - 10x - 8$

(25) $5y^2 + 9y + 4$ (26) $6x^2 + 35x - 6$

(27) $6a^2 + 11ab + 5b^2$ (28) $6p^2 - 5pq - 6q^2$

(29) $st - sq + rt - rq$ (30) $ax + by - ay - bx$

(31) $6st - 4s + 3t - 2$ (32) $2xz - 3y - 6x + yz$

8.6 ALGEBRAIC FRACTIONS. Level 3

The rules for numerical fractions apply to algebraic fractions. When fractions are multiplied, their numerators are multiplied and their denominators are multiplied. When fractions are added, they must first be put over a common denominator.

8.6.1 Examples for level 3

(1)   Express as a single fraction: (a) $\left(\frac{2a}{5}\right) x \left(\frac{3}{x}\right)$

Solution   Multiply together the tops and bottoms.

$$\frac{6a}{5x}$$

(2)   Simplify $2a^2/3 \div {}^{4a}/9$

Solution   To divide by ${}^{4a}/9$ is the same thing as to multiply by ${}^9/4a$. The expression can be written:

$$2a^2/3 \times {}^9/4a = 18a^2/12a$$

6a divides both top and bottom. Cancel to obtain:

$$3a/2$$

(3)   Simplify ${}^{2p}/3 + {}^{5p}/2.$

Solution   Here both terms must be put over the common denominator of 6.

$$4p/6 + {}^{15}p/6$$

$$(4p + 15p)/6 = {}^{19}p/6$$

(4)   Express as a single fraction: $1/(x-1) + 2/(x + 3)$

Solution   The common denominator of this fraction is $(x - 1)(x + 3)$. Put both terms over this:

$$\frac{(x + 3) + 2(x - 1)}{(x - 1)(x + 3)}$$

$$\frac{3x + 1}{(x - 1)(x + 3)}$$

8.6.2 Exercises for level 3

Express the following as single fractions:

(1) ${}^a/b \times {}^c/d$

(2) $\left({}^{3x}/5\right) \times \left({}^{4x}/6\right)$

(3) ${}^{2x}/3y + {}^y/2x$

(4) ${}^x/y \div {}^z/w$

(5) ${}^{7ab}/4c \times {}^{2c}/ab$

(6) ${}^{5pq}/r \div {}^{10p}/qr$

(7) ${}^x/3 + {}^{2x}/5$

(8) ${}^{3a}/2 - {}^a/4$

(9) ${}^r/4 + (r - 1)/3$

(10) $3(x - 2)/4 - {}^{5x}/3$

(11) $(7b - 1)/2 + (3b + 1)/3$

(12) $4(c - 1)/3 - 3(c + 1)/4$

(13) ${}^2/x + {}^3/xy$

(14) ${}^3/a + {}^4/bc$

(15) $\left({}^1/3x\right) \times \left(1/(x - 1)\right)$

(16) $3z/(z - 1) \times 2z/(z + 1)$

(17) $(x + 2)/3 \div (x - 1)/2$

(18) $2/(1 - y) \div 3/(2 - y)$

(19) ${}^3/x + 4/(x + 1)$

(20) ${}^5/z - 2/(1 - z)$

(21) $z/3 + 2/(z + 1)$ (22) $5/(3y + 2) + {}^4/y$

(23) $2/(y - 1) + 1/(y + 3)$ (24) $1/(3w + 1) - 1/(w + 2)$

(25) $3/(2p - 1) + 5/(2p + 3)$ (26) $7/(3 - 2q) - 2/(1 - q)$

COMMON ERRORS.

(1) Collecting.

It is incorrect to add together terms unless they are of the same sort, with the same letters occurring.

$5a^2b$ and $7a^2b$ can be added to give $12a^2b$.

But $5a^2b + 7ab^2$ cannot be simplified in this way.

(2) Brackets.

(a) Do not ignore brackets:

$$a(b + c) \neq ab + c$$

(b) Do not forget to put brackets in when they are necessary.

From $y/4 = x + 3$ go to $y = 4(x + 3)$.

Do not go to $y = 4x + 3$. This gives incorrect values.

(3) Multiplying negative terms.

Remember the basic rules of:

$$minus\ times\ minus\ is\ plus \quad - \times - = +$$
$$plus\ times\ plus\ is\ plus \quad + \times + = +$$
$$minus\ times\ plus\ is\ minus \quad - \times + = -$$

It is quite common to get these wrong, when expanding out something like $(x - y)(p - q)$. The yq term is positive.

(4) Expansion.

When multiplying out brackets, *all* the terms in the first pair must multiply *all* the terms in the second pair. Do not just multiply together the first two and the last two.

$$(a + b)(x + y) \neq ax + by.$$

The correct expansion is $ax + ay + bx + by$

In particular, be careful when expanding $(a + b)^2$ or $(a - b)^2$

$$(a + b)^2 = a^2 + 2ab + b^2 \neq a^2 + b^2$$

$$(a - b)^2 = a^2 - 2ab + b^2 \neq a^2 - b^2$$

(5) **Factorizing.**

(a) The following are common mistakes when factorizing:

$$49x^2 - 1 \neq (49x - 1)(49x + 1)$$

$$x^2 + 5x + 6 \neq (x + 2x)(x + 3x)$$

(b) $x^2 + a^2$ is not the difference of two squares. It cannot be factorized at all.

(c) Make sure you factorize fully. It is true that:

$$6st + 3sr = s(6t + 3r)$$

But a further factor of 3 can be taken out.

(6) **Changing the subject of a formula.**

(a) Frequently the wrong operation is done when making x the subject of a formula.

If a term has been *added* to x, we must *subtract* it away.

If a term is *multiplying* x, then we must *divide* by it.

In each case we do the opposite operation to get x on its own. We must also be sure to do them in the correct order.

(b) When x is made the subject of a formula, it must be expressed as a function of the other terms. It cannot occur on both sides of the equation. If the answer is of the form $x = 5c + 7xy$, then x has not been made the subject.

(7) **Algebraic Fractions.**

(a) When multiplying a fraction by another term, it is incorrect to multiply both numerator and denominator by that term.

$$5 \times {}^a/_b \neq {}^{5a}/_{5b}$$

It may be helpful to think of 5 as ${}^5/_1$, which gives the correct expression:

$${}^5/_1 \times {}^a/_b = (5xa)/(1xb) = {}^{5a}/_b$$

(b) Wrong addition of fractions. When adding fractions, we cannot simply add the numerators and the denominators.

$${}^a/_b + {}^c/_d \neq (a + c)/(b + d)$$

(c) Sums of fractions cannot be simply inverted.

From ${}^1/_f = {}^1/_h + {}^1/_k$ it does not follow that $f = h + k$.

9 Equations

An *equation* states that one mathematical expression is equal to another. Examples of equations are:

$$x + 3 = 56; \quad y^2 + 3y - 2 = 0; \quad c = b + 4d^2$$

Usually an equation contains an *unknown* . To solve the equation is to find the value or values of this unknown.

There is one basic rule to obey when solving equations.

Do to the left what you do to the right.

So if 5 is added to the left hand side of an equation, 5 must be added to the right also. If the left is divided by 4.7, the right also must be divided by 4.7.

9.1 EQUATIONS WITH ONE UNKNOWN. Levels 2 and 3

9.1.1 Examples for level 2

(1)   Solve the equation $x - 2 = 4$

Solution   Add 2 to both sides.

$$x = 6$$

Check the answer by substituting it into the original equation.

$$x - 2 = 6 - 2 = 4$$

So the result is correct.

(2)   Solve the equation $3x + 13 = 40$

Solution   Subtract 13 from both sides:

$$3x = 27$$

Divide both sides by 3:

$$x = 9$$

9.1.2 Exercises for level 2

Solve the following equations:

(1) x + 3 = 5

(2) y - 4 = 13

(3) 12 - x = 5

(4) 7 - x = 9

(5) 13 = 7 + x

(6) 9 = 13 - z

(7) 2x + 5 = 11

(8) 5z - 3 = 7

(9) 5 - 3a = 2

(10) 7 + 3b = 10

(11) 9 = 3 + 2x

(12) 5 = 17 - 3z

(13) 5(x - 2) = 15

(14) 4(b + 3) = 16

(15) $z/7 = 21$

(16) $^{12}/y = 4$

9.1.3 Examples for level 3

(1)   Solve the equation 3y - 4 = 2y + 1

Solution   Subtract 2y from both sides.

$$3y - 2y - 4 = 1$$

$$y - 4 = 1$$

Add 4 to both sides,

$$\mathbf{y = 1 + 4 = 5}$$

Check the answer by putting y = 5 in the original equation.

Left hand side = 3x5 - 4 = 11
Right hand side = 2x5 + 1 = 11

(2)   Solve the equation $p/3 + (p - 1)/5 = 11$

Solution   Multiply both sides by 15.

$$5p + 3p - 3 = 165$$

Collect together terms.

$$8p = 168$$

$$\mathbf{p = 21}$$

9.1.4 Exercises for level 3

Solve the following equations:

(1) $2x - 10 = 18$

(2) $3y/4 + 1 = 28$

(3) $5m + 1 = 19 - 4m$

(4) $n/7 - n/5 = 4$

(5) $3z + 5 = z - 7$

(6) $8(2 - x) = 64$

(7) $5g = 4(18 - g)$

(8) $x/2 + x/3 + x/4 = 26$

(9) $3(x + 4) = 5(x - 2)$

(10) $w/4 - 2w/5 = 9$

(11) $y/2 + (y + 1)/3 = 25$

(12) $(z - 3)/4 = (2 + z)/3$

(13) $(3q + 6)/8 - (2q - 5)/4 = 6$

(14) $y/(y + 1) = 1/y + 1$

(15) $(p - 1)/p = p/(p - 3)$

(16) $v/(v + 2) - 1/(v - 2) = 1$

9.2 SIMULTANEOUS EQUATIONS. Level 3

Simultaneous equations consist of two equations with two unknowns.

9.2.1 Examples for level 3

(1)
$$[1]\ \ x + 4y = 8$$
$$[2]\ \ x + 5y = 12$$

Solution   The way to solve this is to get rid of the x term.

Subtract [1] from [2]:

$$[2] - [1] = [3]\ \ \ y = 12 - 8 = 4$$

The value of y can be put into either of the original equations to find x.

$$x + 4 \times 4 = 8$$

$$x = 8 - 16 = -8$$

Equation [1] was used to find the value of x. The working can be checked by putting the values in [2]

$$-8 + 5 \times 4 = -8 + 20 = 12.\ \ \ \text{Correct.}$$

(2)
$$[1]\ \ 3x - 5y = 1$$
$$[2]\ \ 2x + 7y = 11$$

Solution Here both equations must be multiplied by a number in order to match either the x terms or the y terms. If the y terms are matched:

$$[3] = 7 \times [1] \quad 21x - 35y = 7$$

$$[4] = 5 \times [2] \quad 10x + 35y = 55$$

Eliminate y by adding.

$$[5] = [3] + [4] \quad 31x = 62$$

$$x = 2$$

Substitute in either [1] or [2] to find that

$$y = 1.$$

9.2.2 Exercises for level 3

Solve the following pairs of simultaneous equations.

(1) $x + y = 3$
     $x - y = 7$

(2) $2z + w = 3$
     $3z + w = 9$

(3) $3x + 2y = 4$
     $5x - 2y = 12$

(4) $3x + y = 7$
     $3x - 4y = -13$

(5) $5q - 3p = 11$
     $q + p = 7$

(6) $7x + 5y = 4$
     $3x + 2y = 7$

(7) $3z - 5w = 8$
     $z + 3w = 12$

(8) $4q - 3p = 5$
     $7q - 6p = 5$

(9) $3x + 2y = 14$
     $x - 3y = 1$

(10) $2x - 4y = 14$
      $3x + y = 7$

(11) $3x + 2y = 20$
      $-5x + 3y = 11$

(12) $4s + 5r = 1$
      $3r - 2s = -17$

(13) $2x + 3y = 13$
      $7x - 5y = -1$

(14) $6p - 7q = -1$
      $5p - 4q = 12$

(15) $x/6 + y/3 = 8$
      $x/4 - y/9 = 1$

(16) $m/6 + 2n/3 = 6$
      $2n/5 - m/10 = 2$

9.3 QUADRATIC EQUATIONS. Level 3

If an equation contains the square of the unknown then it is called a *quadratic equation*.

Sometimes the equation can be solved by factorizing. Sometimes the *formula* must be used.

$$\text{For } ax^2 + bx + c = 0,$$

$$x = \frac{-b +/- \sqrt{b^2 - 4ac}}{2a}$$

9.3.1 Examples for level 3

(1) Solve the equation $x^2 - 3x + 2 = 0$

Solution: This quadratic factorizes, to become:

$$(x - 2)(x - 1) = 0.$$

Either $x - 2 = 0$ or $x - 1 = 0$. There are the two solutions:

$$x = 2 \text{ or } x = 1$$

(2) Solve the equation $y^2 + 17y = 60$

Solution The left hand side should not be factorized until there is a 0 on the right hand side. So subtract 60 from both sides:

$$y^2 + 17y - 60 = 0.$$

This factorizes, to become:

$$(y + 20)(y - 3) = 0$$

$$y = -20 \text{ or } y = 3$$

(3) Solve the equation $z^2 + 13z + 2 = 0$, giving your answer to 3 decimal places.

Solution: This equation does not factorize. So the formula must be used. Here a = 1, b = 13, c = 2. Substitute these values in to obtain:

$$x = \frac{-13 +/- \sqrt{13^2 - 4 \times 1 \times 2}}{2 \times 1}$$

$$x = \frac{-13 +/- \sqrt{161}}{2}$$

$$z = -0.156 \text{ or } z = -12.844$$

9.3.2 Exercises for level 3

Solve the following equations:

(1) $x^2 + 5x - 6 = 0$

(2) $y^2 + 7y + 12 = 0$

(3) $x^2 - 10y + 9 = 0$

(4) $x^2 + 19x + 60 = 0$

(5) $x^2 - 28x - 60 = 0$

(6) $4x^2 - 4x - 3 = 0$

(7) $z^2 - 3z = 10$

(8) $p^2 = 8p - 15$

(9) $(x - 2)(x + 1) = 18$

(10) $y/(y - 3) = y - 4$

(11) $^6/y = y + 5$

(12) $z + {}^{12}/z = 7$

(13) $4x^2 + 10x + 3 = 0$

(14) $y^2 - 5y + 2 = 0$

(15) $z^2 + 3z - 5 = 0$

(16) $q^2 - q - 3 = 0$

(17) $x + {}^1/x = 3$

(18) $(y + 1)(y + 2) = 28$

9.4 PROBLEMS WHICH LEAD TO EQUATIONS. Levels 2 and 3

9.4.1 Examples for level 2

(1)   The rental for a telephone is £12 per quarter, and the call charges are 2 p per unit. A bill was £25. Let x be the number of units used. Form an equation in x and solve it.

Solution   The total cost is $2x + 1200$ pence. The bill is £25, so this gives the equation:

$$2x + 1200 = 2500$$

This can be easily solved to give $x = 650$.

(2)   A man is three times as old as his son. In 12 years time he will be twice as old. Let the son be aged y years. Find an equation in y and solve it.

Solution   The father is 3y years. In 12 years time the son will be $y + 12$, and the father will be $3y + 12$.

The father will then be twice as old. This gives the equation:

$$3y + 12 = 2(y + 12)$$

Expand out and solve to obtain $y = 12$ years.

9.4.2 Exercises for level 2

(1)   The hire of a car is £20 per day plus a charge of 10 p per mile. Letting x be the number of miles covered, find an expression giving the charge in terms of x.
If the total cost of hiring was £37, how many miles were covered?

(2)   I think of a number x, double it and then add 5. If the result is 37, what was the original number?

(3)   Two partners in a business agree that the first partner shall receive £1500 more of the profits than the second partner.
    Letting x be the amount received by the second partner, find the total profit in terms of x.
    If the total profit was £27,700, how much did each partner get?

(4)   The three angles of a triangle are x°, 3x - 30°, 60 - x°. Find x, given that the sum of the angles in a triangle is 180°.

(5)   Jane tells John: "Think of a number, multiply it by 3, subtract 7 and then take away the number you first thought of."
    If John thought of x, what number has he ended with?
    If John ends with 11, what was his original number?

(6)   A car is driven at x mph for $1^1/2$ hours, then at x + 20 mph for $^1/2$ hour. The total distance covered is 90 miles. Find x.

(7)   A woman is six times as old as her daughter. 3 years ago she was 11 times as old. Let the daughter's age be y years. Form an equation in y.
    Solve the equation to find how old the woman is.

(8)   The sum of three consecutive numbers is 441. Letting n be the smallest of the three, write down the other numbers in terms of n. Find an equation in terms of n, and solve it.

(9)   A girl buys x magazines at 50p each and x - 3 magazines at 40p each. She spends £6 in all. Find x.

(10)   Money is left in a will, so that Anne gets twice as much as Brian, and Brian gets three times as much as Christine. If Christine gets £x, express the amounts received by Brian and Anne in terms of x. If the total amount left was £1800, find how much each received.

(11)   Adrian takes a holiday job at £x per week. After 4 weeks his pay increases by £10 per week. He then works for 5 more weeks. Write in terms of x the total amount he has earned.
    If the total is £617, find x.

(12)   11 children have an average age of 10 years. Write down the total of all their ages.
    Another child aged x years joins them, and the new average is 10 years 1 month. How old is the new child?

9.4.3 Examples for level 3

(1)   3 kg of plums and 2 kg of apples cost £3.40, while 1 kg of plums and 4 kg of apples cost £2.80. Find the cost of 1 kg of apples.

Solution   Here there are two unknown. Let x p. be the price of 1 kg of apples, and y p. be the price of 1 kg of plums. There are two bits of information, which lead to the simultaneous equations;

$$3y + 2x = 340$$
$$y + 4x = 280$$

Solve to obtain x = 50.

1 Kg of apples costs 50p

(2)   A rectangle is 3 cm longer than it is wide. The area is 108 cm^2. Find the length.

Solution   Let the length be x cm. The width must be x - 3 cm. The area is 108, which gives the equation:

$$x(x - 3) = 108$$

Expand and re-arrange to get:

$$x^2 - 3x - 108 = 0$$

This has solutions x = 12 or x = -9. The negative answer is not meaningful here. The answer is;

$$x = 12 \text{ cm}$$

9.4.4 Exercises for level 3

(1)   £1200 is invested, £x at 9% and £y at 11%. If the total income is £122, find how much was invested at 9%.

(2)   At a concert 500 tickets were sold: the cheaper ones cost £2.50 and the more expensive ones £4.50. The total receipts were £1,610. Let x be the number of cheap tickets, and y the number of expensive tickets. Form two equations in x and y, and hence find how many cheap tickets were sold.

(3)   In a pub, brandy is 10 p. more expensive than whisky. A round of 3 brandies and 5 whiskies cost £5.26. Let x be the cost of a whisky, and y the cost of a brandy. Find the cost of a whisky.

(4)   3 pounds of butter at x p per pound and 4 pints of milk at y p per pint cost £3.84. 5 pounds of butter and 7 pints of milk cost £6.48. Form two equations in x and y and solve them.

(5)   Chuck saves money by putting every 50 p and every 20 p coin he receives in a box. After a while he find that he has 54 coins, amounting to £17.10. How many 50 p coins does he have?

(6)   The manager of a shoe shop spends £S on shoes at £10 a pair, and £(S + 100) on boots at £15 a pair. If she obtains 50 pairs in all, find S.

(7)   A train runs for x + 10 miles at 40 mph, then for x - 60 miles at 60 mph. The total time for the run was 3 hours. Find x.

(8)   The sum of two numbers is 27, and the difference is 17. Find the numbers.

(9)   A two digit number xy is increased by 27 when its digits are reversed to yx. The sum of the digits is 13. Find the number.

(10)   From Leeds to London is 180 miles. A car travels from London to Leeds at 60 m.p.h., and starting at the same time a lorry travels from Leeds to London at 30 m.p.h. How far from London do they pass each other?

(11)   A positive number x is 156 less than its square. Find the number.

(12)   The area of a circle is 3π greater than the perimeter. Find the radius.

(13)   The sides of a right-angled triangle are 2x + 1 cm, 2x - 1 cm, and x cm. Find x.

(14)   The area of a rectangle is 72 cm^2, and its perimeter is 34 cm. Find the lengths of the sides.

(15)   A piece of wire of length 80 cm is cut in two, and the segments are each bent into a square. The total area is 208 cm^2. Find the sides of the squares.

(16)   A boy bats in 6 matches. In the 7th match he makes 8 runs and his average decreases by 1. How many runs has he scored in all?

(17)   A man can row in still water at 4 m.p.h. He rows upstream for 5 miles, then downstream for the same distance. The total time taken was 3 hours. Find the speed of the current.

COMMON ERRORS

(1)   Arithmetic

All the common errors which arise in basic algebra are very easy to make when solving equations. Be very careful when multiplying out brackets, or when multiplying two negative numbers together. See Chapter 8 for a list of some common algebraic errors.

(2)   Equations in one unknown

The unknown cannot appear on both the sides of the answer. If your solution is of the form $x = 3 - 2x$ then you have not solved the equation at all, you have merely re-written it in another form. The solution of the equation must be a statement which gives us the *numerical value* of the unknown.

(3)   Solving equations

It is common to do the wrong operations when solving equations. The following examples show which operations should be used:

For $x + 3 = 5$, *subtract* 3 from both sides.
For $y - 4 = 7$, *add* 4 to both sides.
For $3z = 9$, *divide* by 3.
For $^w/_8 = 10$ *multiply* by 8.

(4)   Simultaneous equations

(a) Do not try to get by with one equation for two unknowns. Nearly always you must have two equations for two unknowns.

(b) Incorrect algebra. One common error is to disobey the basic rule of equations. In other words, to go from:

$$2x + 3y = 4$$
$$\text{to} \quad 4x + 6y = 4$$

Here the right hand side was not multiplied by the same thing as the left hand side. The correct step would be:

$$4x + 6y = 8$$

(c) Another common error occurs when adding or subtracting equations. Remember the basic rule that *two minuses make a plus* . So subtracting a negative number is the same as adding a positive one. For example, in the situation:

[1] $3x + 4y = 7$
[2] $3x - 5y = 2$

When we subtract [2] from [1] we obtain:

$$4y - (-5y) = 5$$
$$4y + 5y = 5$$

(5)   Quadratic equations

(a) Incorrect assignment of a, b, c. For the equation:

$x^2 - 2x + 2 = 0$, then a = 1 (not 0) and b = -2 (not 2).

(b) Incorrect algebra. Mistakes are often made when b or c are negative. Remember that "minus times minus is plus". For the equation:

$x^2 - 3x - 8 = 0$, both b and c are negative.

The formula gives:

$$x = \frac{+3 +/- \sqrt{9 - 4 \times 1 \times (-8)}}{2}$$

$$x = \frac{3 +/- \sqrt{41}}{2}$$

$$x = 4.70 \text{ or } x = -1.70.$$

(c) Errors in calculation. When using the formula for a quadratic, remember that the 2a term must divide the whole of the top line. So press the ⊟ button on your calculator before you divide by 2a.

(6)   Problem solving

(a) It is important to state clearly what your letters stand for. It is not enough to say: *Let x be the apples*. We must make it clear whether x is the price, or the weight, or the number of the apples. In the example above we stated that *1 kg of apples costs x pence*.

(b) It is very common to make mistakes with the quantities *Distance, Time* and *Speed*.

Travel for t hours at s m.p.h., the distance covered is ts miles.
Travel for d miles at s m.p.h., the time taken is d/s hours.

If in doubt, think of an actual journey. If we travel 100 miles at 40 m.p.h., it will take us $2^1/2$ hours.

10 Inequalities

10.1 ORDER

10.1.1 Examples for level 1

(1)   Find the largest of 0.12, $^1/_8$, $^2/_{15}$.

Solution   Write all three as decimals.

0.12: $^1/_8 = 0.125$: $^2/_{15} = 0.13333$.

The largest is $^2/_{15}$.

(2)   Which of the sizes of soap, shown in fig 10.1, offers better value?

Solution   The cost of soap in the economy size packet is:

$^{40}/_{150} = 0.267$ p. per gram.

The cost of soap in the family size packet is:

$^{60}/_{250} = 0.24$ p. per gram.

The Family size offers better value.

10.1.2 Exercises for level 1

(1)   Write down the larger of $6^3/_4$ and 6.7.

(2)   Write down the smaller of 4.34 and $4^1/_3$.

(3)   Arrange in increasing order:

$$1.3, 1^1/_4. 1.42, ^9/_7.$$

(4)   Arrange in decreasing order:

$$0.0011, 0, ^1/_{100}, \text{a millionth, a thousandth.}$$

(5)   Which of the jam-jars of fig 10.2 offers better value?

Fig 10.2

92

(6)   Jason has a roll of 28 holiday snaps he wishes to print. Which is the cheaper way of paying for them? (See fig 10.3).

| £3.20 per roll |
| or 12p per print |

Fig 10.3

(7)   Which is the cheaper way for Linda to buy her motor-cycle? How much will it cost her? (See fig 10.4)

| 1. £50 deposit, and 24 payments of £15 |
| 2. £100 deposit, and 12 payments of £25 |

Fig 10.4

(8)   The first half of a journey took $1/2$ hour and was 20 miles in distance: the second half took 20 minutes and was 17 miles long. During which half was the average speed greater?

(9)   Five days a week Kevin takes the tube to and from work. His fares are 60 p. each way. Is it worth his while to buy a weekly ticket at £5.40?

(10)   Samantha sees two jobs advertised. One pays £73 per week, the other £3,800 per year. Allowing 52 weeks in the year, which job is better paid?

(11)   George is paid £125 per week. He is offered either an extra £17 per week or a 12% pay rise. Which should he take?

(12)   While on holiday in Holland, Mary sees perfume advertised at 50 Guilders for 20 c.c. She knows that in England the same perfume will cost £16 for 25 c.c. If there are 3.9 guilders to the pound, is the perfume cheaper in Holland or in England?

(13)   A tool set contains spanners whose sizes go up in steps of an eighth of an inch up to 1 inch. Arrange the sizes in order.

$$1, \ ^7/_8, \ ^1/_4, \ ^1/_2, \ ^3/_8, \ ^5/_8, \ ^3/_4, \ ^1/_8.$$

10.2 INEQUALITIES IN ONE VARIABLE. Levels 2 and 3

a < b means that a is *less than* b.

a≤ b means that a is *less than or equal to* b.

a > b means that a is *greater* than b.

a≥b means that a is *greater or equal to* b.

10.2.1 Exercises for level 2

(1)   Which of the following are true?

(a) 3 < 5 (b) 5 < 3.6 (c) $^1/_3 \leqslant 0.3$ (d) $10^2 > 90$ (e) $^1/_8 \leqslant 0.125$.

(2)   Arrange the following in increasing order, using the < sign.

$$4.6, \ 4.45, \ ^{19}/_4, \ ^{14}/_3.$$

(3)   Arrange the following in decreasing order, using the > sign.

$^1/8$, 0, 0.2, $-^1/2$, $-^1/8$, $-^3/17$, $-^1/4$.

(4)   Find a whole number x for which x > 3 and 2x < 12.

(5)   Find an integer y for which y ⩽ -3 and y > -5.

(6)   List the positive integers for which:

(a) $0 < x < 6$ (b) $2 \leqslant x \leqslant 7$ (c) $13 < x \leqslant 17$ (d) $7 \leqslant x < 10$

(7)   List the whole numbers for which:

(a) $-2 < x < 2$ (b) $-7 \leqslant x \leqslant -3$ (c) $-1 \geqslant x > -5$ (d) $4 > x \geqslant -2$

For level 3

An *inequality* looks like an equation, with the = sign replaced by one of the order signs. (<, >, ⩽, ⩾).

When solving inequalities, the same basic rule applies as to equations.

Do to the left what you do to the right

There is one additional rule for the solving of inequalities. When multiplying or dividing by a *negative* number, the inequality sign changes round.

If a < b, then -2a > -2b

If c ⩽ d, then $-^1/2c \geqslant -^1/2d$

10.2.2 Examples for level 3

(1)   Find the solution set of the inequality 3x - 2 < 4, giving your answer in the form {x : x ... }.

Solution   Add 2 to both sides.

$$3x < 6$$

Divide both sides by 3. As 3 is a positive number, the inequality sign stays the same way round.

$$x < 2$$

The solution set is:

$$\{x : x < 2\}$$

(2)   Solve the inequality $2 - 3x \leqslant 2x + 17$, illustrating your answer on a number line.

Solution   Subtract 2x from both sides, and subtract 2 from both sides.

$$-5x \leqslant 15$$

Divide by -5, being sure to reverse the inequality sign.

$$x \geqslant \textbf{-5}$$

This is shown on the number line as follows:

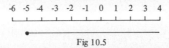

Fig 10.5

(The dot at the left of the line is filled in, to indicate that -5 is included.)

(3)   The hire of a car is £20 per day and 10 p per mile. A woman drives for x miles and her bill is less than £35. Form an inequality in x and solve it.

Solution   The total bill, of fixed charge plus mileage charge, comes to:

$$20 + {}^{x}/10.$$

This must be less than £35.

$$20 + {}^{x}/10 < 35$$

$${}^{x}/10 < 15$$

$$x < \textbf{150}$$

10.2.3 Exercises for level 3

(1)   Find the solution sets for the following inequalities, giving your answer in the form $\{x : x \dots \}$:

(a) $x - 3 < 4$ (b) $2x + 5 > 7$

(c) $3x + 1 \geqslant 10$ (d) $2x + 11 < 3$

(e) $12 - x < 5$ (f) $13 - 3x \geqslant 4$

(2)   Solve the following inequalities, illustrating your answers on number lines:

(a) $x + 2 < 2x - 10$ (b) $4x + 5 < x + 14$

(c) $x + 1 < {}^{1}/2x + 4$ (d) $(x-3)/2 \leqslant 5$

(e) $(x + 2)/3 \geqslant (1-x)/2$ (f) ${}^{1}/2(3x + 5) < 2(x - 7)$

(3) Find the solution sets for the following inequalities, both in the form {x : ... } and on the number line:

(a) $3 < x + 1 < 7$

(b) $7 < 2x - 1 < 11$

(c) $-4 \leqslant 3x + 2 \leqslant 20$

(d) $1 \geqslant 3 - x \geqslant -8$

(e) $7 \leqslant 3 - 2x < 13$

(f) $8 > 2 - 3x \geqslant -10$

(4) Express in the form {x : ... } the inequalities represented by the following number lines:

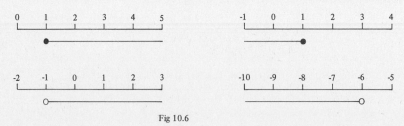

Fig 10.6

(5) A boy buys x apples at 10 p. each. If he has at most £1 to spend, form an inequality in x and solve it.

(6) Cakes cost 25 p. each. Jane buys y of them, and spends less than £1.50. Form an inequality in y and solve it.

(7) A car does 30 miles per gallon, and there are z gallons of petrol in the tank. If the car goes for 100 miles without re-fuelling, find an inequality in z and solve it.

(8) I have hired a car for 5 hours, and the maximum speed is 50 m.p.h. If I drive for m miles, form an inequality in m and solve it.

(9) A man buys x stamps at 17 p. and 2x stamps at 12 p. He spends less than £5. Form an inequality in x and solve it.

(10) The rental charge of a phone is £15 per quarter, and it costs 5 p. per unit. A man uses x units, and his bill is less than £50. Form an inequality in x and solve it.

(11) A room is 2 metres longer than it is wide. Its width is x metres, and the perimeter is less than 60 m. Form an inequality in x and solve it.

10.3 INEQUALITIES IN 2 VARIABLES. Level 3

The solution set for simultaneous inequalities in 2 variables consists of a region of a plane. Hence the solution set can be illustrated on graph paper.

10.3.1 Examples for level 3

(1)   Illustrate on graph paper the solution set of the inequality $2x + 3y \leqslant 6$.

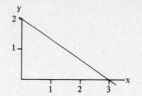

Fig 10.7

Solution   First draw the line $2x + 3y = 6$.

The line goes through the points $(0,2)$ and $(3,0)$. Fig 10.7.

Because the inequality requires $2x + 3y$ to be less than 6, the solution set is the region below the line. Fig 10.8.

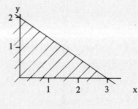

Fig 10.8

(2)   Illustrate on graph paper the solution set of the simultaneous inequalities

$x > \frac{1}{2}, y > \frac{1}{2}, 4x + 3y < 12, 3x + 4y < 12$.

Find the pairs of whole numbers which satisfy these inequalities.

Solution   First draw the lines corresponding to the 4 equalities. Fig 10.9.

The region required is above the line $y = \frac{1}{2}$, to the right of the line $x = \frac{1}{2}$, and below the two slanting lines.

The pairs of point within the required region are:

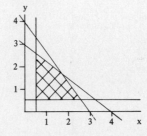

Fig 10.9

(1,1), (1,2), (2,1)

(3)   A man has £10 to spend on cigars. Superbas cost 50 p each, and Grandiosos cost 75 p. If he buys x Superbas and y Grandiosos, obtain an inequality in x and y. Illustrate this inequality on a graph.

Solution   The total amount he spends on cigars is given by:

$50x + 75y$ p.

This must be less than 1,000 p.

$50x + 75y \leqslant 1000$

$2x + 3y \leqslant 40$

This is illustrated in fig 10.10.

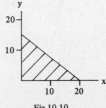

Fig 10.10

10.3.2 Exercises for level 3

(1) Illustrate on graph paper the regions corresponding to the solution sets of the following:

(a) $x + y \leq 1$ (b) $2x + 4y \leq 8$

(c) $3x + 5y \leq 15$ (d) $4x + 3y \leq 12$

(e) $x + 2y \geq 4$ (f) $3x + 2y \geq 12$

(g) $x - y \geq 0$ (h) $2x - 3y \leq 6$

(i) $y - 3x \leq 6$ (j) $5y - 4x \geq 20$

(2) Illustrate the following inequalities on graph paper. Each group should be illustrated simultaneously.

(a) $x \geq 0, y \geq 0, x + y \leq 6$ (b) $x + y \geq 3, x \leq 7, y \leq 8$

(c) $x \geq 0, y \geq 0, 2x + 3y \leq 1$ (d) $x \geq 0, y \geq 0, x + 3y \leq 9, 2x + y \leq 10$

(e) $y \geq 0, y \leq x, 3x + 4y \geq 12, 3x + 4y \leq 18$

(f) $x - y \geq 1, x - y \leq 3, x + y \leq 8, x + y \geq 4$

(3) Describe the shaded regions of fig 10.11 by inequalities in x and y.

(a) 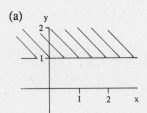   (b)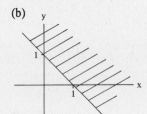

(c) 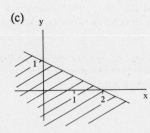   (d)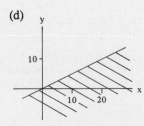

Fig 10.11

(4) A boy has £1 to spend on sweets. Liquorice bars cost 10 p each, and toffee bars cost 15 p. Say he buys x liquorice bars and y toffee bars. Find an inequality in x and y, and illustrate it on a graph.

(5)   If, in exercise 4, the boy's mother tells him not to buy more than 8 bars, find another inequality in x and y. Illustrate the new inequality on the same graph.

(6)   A patient can be prescribed either pills or tablets. Tablets contain 3 units of vitamin A, and pills contain 4 units. The daily intake must be at least 15 units. If the patient takes x tablets and y pills, obtain an inequality in x and y and illustrate it on a graph.

(7)   A factory manager is to buy two sorts of machines. Type 1 occupies 3 m^2 and costs £200, type 2 occupies 2 m^2 and costs £1000. There is 40 m^2 of space available, and £9000 to spend. Let x be the number of type 1 machines, and y the number of type 2 machines
    Show that the restrictions can be expressed by the inequalities:

$$3x + 2y \leqslant 40 \text{ and } x + 5y \leqslant 45.$$

    Illustrate these inequalities on a graph.

(8)   In a car park there is 6000 m^2 of land available. A car space occupies 15 m^2, and a lorry space 30 m^2. There must be space for at least 50 lorries, and there must be at least twice as many car spaces as lorry spaces. Let c be the number of car spaces, and l the number of lorry spaces.
    Show that the restrictions can be expressed by the inequalities:

$$c + 2l \leqslant 400, l \geqslant 50 \text{ and } c \geqslant 2l.$$

    Illustrate these inequalities on a graph.

(9)   Food A contains 4 units of protein and 5 units of starch per kg, and food B contains 6 units of protein and 3 units of starch per kg. The minimum daily intake of protein is 16 units, and the minimum daily intake of starch is 11 units. Let X be the number of kg of A to be eaten, and Y the number of kg of B. Obtain inequalities in X and Y, and illustrate them on a graph.

(10)   A mail order firm must deliver 900 parcels using a lorry which takes 150 at a time or a van which takes 80 at a time. Each journey costs £5 for the lorry and £4 for the van. The total cost cannot exceed £44 and the van must make more journeys than the lorry. Let x be the number of van journeys, and y the number of lorry journeys. Find inequalities in x and y and illustrate them on a graph.

(11)   An aircraft has 600 m^2 of cabin space, and can carry 5000 kg of luggage. An economy passenger gets 3 m^2 of space, and is allowed 20 kg of luggage. A first class passenger gets 4 m^2 of space, and is allowed 50 kg of luggage. There must be space for at least 50 economy passengers. Let x be the number of first class seats, and y the number of second class seats. Obtain inequalities in x and y, and illustrate them on a graph.

(12)   A grocer has 240 kg of Indian tea, and 250 of China tea. He can make two blends, which are mixed in the ratio of 1:4 and 5:1. Let x be the weight of the first blend, and y the weight of the second blend. Obtain inequalities in x and y, and illustrate them on a graph.

COMMON ERRORS

(1)   **Solving inequalities**
        Do not forget the rule that when multiplying or dividing by a negative number the inequality sign must be reversed. If you ignore this rule you will get exactly the wrong answer, i.e. you will get:

$$x < 3 \text{ instead of } x > 3.$$

(2) Two-dimensional inequalities

(a) Be sure to draw the correct line. The line $2x + 3y = 6$ goes through $(3,0)$ and $(0,2)$.

But do not draw the vertical and horizontal lines through these points. Fig 10.12 shows the correct and incorrect diagrams.

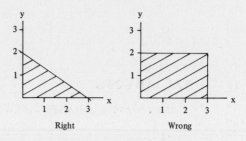

(b) Be sure that you take the correct side of the line. If in doubt, take a test point. If your inequality is $2x + 3y \leqslant 6$, then test by putting $x = 1$ and $y = 1$.

Right   Wrong

Fig 10.12

$2x1 + 3x1 \leqslant 6$, so the point $(1,1)$ must lie on the shaded part of your diagram.

(c) When dealing with a problem, make sure that you obtain the full inequalities.

If a boy buys x items at 20 p, and y items at 30 p, out of a budget of 60 p, then the inequality is $20x + 30y \leqslant 60$.

Do not just put $x \leqslant 3$ and $y \leqslant 2$. These do not express the full situation.

11 Plane Figures

11.1 ANGLES AND LINES

The angles round a point add up to 360°.
Note that in fig 11.1:

$$130° + 140° + 90° = 360°$$

Fig 11.1

The angles along a line add up to 180°.
Note that in fig 11.2:

$$75° + 105° = 180°$$

An angle of 90° is a *right-angle*. Lines which meet at a right-angle are *perpendicular*.

An angle between 0° and 90° is *acute*.

An angle between 90° and 180° is *obtuse*.

Lines which are in the same direction are *parallel*.

Fig 11.2

right angle

acute

11.1.1 Examples for level 1

(1)   Find the angle labelled x° in fig 11.4.

Solution The angles must add up to 360°.
Subtract the other angles from 360°:

$$x° = 360° - 140° - 100° = 120°$$

(2)   The lines of fig 11.5 are perpendicular. Find the angle labelled x.

Solution   The angles along the line must add up to 180°. The two angles labelled x must add up to 90°.

$$x = {}^{90°}/_2 = 45°$$

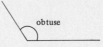

obtuse

Fig 11.3

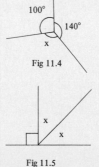

100°

140°

x

Fig 11.4

x

x

Fig 11.5

11.1.2 Exercises for level 1

(1)   Find the unknown angles of the diagrams in fig 11.6.

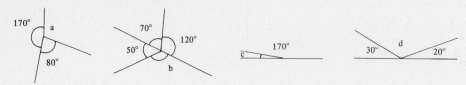

Fig 11.6

(2)   Find the values of the unknowns in the diagrams in fig 11.7.

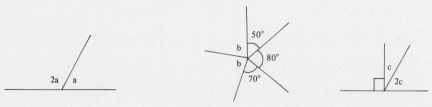

Fig 11.7

(3)   One turn corresponds to 360°. How many degrees correspond to:

(a) Half a turn (b) A quarter turn (c) 3 turns.

(4)   How many turns correspond to:

(a) 720° (b) 3,600° (c) 540° (d) 120° ?

(5)   A record turns at 45 revolutions per minute. How many degrees does it turn through in 1 second?

(6)   What is the angle between the hands of a clock at the following times:

(a) Three o'clock (b) Six o'clock (c) Five o'clock ?

(7)   In ten minutes, through what angle does the minute hand of a clock pass?

(8)   If the hour hand of a clock passes through 120°, how much time has passed?

(9)   Find the unknown angles in fig 11.8.

(10)   Two lines meet as shown in fig 11.9. Write down the acute angle between them and the obtuse angle between them.

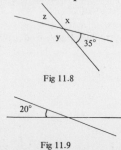

Fig 11.8

Fig 11.9

102

For level 2

When two lines cross, the *opposite* angles are equal.

A line which crosses two parallel lines is called a *transversal*. The angles on either side of the transversal are equal, and are called *alternate* angles.

The angles on the same side of the transversal are equal and are called *corresponding* angles.

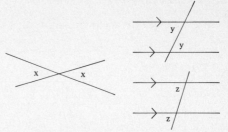

Fig 11.10

11.1.3 Examples for level 2

(1)   Find the angles labelled x, y and z in fig 11.11.

Solution   The angle labelled x is opposite to the 50° angle.

$$x = 50°$$

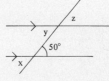

Fig 11.11

The angle labelled y is alternate to the angle of 50°.

$$y = 50°$$

The angle labelled z is corresponding to the angle of 50°.

$$z = 50°$$

(2)   In fig 11.12 AC is extended to D, and CE is parallel to AB. Find the sum of the angles in △ABC.

Solution   <BAC = <ECD (Corresponding angles)

<ABC = <BCE (Alternate angles)

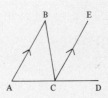

Fig 11.12

So the sum of the angles of △ ABC is the sum of the angles BCA, BCE and ECD. These are the angles along a straight line.

The angles of △ABC add up to 180°

11.1.4 Exercises for level 2

(1)   Find the angles labelled a, b, c, d, e, f, g in fig 11.13.

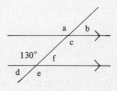

Fig 11.13

(2)   Find the unknown angles in fig 11.14.

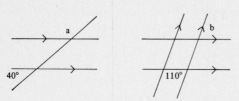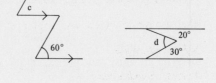

Fig 11.14

(3)   AB and CD are parallel lines, and BD is a transversal. The bisectors of the angles ABD and CBD meet at E. If <BDC = 80°, find <BED.

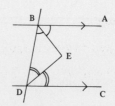

Fig 11.15

(4)   As far as possible label the angles of fig 11.16. What is the relationship between a and b?

Fig 11.16

(5)   Find the angle x of fig 11.17.

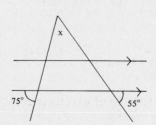

Fig 11.17

(6)   In fig 11.18, which of the diagrams contain a pair of parallel lines?

(a)   (b)   (c)   (d)

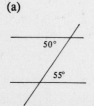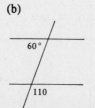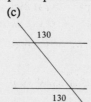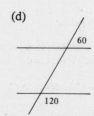

Fig 11.18

11.2 POLYGONS

A plane figure with straight sides is called a *polygon*.

The names of polygons are as follows:

3 sides: a *triangle*

4 sides: a *quadrilateral*

5 sides: a *pentagon*

6 sides: a *hexagon*

If all the angles and sides of a polygon are equal, then it is *regular*.

A pattern of polygons is called a *tessellation* or *tiling*. Fig 11.19.

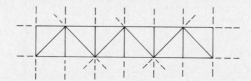

Fig 11.19

The angles of a triangle add up to 180°.

A triangle with two equal sides is *isosceles*. The base angles of an isosceles triangle are also equal.

A triangle with all its sides equal is *equilateral*. All the angles of an equilateral triangle are equal to 60°.

A triangle, one of whose angles is 90°, is a *right-angled* triangle.

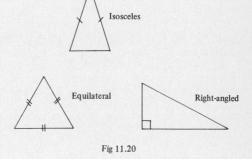

Fig 11.20

The angles of a quadrilateral add up to 360°.

A quadrilateral with one pair of parallel sides is a *trapezium*.

A quadrilateral with two pairs of parallel sides is a *parallelogram*.

A quadrilateral with all sides equal is a *rhombus*.

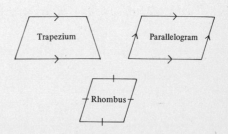

A quadrilateral with all angles equal to 90° is a *rectangle*.

A quadrilateral with all sides equal and all angles equal is a *square*.

A quadrilateral with two pairs of adjacent sides equal is a *kite*.

Fig 11.21

If a figure is identical on both sides of a line, then it is *symmetrical* about that line.

Fig 11.22

11.2.1 Examples for level 1

(1)   Find the angle labelled x in fig 11.23.

Solution The two base angles add up to 140°. The triangle is isosceles, so they must both be equal to x.

$$x = {}^{140°}/_2 = 70°$$

Fig 11.23

(2)   Find the angle labelled y in fig 11.24.

Solution To find the fourth angle of the quadrilateral, subtract the other three angles from 360°:

$$360° - 110° - 70° - 80° = 100°.$$

Subtract this angle from 180°.
$$y = 180° - 100° = 80°$$

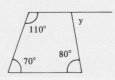

Fig 11.24

11.2.2 Exercises for level 1

(1)   Find the unknown angles in the triangles of fig 11.25.

Fig 11.25

(2)   Find the unknown angles of the quadrilaterals of fig 11.26.

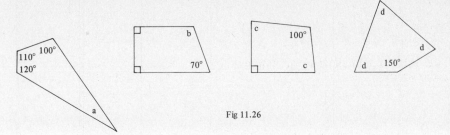

Fig 11.26

(3)   Find the exterior angles in the diagrams of fig 11.27.

Fig 11.27

(4)   Find the unknown lengths in the triangles of fig 11.28.

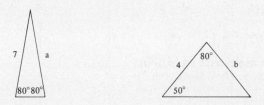

Fig 11.28

(5)   Find the angles x and y in fig 11.29.

(6)   Find the angle z in fig 11.30.

(7)   ABC is a triangle in which AB = AC and <BCA = 75°. Find <BAC.

(8)   DEF is a triangle in which <DFE = <DEF and <FDE = 58°. Find the unknown angles.

(9)   In the triangle PQR, PQ = PR and <PQR = 60°. What sort of triangle is Δ PQR?

(10)   Draw dotted lines to show the lines of symmetry of the shapes in fig 11.31.

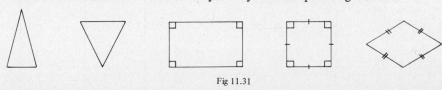

Fig 11.31

107

(11)   Complete the tessellations of fig 11.32.

(a) 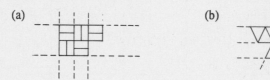   (b)

Fig 11.32

(12)   Draw a quadrilateral which is both a rectangle and a rhombus. What is the name for this figure?

(13)   Draw a quadrilateral which is both a parallelogram and a kite. What is the name for this figure?

For level 2

If a polygon has n sides, the sum of its *interior* angles is 180n - 360°.

The sum of its *exterior* angles is 360°.

Fig 11.33

11.2.3 Examples for level 2

(1)   Find the sum of the interior angles of a decagon. (10 sides). If the figure is regular, how big is each exterior angle?

Solution   Put n = 10 into the formula above.

The sum of the interior angles is 180x10 - 360 = 1,440°

If the figure is regular, then the exterior angles are equal.

Each exterior angle is $^{360}/_{10} = 36°$

(2)   A tesselation is made out of regular n-sided polygons. Three tiles meet at each vertex. Find n.

Solution   A vertex is shown in fig 11.34.
The interior angle of each polygon is $^{360}/_3 = 120°$.

The internal angle of a regular *hexagon* is equal to 120°.

The number of sides is 6

11.2.4 Exercises for level 2

Fig 11.34

(1)   Find the sum of the interior angles of:

(a) A pentagon (b) a hexagon (c) an octagon (8 sides) (d) a 20 sided figure.

(2)   Find the value of each interior angle of a regular figure with:

(a) 5 sides (b) 10 sides (c) 12 sides (d) 30 sides.

(3)   Find the value of each exterior angle of a regular figure with:

(a) 8 sides (b) 12 sides (c) 18 sides (d) 30 sides.

(4)   The sum of the interior angles of a polygon is 900°. How many sides does it have?

(5)   The sum of the interior angles of a polygon is 1260°. How many sides does it have?

(6)   Each exterior angle of a regular polygon is 72°. How many sides does it have?

(7)   Each interior angle of a regular polygon is 150°. How many sides does it have?

(8)   The angles of a pentagon are $x°$, $x° + 10°$, $x + 20°$, $x° + 30°$, $x° + 40°$. Find x.

(9)   The angles of a hexagon are $y°$, $2y°$, $4y°$, $y°$, $y°$, $3y°$. Find y.

(10)   Three of the angles of a quadrilateral are equal, and the fourth is 30°. Find the other angles.

(11)   Which of the following angles could be the interior angles of a regular polygon? In each possible case give the number of sides of the polygon.

(a) 170° (b) 160° (c) 145° (d) 150° (e) 130°.

(12)   The interior angle of a regular polygon is 100° greater than the exterior angle. Find the number of sides of the polygon.

(13)   The interior angle of a regular polygon is 4 times as great as the exterior angle. Find the number of sides of the polygon.

(14)   ABCDE is a regular pentagon. Find the angles ABC, ADE, ACD.

Fig 11.35

(15)   ABCDEF is a regular hexagon. Find the angles ABC, ACD, ADE, ADF. What can you say about the quadrilateral ACDF?

Fig 11.36

(16) ABCD is a parallelogram. <ADB = <ABD = 40°. Find <BCD. What can you say about ABCD?

Fig 11.37

109

(17)   Find the interior angle of a regular pentagon. Show that it is not possible to tile a floor with regular pentagons.

(18)   Suppose we wish to tile a floor with regular n-sided polygons. What values of n are possible?

(19)   A floor is tiled with equilateral triangles and with certain regular polygons. Part of the tiling is shown. Find the internal angle of the polygon, and hence find the number of its sides.

(20)   Extend the following tessellations:

Fig 11.38

(a)

(b)

Fig 11.39

11.3 SIMILARITY AND CONGRUENCE. Levels 2 and 3

Two figures are *congruent* if they have the same shape and size.

Two triangles are congruent when they obey any of the following sets of conditions.

I. The sides of one triangle are equal to the sides of the other triangle. (*SSS*).

II. One side and two angles of one triangle are equal to one side and two angles of the other triangle. (ASA).

III. Two sides and the enclosed angle of one triangle are equal to two sides and the enclosed angle of the other triangle. (*SAS*).

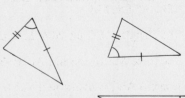

IV. Both triangles are right-angled, with the hypoteneuse and one other side of one triangle equal to the hypoteneuse and one other side of the other triangle. (RHS).

Fig 11.40

110

Note that if Δ ABC is congruent to Δ DEF, that means that <A = <D, <B = <E, <C = <F. So the order in which the letters appear is important.

Two figures are *similar* if they have the same shape. (But not necessarily the same size).

If two triangles ABC and DEF are similar then there is a fixed ratio between the sides of Δ ABC and the sides of Δ DEF.

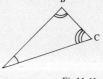

AB/DE = BC/EF = AC/DF

The ratio of sides is the same for both triangles.

AB/BC = DE/EF : AB/AC = DE/DF : AC/BC = DF/EF.

Fig 11.41

Two triangles are similar if any of the following sets of conditions are true:

I. The angles of one triangle are equal to the angles of the other triangle.

II. One pair of angles are equal and the ratios of the enclosing sides are equal

III. The ratios of corresponding sides are equal.

11.3.1 Examples for level 2

(1)   ABCD is a parallelogram. Join A to C; Find a pair of congruent triangles. What can be said about the sides of the parallelogram?

Solution   Consider the two triangles ABC and ADC.

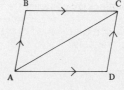

They have the line AC in common.

Angles BCA and DAC are alternate

Angles BAC and DCA are alternate.

Fig 11.47

Hence by condition ASA the triangles ABC and CDA are congruent.

From congruence it follows that AB = DC and AD = BC.

(2)   L and M are the midpoints of the sides AB and AC of the triangle ABC. Find a pair of similar triangles. If BC = 8 find LM.

Solution   The triangles ABC and ALM have angle A in common.

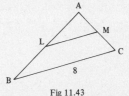

The ratios AL:AB and AM:AC are both equal to $^1/_2$.

Hence Δ ABC is similar to Δ ALM.

It follows that LM:BC is also equal to $^1/_2$.

Fig 11.43

LM = 8x$^1/_2$ = 4.

11.3.2 Exercises for level 2

(1)   Which of the following pairs of triangles are congruent?

(a)   (b)   (c)

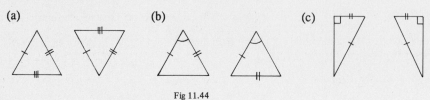

Fig 11.44

(2) ABD and CBE are straight lines meeting at B. CB = BD and AB = BE. Find two congruent triangles. Write down two pairs of equal angles.

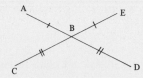

Fig 11.45

(3) ABD and CBE are straight lines meeting at B. CB = BE and AB = BD. Find two congruent triangles. Write down two pairs of equal angles. What can you say about AC and DE?

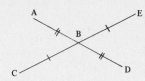

Fig 11.46

(4) PQ and RS are equal and parallel lines. PR and SQ cross at X, as in fig 11.47. Find two congruent triangles. Write down two pairs of equal sides.

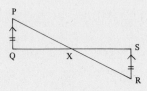

Fig 11.47

(5)   ABCD is a quadrilateral in which AB = CD and AD = BC. Draw the diagonal AC. Find a pair of congruent triangles. Write down which pairs of angles are equal.

What can you say about AB and CD, and about AD and BC? What sort of figure is ABCD?

(6)   ABCD is a parallelogram. The diagonals AC and BD meet at X. Write down as many pairs of congruent triangles as you can.

(7)   ABCD is a kite in which AB = AD and CB = CD. The diagonals AC and BD meet at X.

    What is the line of symmetry of the figure? Write down 3 pairs of congruent triangles. Which angles are equal to each other?

(8)   Which of the following triangles are similar to each other?

(a)   (b)   (c)

 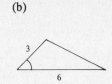 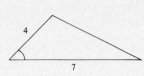

Fig 11.48

(9) L and M lie on the sides AB, AC of the triangle ABC, so that LM is parallel to BC. Write down a pair of similar triangles.

Fig 11.49

(10) L, M and N lie on the sides AB, BC, CA of Δ ABC. LM, MN, NL are parallel to CA, AB, BC respectively. Find as many pairs of similar triangles as you can.

Fig 11.50

(11) Fig 11.51 shows a 3 barred gate in which AB, EF, DC are horizontal and AD, BC are vertical. AGC is a straight line. Write down as many similar triangles as you can.

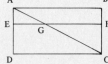

Fig 11.51

(12) The triangles ABC and DEF of fig 11.52 are similar. They are not drawn to scale.

        (a) If AB = 4, DE = 2 and AC = 6, find DF.

        (b) If AB = 3, AC = 2 and DE = 6, find DF.

Fig 11.52

        (c) If BC = 9, EF = 12 and DE = 8, find AB.

        (d) If AB = ³/4, DE = ¹/2 and BC = 1¹/2, find EF.

(13) In the camp-stool of fig 11.53 the lengths of wood above and below the crossing are 20 cm and 30 cm respectively. If the feet are 45 cm apart, how wide is the seat?

Fig 11.53

113

(14)   In fig 11.51, AE = 40 cm, ED = 80 cm, AB = 240 cm. Find the lengths of EG and GF.

(15) In fig 11.54 a ladder AB is laid against two walls as shown. The higher wall is 3 m high, and the lower wall is 1 m. high. If the foot of the ladder is 0.5 m. from the lower wall, how far is it from the higher wall? How far apart are the walls?

Fig 11.54

(16)   You are given an isosceles triangle ABC, which is not equilateral. By drawing at least 4 more triangles show that it is possible to tessellate the plane with triangles congruent to △ ABC.

Fig 11.55

(17)   You are given a right-angled triangle ABC, which is not isosceles. By drawing at least 4 more triangles show that it is possible to tessellate the plane with triangles congruent to △ ABC.

Fig 11.56

11.3.3 Examples for level 3

(1)   ABC is an equilateral triangle. L, M, and N are points on AB, BC, CA respectively such that AL = BM = CN. Show that △ LMN is also equilateral.

Solution   Consider the three triangles ALN, BML, CNM.

Fig 11.57

As △ ABC is equilateral, it follows that the angles <A, <B, <C are equal.

It is given that AL, BM and CN are equal.

The three sides of △ ABC are equal, hence it follows that AN, CM, and BL are equal.

The three triangles are congruent, by SAS. It follows that:

LN = NM = ML. △ LMN is equilateral.

(2)   L and M are on the sides AB and AC of the triangle ABC. AM = 3, MC = 9, AL = 4, LB = 5.

Find the ratio LM:BC. If △ ABC has area 54, what is the area of the quadrilateral BLMC?

Solution   AB = 9 and AC = 12.

The triangles ABC and AML have <A in common.

The ratio AL:AM = 4:3 and the ratio AC:AB = 4:3.

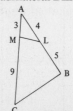

Fig 11.58

114

Hence △ ABC is similar to △ AML. It follows that:

LM:BC = AM:AB = 1:3

The sides of the triangles are in the ratio 1:3. Hence the areas of the triangles are in the ratio $1^2:3^2 = 1:9$.

It follows that the area of △ ALM is 54÷9 = 6.

The area of BLMC is 54 - 6 = 48

11.3.4 Exercises for level 3

(1)   ABCD is a rectangle. Show that the triangles ABC and BAD are congruent. Deduce that the diagonals AC and BD are equal.

Fig 11.59

(2)   ABCD is a parallelogram, in which the diagonals AC and BD are equal. Show that △ ABC is congruent to △ BAD. Deduce that ABCD is a rectangle.

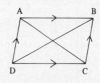

Fig 11.60

(3)   ABCD is a rhombus. Let the diagonals AC and BD meet at X. Show that △ ABX is congruent to △ ADX. Deduce that AC and BD are perpendicular.

(4)   ABCD is a parallelogram in which the diagonals AC and BD are perpendicular. Show that ABCD is a rhombus.

(5)   ABCDE is a regular pentagon. X and Y are on BC and CD respectively, and CX = CY. Show that △ ABX ≡ △ EDY. Deduce that AX = EY.

(6) In fig 11.61 ACDE and BCFG are squares on the equal sides of the isosceles triangle ABC. Show that AF = BD.

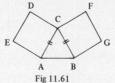

Fig 11.61

(7)   M is the midpoint of the side BC of △ ABC. AM is extended to N, where AM = MN. Show that ABNC is a parallelogram.

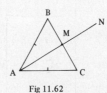

Fig 11.62

(8)   ABCDE... are adjacent points of a regular 17 sided figure. Show that △ ABD ≡ △ BCE.

(9) You are given a triangle ABC which is neither isosceles nor right-angled. By drawing at least 4 more triangles show that it is possible to tessellate a plane with triangles congruent to ABC.

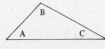

Fig 11.63

(10) X and Y are on the sides AB and AC of Δ ABC, such that AX:XB and AY:YC are both equal to 1:3. Find the ratio XY:BC. If Δ ABC has area 32, find the area of Δ AXY and of XYCB.

Fig 11.64

(11) ABCD is a parallelogram, and E lies on AB so that AE:EB = 1:2. ED and AC meet at F. Find a pair of similar triangles. What is the ratio of their areas?

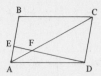

Fig 11.65

(12)   X and Y lie on the sides AB and AC of Δ ABC. AX = 4, XB = 8, AY = 5, YC = 10.

Write down a pair of similar triangles. Find the ratio XY:BC. If Δ AXY has area 8, find the area of XYCB.

(13)   P and Q lie on the sides AB and AC of Δ ABC. AP = 12, PB = 3, AQ = 10, QC = 8.

Write down a pair of similar triangles. Which pairs of angles are equal? If Δ ABC has area 81, find the area of Δ APQ and of PQCB.

(14) In fig 11.66 <BAD = <ABC = 90°, and <BCA = <ABD. Write down a pair of similar triangles.

(a) If AB = 10 and AD = 15, find BC.

(b) If BC = 2 and AD = 8, find AB.

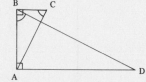

Fig 11.66

(15) Δ ABC is right-angled at A. The perpendicular from A meets BC at D.

Write down three similar triangles.

Complete the equations: $^{BD}/_{AB} = {^{AB}}/_{-} : {^{CD}}/_{AC} = {^{-}}/_{BC}$.

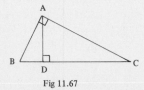

Fig 11.67

Show that BC = $AB^2/BC + AC^2/BC$. Multiply this equation by BC. What theorem have you now proved?

COMMON ERRORS

(1)   **Angles and lines**
Do not assume facts about geometrical diagrams. In particular, do not assume the following unless you are told they are true:

> That a certain line is straight.
> That two lines are parallel.
> That a certain angle is a right-angle.

(2)   Polygons

(a) Make sure that you label polygons correctly. The letters must follow round the figure, either clockwise or anti-clockwise. They must not jump across the diagonal.

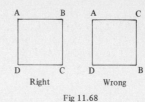

Fig 11.68

(b) If you are told that a polygon is a rectangle, then do not assume that it is not a square. It might be. Similarly, a parallelogram may be a rhombus, and an isosceles triangle may be equilateral.

(c) Do not assume a polygon is regular unless you are told so. The interior angles of a hexagon add up to 720°. But we can only say that each interior angle is 120° if the hexagon is regular.

(3)   Congruence and similarity

(a) If Δ ABC is congruent to Δ DEF, then the triangles are congruent in that order. So <A = <D etc., AB = DE etc. It would then be wrong to state that Δ ABC is congruent to Δ EFD.

(b) If you are told that L is on AB, with AL:LB = 1:3, then the ratio AL:AB is 1:4. Do not think that AL:AB is also 1:3.

12 Circles

12.1 CENTRE, RADIUS, CIRCUMFERENCE

A *circle* consists of all the points which are the same distance from a fixed point.

The fixed point is the *centre*.

The distance is the *radius*.

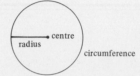

Fig 12.1

The greatest distance across the circle is the *diameter*. The diameter is twice the radius.

The length round the circle is the *circumference*. The ratio of the circumference to the diameter is the same for every circle. The ratio is called π (pronounced *pie*). π can never be written down exactly. It is sometimes approximated by $^{22}/_7$ or by 3.14.

12.1.1 Examples for level 1

(1)   The radius of a big-wheel at a fair is 3 m. How far does a point on the rim travel during one revolution? (Take π to be 3.14).

Solution   The diameter of the wheel is twice the radius, i.e. 6 m.

The distance travelled is the circumference of the wheel, which is π times the diameter.

Distance travelled is 3.14 x 6 = 18.8 m.

Fig 12.2

(2)   A and B are points on a circle with centre O. <AOB = 40°. Find the other angles of the triangle AOB.

Solution   OA and OB are both radii of the circle. Hence they are equal.

The angles <OAB and <OBA are equal. Together they must come to 180° - 40° = 140°.

$$<OAB = <OBA = {}^{140°}/_2 = 70°$$

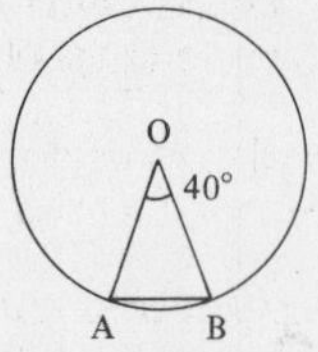

Fig 12.3

12.1.2 Exercises for level 1

Throughout these exercises either use the π button on your calculator or take π to be 3.14.

(1) On the circles below mark (a) the centre (b) a radius (c) a diameter.

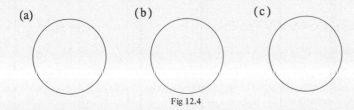

(a)　　　　　　(b)　　　　　　(c)

Fig 12.4

(2)　A circular running track has radius 50 m. What is the diameter? If an athlete runs round once, how far has she gone?

(3)　The minute hand of a clock is 4 cm. long. How far does the tip of the hand travel during one hour?

(4)　The hour hand of a clock is $2^1/2$ cm. long. How far does it travel in 6 hours?

(5)　A coin of diameter 3 cm. rolls along the ground. How far has the coin travelled when it has made a complete revolution?

(6)　The driving wheel of a car is 6 inches in radius. How far does a point on the rim move when the wheel goes through a quarter turn?

(7)　A running track is to be made in the form of a circle with circumference 400 metres. What should the diameter be? What should the radius be?

(8)　The circumference of a circle is 20 cm. What is the radius?

(9) A square is drawn inside a circle of radius 5 cm. What is the length of the circle between two adjacent corners of the square?

Fig 12.5

(10)　A regular pentagon is drawn inside a circle of radius 8 cm. What is the length of the circle between adjacent corners of the pentagon?

(11)　A record of diameter 12 inches rotates at 33 revolutions per minute. How far does a point on the rim travel in one minute? How far does it travel in one second?

(12) The diameter of a cartwheel is 120 cm., and the boss in the centre of the wheel is of diameter 15 cm. How long are the spokes of the wheel? If the cart travels 100 m., how many times has the wheel turned?

Fig 12.6

(13)   Find the unknown angles in the following diagrams:

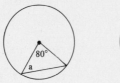

Fig 12.7

12.2 CHORDS, ARCS, SECTORS. Level 2

The straight line joining two points A and B on a circle is a *chord*.

The part of the circle between A and B is an *arc*.

The region of a circle between two radii is a *sector*.

A diameter splits a circle into two *semi-circles*.

If AB is a diameter of a circle and C a point on the circumference, then <ACB = 90°.

A chord is perpendicular to the radius through its centre.

Fig 12.8

12.2.1 Examples for level 2

(1)   AB is a diameter of the circle in fig 12.9. If <CAB = 25° find the other angles.

Solution   Because AB is a diameter it follows that <ACB is a right-angle. The third angle can be found by subtraction:

<ACB = 90° and <ABC = 180° - 90° - 25° = 65°

Fig 12.9

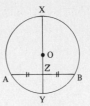

(2)   The chord AB of the circle in fig 12.10 is bisected by the diameter XY. Show that AX = BX and that AY = BY. What sort of figure is XAYB?

Fig 12.10

Solution   The radius OY of the circle goes through the midpoint of AB. Hence XY and AB are perpendicular. Hence <AZX = <BZX. It follows that Δ AXZ is congruent to Δ BXZ.

This implies that AX = BX.

By exactly similar reasoning AY = BY. The quadrilateral XAYB has two pairs of adjacent sides equal.

XAYB is a kite

12.2.2 Exercises for level 2

(1)   On the circles below mark (a) the chord AB (b) the arc AB.

(a)

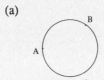

(b)

Fig 12.11

(2)   Find the unknown angles in the diagrams below.

(a)

(b)

(c)

Fig 12.12

(3)   AB and CD are diameters of a circle. Show that ACBD is a rectangle.

Fig 12.13

(4)   AB is a diameter of a circle with centre O. C is a point of the circumference. <COB = 40°. Find the angles OCB, OCA. What is angle ACB?

(5)   AB is a chord of a circle centre O and radius 5, and <AOB = 30°. What is the length of the arc AB?

(6)   An athlete runs 100 m. round a running track, which is a circle centre O and radius 70 m. If A is the starting point and B the finishing point, what is the angle between OA and OB?

(7)   Find the ratio of the arcs AB, BC, CA in fig 12.14.

Fig 12.14

(8)   AB is a chord of a circle centre O, and C is the midpoint of AB. What can you say about <ACO? What is the relationships between the triangles OCA and OCB? If <OAB = 50° find <COB.

(9)   AB and AC are equal chords of a circle centre O. Draw a diagram of the circle. What is the line of symmetry of the figure? If <OBA = 20° find <BAC.

(10)   AB is a chord of a circle, and CD is a diameter going through the midpoint of AB. Name 3 pairs of equal sides in fig 12.15. Name a pair of congruent triangles. What sort of quadrilateral is ACBD?

Fig 12.15

(11)   AB and CD are parallel chords of the circle centre O in fig 12.16. Draw the line of symmetry of the figure. If <A = 80° find the other three angles of the quadrilateral.

Fig 12.16

(12)   In fig 12.17 AB and CD are equal chords of the circle centre O. What is the relationship between Δ OAB and Δ OCD? Draw the perpendiculars from O to the two chords. What can you say about the two perpendicular distances?

Fig 12.17

12.3 CIRCLE THEOREMS. Level 3

A chord divides a circle into two *segments*.

A quadrilateral inscribed inside a circle is *cyclic*.

An angle greater than 180° is a *reflex* angle.

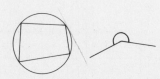

Fig 12.18

I.   Let AB be a chord of a circle with centre O, and let C be a point on the circumference in the same segment as O. Then.

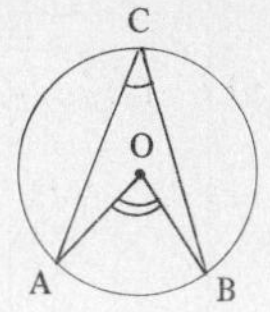

<AOB = 2 x <ACB

Fig 12.19

(The angle at the centre is twice the angle at the circumference.)

II.   Let AB be a chord of a circle, and let C and D be points in the same segment. Then:

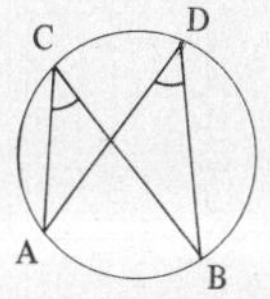

<ACB = <ADB

(Angles in the same segment are equal.)

Fig 12.20

III.   The opposite angles of a cyclic quadrilateral add up to 180°.

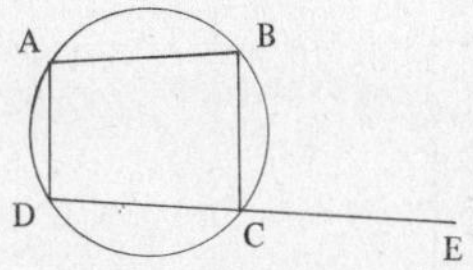

<A + <C = 180°

IV. The exterior angle of a cyclic quadrilateral is equal to the opposite angle.

<A = <BCE

Fig 12.21

12.3.1 Examples for level 3

(1)   O is the centre of the circle in fig 12.22, and <AOB = 80°. Find the reflex angle <AOB, <C and <D.

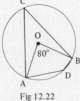

Solution   The angles round O add up to 360°.

Fig 12.22

The reflex angle <AOB is 360° - 80° = 280°

C is in the same segment as O. Hence Theorem I applies.

<C = ¹/₂ <AOB = 40°

ACBD is a cyclic quadrilateral. Hence Theorem III applies.

<D = 180° - <C = 140°

(2)   The chords AB and CD of a circle cross at X. Show that Δ AXC is similar to Δ CXB.

Fig 12.23

123

Solution   C and B are in the same segment of the circle defined by the chord AD. Theorem II applies.

$$<ACD = <ABD.$$

A and D are in the same segment of the circle defined by the chord BC. Theorem II applies.

$$<CDB = <CAB.$$

The angles <AXC and <BXD must also be equal.

△ AXC is similar to △ DXB

12.3.2 Exercises for level 3

(1) O is the centre of the circle of fig 12.24. The diagram is not to scale.

Fig 12.24

   (a) If <ACB = 25° find <AOB.
   (b) If <AOB = 66° find the reflex angle <AOB.
   (c) If <AOB = 122° find <ACB.

(2) The diagram of fig 12.25 is not to scale.

Fig 12.25

   (a) If <ADB = 43° find <ACB.
   (b) Find an angle equal to <CAB.
   (c) Find two pairs of similar triangles.

(3) The diagram of fig 12.26 is not to scale. ADE is a straight line.
   (a) If <A = 56° find <C.
   (b) If <B = 74° find <ADC and <CDE.

Fig 12.26

(4)   Find the unknown angles in the following figures. Throughout O denotes the centre of the circle.

(a)   (b)   (c)   (d)   (e)

Fig 12.27

(5)   AB is a chord of a circle centre O. C is a point on the circumference, so that O and C lie in different segments of the circle. Show that <ACB is equal to half the reflex angle <AOB.

(6) In fig 12.28 O is the centre of the circle. <AOC = 108 and <CBO = 56°. Find <AOB.

Fig 12.28

124

(7) In fig 12.29 the chords AB and DC are parallel. <BAX = 35°. Find <AXD, <CXB, <ABD.

(8)   In fig 12.29 the chords AB and DC are parallel. If <CXB = 88° find <CAB and <ACD.

Fig 12.29

(9) In fig 12.30 ABC and FED are straight lines. <A = 110°. Find <BEF, <BED, <C. What can you say about the lines AF and CD?

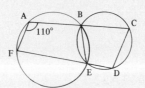

Fig 12.30

(10) In fig 12.31 ABCD is a parallelogram. The circle through ADC cuts BA at E. Prove that Δ CBE is isosceles.

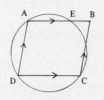

Fig 12.31

(11)   ABCD is a cyclic parallelogram. Draw a diagram of ABCD and the circle. What else can be said about ABCD?

(12)   Draw pairs of circles which meet:

   (a) At two points (b) at one point (c) at no points.

12.4 TANGENTS. Levels 2 and 3

A line which meets a circle at one point only is a *tangent* to the circle.

A tangent is perpendicular to the radius at that point.

12.4.1 Examples for level 2

Fig 12.32

(1)   In fig 12.33 XP is a tangent to the circle with centre O. OX cuts the circle at Q. If <PXO = 24° find <POQ and <OPQ.

Solution   OP is a radius, and PX is a tangent. Hence they are perpendicular.

$$<POQ = 180° - 90° - 24° = 66°.$$

OP and OQ are equal radii. Hence <OPQ = <OQP.

$$<OPQ = (180° - 66°)/2 = 57°.$$

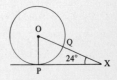

Fig 12.33

(2)   Two balls touch each other. Show that the two centres and the point of contact lie in a straight line.

Solution   As the balls touch, they have a common tangent, shown by the dotted line in fig 12.34.

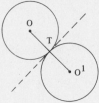

        Both the radii to the point of contact are perpendicular to the tangent. OTO' = 90° + 90° = 180°.

                    OTO' is a straight line.   Fig 12.34

12.4.2 Exercises for level 2

(1)   Find the unknown angles in the diagrams below.

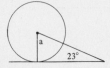
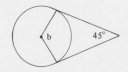
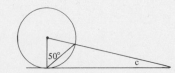

Fig 12.35

(2) In fig 12.36 AT is a diameter of a circle and TB is a tangent. If <A = 55° find <B.

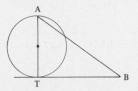

Fig 12.36

(3) In fig 12.37 the small circle touches the large circle. Show that the two centres and the point of contact are in a straight line.

Fig 12.37

(4) In fig 12.38 TA and TB are tangents to the circle with centre O. If <T = 72° find <O.

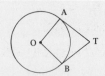

Fig 12.38

(5) In fig 12.39 AB and CD are diameters, and the tangents at A, B, C, D meet at P, Q, R, S. Show that PQRS is a parallelogram.

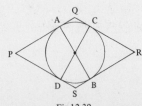

Fig 12.39

126

(6) If in question 5 we know that AB and CD are perpendicular, what more can we say about PQRS?

For level 3

If P lies outside a circle, then the two tangents from P to the circle are equal in length.

PA = PB.

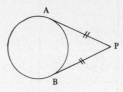

Fig 12.40

Let a chord AB meet a tangent AX. Let C be in the opposite (alternate) segment. The *Alternate Segment* theorem states:

The angle between a chord and a tangent is equal to the angle in the opposite segment.

<XAB = <ACB

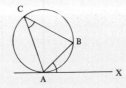

Fig 12.41

12.4.3 Examples for level 3

(1) XA and XB are tangents to a circle. If <AXB = 28°, find <XAB.

Solution As XA and XB are tangents, they are equal. Hence Δ AXB is isosceles. It follows that:

<XAB = ¹/₂(180° - 28°) = 76°

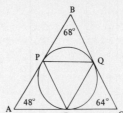

Fig 12.42

(2) The angles of a triangle are 48°, 68°, 64°. A circle is inscribed in the triangle. What are the angles of the triangle formed by the three points of contact?

Solution In fig 12.43, Δ APR is isosceles. It follows that <APR = 66°.

AP is a tangent to the circle, and PR is a chord. By the Alternate Segment theorem:

<PQR = <APR = 66°

By similar calculations:

<PRQ = 56° and <RPQ = 58°

12.4.4 Exercises for level 3

(1) In the diagram of fig 12.44 mark in pairs of equal sides and pairs of equal angles.

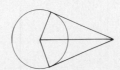

Fig 12.44

127

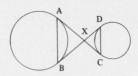

Fig 12.45

(2) Fig 12.45 shows the tangents AXC and BXD to two circles. Mark in pairs of equal sides and pairs of equal angles. What is the relationship between Δ AXB and Δ CXD?

(3)   In each of the diagrams of fig 12.46 draw the common tangents to both circles. Draw the lines of symmetry of each diagram.

(a)   (b)   (c)   (d)   (e)

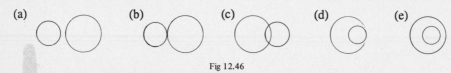

Fig 12.46

(4) In fig 12.47 a quadrilateral ABCD is drawn round a circle. Show that AB + CD = BC + AD.

Fig 12.47

(5) Fig 12.48 shows three circles, each of radius 10 cm, touching each other at A, B, C. Find the side of the triangle ABC.

Fig 12.48

(6)   Find the unknown angles in the diagrams of fig 12.49.

Fig 12.49

(7)   A circle is drawn inside a triangle with angles 60°, 66°, 54°. Find the angles of the triangle made by the points of contact.

(8)   A circle is drawn inside a triangle, and the triangle made by the points of contact has angles 55°, 63°, 62°. Find the angles of the original triangle.

(9) In fig 12.50 AB is a diameter of the circle, and CT is the tangent at C. If <BCT = 70°, find <ABC.

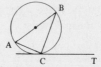

Fig 12.50

128

(10) The quadrilateral ABCD has angles 100°, 95°, 70°, 95° at A, B, C, D. A circle is drawn inside the quadrilateral, touching AB, BC, CD, DA at P, Q, R, S. Find the angles of PQRS.

(11) In fig 12.51 O is the centre and TA and TB are tangents. <ATB = 40° and <TBY = 25°. Find <AOB, <AXB, <YAT.

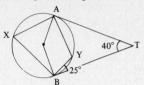

Fig 12.51

(12) Find the unknown angles in the diagrams of fig 12.52.

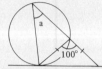

Fig 12.52

12.5 INTERSECTING CHORDS. Level 3

AB and CD are chords of a circle meeting at X. (Either inside or outside the circle). The *Intersecting Chords* theorem states:

$$XA \times XB = XC \times XD$$

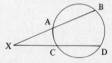

Fig 12.53

If AB meets a tangent XT at X, then:

$$XA \times XB = XT^2.$$

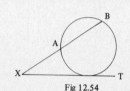

Fig 12.54

12.5.1 Examples for level 3

(1) The chords AB and CD to a circle meet at X inside the circle. If XB = 3, XC = 2 and AB = 7 then find XD.

Solution Note that XA = 7 - 3 = 4. This gives the equation:

$$3 \times 4 = 2 \times XD$$

$$XD = 6$$

Fig 12.55

(2) A circular mirror with radius 30 cm hangs with its centre 50 cm. below a nail, held by a string wrapped round the mirror. What length of string is not in contact with the mirror?

Solution XT is a tangent to the circle. If A and B are the highest and lowest points of the mirror, then XA = 20 cm. and XB = 80 cm. This gives:

$$XT^2 = 20 \times 80 = 1600 \text{ cm}^2.$$

$$XT = 40 \text{ cm}.$$

The length of free string is 2 x 40 = 80 cm.

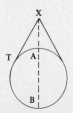

Fig 12.56

129

12.5.2 Exercises for level 3

(1)   Find the unknown lengths in the diagrams of fig 12.57.

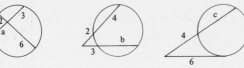

Fig 12.57

(2)   The chords AB and CD of a circle meet at X inside the circle. XA = 3, AB = 8, XC = 2. Find XD.

(3)   The chords AB and CD to a circle meet at X outside the circle. XA = 5, XB = 15, XC = 3. Find CD.

(4)   The chords AB and CD to a circle meet at X outside the circle, C being nearer to X than D. XC = 15, CD = 5, XB = 25. Find XA.

(5)   The chords PQ and RS to a circle meet at O outside the circle. OP = 3, PQ = 13, OS = 12. Find RS.

(6) Fig 12.58 shows a bridge, which is in the shape of an arc of a circle. The width of the river is 12 m., and the highest point of the bridge is 4 m. above the water. What is the radius of the circle?

Fig 12.58

(7)   A, B, C are points on a circle. The chord AB meets the tangent at C at X. XA = 5, XB = 45. Find XC.

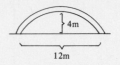

Fig 12.59

(8)   P, Q, R are points on a circle. The chord PQ meets the tangent at R at X. XR = 6, XP = 4. Find PQ.

(9)   The mast of a ship is 40 m. above the level of the sea. How far can a lookout at the top of the mast see? (Take the radius of the Earth to be 6,400,000 m.)

COMMON ERRORS

(1)   **Circumference and radius**

(a) The value of π is only *approximately* 3.14 or 22/7. If you calculate using these values for π, it is misleading to give your answer to more than 3 significant figures.

(b) The circumference of circle is π times the diameter. So if you know the circumference, in order to find the diameter you must *divide* by π. Do not multiply.

(c) Be careful not to confuse the radius and the diameter. Read the question carefully, to see which you are given or which you are required to find.

(2)   Chords

(a) Do not assume that a chord is a diameter unless you are told so.

(b) Do not assume that two chords cross in the centre of a circle unless you are told so.

These two mistakes are often made when a picture like fig 12.60 is used. It is often wrongly assumed that:

AD is a diameter.

X is the centre of the circle.

AB is parallel to CD.

The last error leads one to say that <A = <D, instead of <A = <C.

Fig 12.60

13 Solids

The flat side of a solid is a *face*.

The line on a solid where two faces meet is an *edge*.

The point on a solid where three or more edges meet is a *vertex* or *corner*.

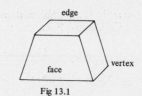

Fig 13.1

A solid with six square faces is a *cube*.

A solid with six rectangular faces is a *cuboid*.

A solid with constant cross-section is a *prism*. If the cross-section is a triangle it is a *triangular* prism.

A solid which tapers to a point from a rectangular base is a *pyramid*.

A perfectly round solid is a *sphere*.

A solid whose cross-section is a circle is a *cylinder*.

A solid which tapers to a point from a round base is a *cone*.

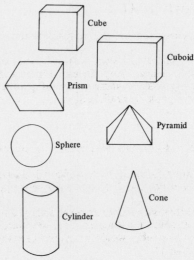

Fig 13.2

A diagram which can be cut out and folded to make a solid is a *net*.

13.1 Examples for level 1

(1)   Fig 13.3 shows a cube. How many faces, edges, vertices does it have? Name a pair of parallel edges and a pair of parallel faces.

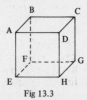

Fig 13.3

Solution   Looking at the diagram, there are 2 horizontal faces and 4 vertical faces.

There are 4 vertical edges and 8 horizontal edges.

There are 4 vertices on the top face and 4 more on the bottom face.

There are 6 faces, 12 edges and 8 vertices.

Any two vertical edges are parallel. The two horizontal faces are parallel.

AE and CG are parallel: ABCD and EFGH are parallel.

(2)   Fig 13.4 shows the net of a solid. What is the name of the solid? Which point will F join?

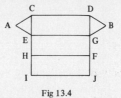

Fig 13.4

Solution   When the solid is assembled, there will be 2 triangular faces CAE and GBD at each end.

The solid is a triangular prism.

IJ is glued to CD, JF is glued to DB.

F will join B.

13.2 Exercises for level 1

(1)   Find the number of faces, edges and vertices of the pyramid of fig 13.2.

(2) Fig 13.5 shows a prism with a triangular cross-section. How many faces, edges and vertices does it have? Name a pair of parallel faces, and two pairs of parallel edges.

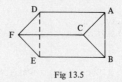

Fig 13.5

(3)   What are the mathematical names for the following common solids?

(a) A matchbox (b) Dice (c) A wedge of cheese (d) A tin can (e) A ping-pong ball (f) A six-sided pencil (unsharpened) (g) The tip of a sharpened pencil (h) A bell-tent.

(4) Fig 13.6 shows a number of cubes arranged to form another solid. How many cubes are there? How many more cubes would be needed to make the solid a cuboid?

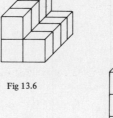

Fig 13.6

(5) Fig 13.7 shows a house which is 3 storeys high, 3 rooms wide and 3 rooms deep. In each room, there is a window in every wall that faces the outside.

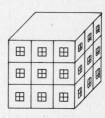

Fig 13.7

How many rooms are there?
How many rooms have no windows?
How many rooms have 1 window?
How many rooms have 2 windows?

Suppose that skylights are put in to the ceilings of all the rooms on the top floor. How many rooms have 3 windows?

(6) Fig 13.8 shows a cuboid formed by 24 smaller cubes. The cuboid is dipped into paint.

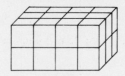

Fig 13.8

(a) How many cubes have one painted face?

(b) How many cubes have 2 painted faces?

(c) How many cubes have 3 painted faces?

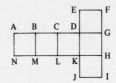

(7) Fig 13.9 shows a net. What solid is formed from the net? What point will join B? What point will join N?

Fig 13.9

(8) A die is marked so that the number of dots on opposite faces always adds up to 7. Fig 13.10 shows the net of a die. Fill in the number of dots on the blank faces.

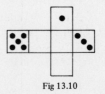

Fig 13.10

(9)   Fig 13.11 shows several possible nets. Which of them will be the net of a cube?

(a)   (b)   (c)   (d)   (e)

Fig 13.11

(10) Fig 13.12 shows a square sheet of cardboard. The shaded bits are removed, and the rest is folded along the dotted lines. What sort of object is made?

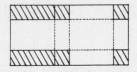

Fig 13.12

(11)   Describe the solids formed from the following nets:

(a)   (b)   (c)   (d)

Fig 13.13

134

For level 2

A solid with four triangular faces is a *tetrahedron*.

Fig 13.14

If a solid is identical on both sides of a plane, then it is *symmetrical* about the plane.

If a solid occupies the same region of space after being rotated about a line, then the line is an *axis of symmetry*.

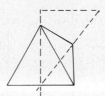

Fig 13.15

13.3 Example for level 2

Fig 13.16 shows an equilateral triangle, with LMN at the midpoints of its sides. What solid is made when it is folded along the dotted lines? Find a plane of symmetry and an axis of symmetry.

Fig 13.16

Solution   Bring the points A, B, C together. Fig 13.17.

Fig 13.17

The solid is a tetrahedron

A plane through AL and the midpoint X of MN will cut the solid into identical halves. Hence this plane is a plane of symmetry.

ALX is a plane of symmetry

Take the line through L and the centre G of $\triangle$ AMN. If the solid is rotated through 120° about this line it will return to the same region of space.

LG is an axis of symmetry

13.4 Exercises for level 2

(1)   For each of the following figures draw a plane of symmetry and an axis of symmetry.

(a) A cuboid which is 3 by 4 by 5.   (b) A cuboid which is 3 by 3 by 5.
(c) A cube which is 3 by 3 by 3.   (d) A cylinder.
(e) A pyramid with square base.   (f) A sphere.
(g) A cone.   (h) A prism, with equilateral triangular
                                        cross-section.

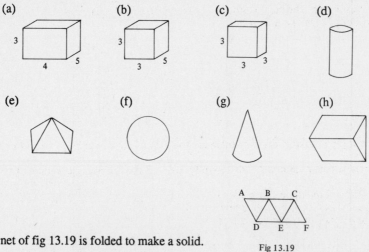

(2) The net of fig 13.19 is folded to make a solid.

Fig 13.19

(a) What is the name of the solid? Which point is joined to F? What point is joined with A?

(b) Mark on the net the triangle which is opposite to A.

(c) How many faces, edges and vertices does the solid have?

(3)   Draw nets for the following solids. Your diagrams should be clearly labelled, but they need not be to scale.

(a) A cuboid 3 by 4 by 5.   (b) A cuboid 3 by 3 by 5
(c) A wedge.
(d) A prism, whose cross-section is a right-angled isosceles triangle.

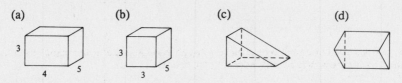

Fig 13.20

(4) Fig 13.21 shows 3 edges of a cube. Complete the diagram.

Fig 13.21

(5) Fig 13.22 shows part of the net for a cube. complete the net.

Fig 13.22

(6) Fig 13.23 shows a cylinder. The curved side is cut along the dotted line. Draw the side when it has been laid flat.

Fig 13.23

13.5 Exercises for level 3

(1)   Find out how many planes of symmetry there are for each of the solids in question 1 of 13.4.

(2)   Find out how many axes of symmetry there are for each of the solids in question 1 of 13.4.

(3)   The curved surface of the cone of fig 13.24 is made out of paper. It is cut along the dotted line. Make a sketch (not to scale) of the paper when it is laid flat.

(4) Fig 13.25 shows a square based pyramid. X and Y are the midpoints of AD and BC.

Fig 13.24   Fig 13.25

    (a) The pyramid is cut in half by the plane through VXY. What is the name of the solid formed by one of the halves? How many faces, edges, vertices does it have?

    (b) The pyramid is cut in half by the plane through VAC. What is the name of the solid formed by one of the halves? How many faces, edges, vertices does it have?

(5) Fig 13.26 shows a cube. W, X, Y, Z are the midpoints of AB, CD, GH, EF.

Fig 13.26

    (a) The cube is cut in half by the plane through WXYZ. What is the name of the solid formed by one of the halves? How many faces, edges, vertices does it have?

    (b) The cube is cut in half by the plane through ACGE. What is the name of the solid formed by one of the halves? How many faces, edges, vertices does it have?

    (c) The cube is cut in two by the plane through BED. What is the name of the smaller part? How many faces, edges, vertices does it have?

(6)   The longest distance within the cube of fig 13.26 is AG. Name 3 other lines in the cube which have the same length.

(7) Make a copy of the cube of fig 13.27. On your diagram mark the midpoints of each of the 6 faces. Join the midpoints together, except when they are directly opposite each other. You should now have a diagram of another solid, called an *octahedron*.

Fig 13.27

(a) How many faces, edges, vertices does this solid have?

(b) Draw the net for this solid.

COMMON ERRORS

(1)   **Diagrams**

Angles and lines get distorted in drawings of solids. Right-angles may not seem to be right-angles, and equal sides may look different. Be careful that you do not misunderstand a diagram.

(2)   **Names of solids**

If a solid is described as a cuboid, then do not assume that it is not a cube. A cube is a special case of a cuboid.

14 Co-ordinates and Graphs

14.1 AXES AND CO-ORDINATES

Axes are two lines, one horizontal, the other vertical. The horizontal line is the *x-axis* and the vertical line is the *y-axis*.

The position of a point along the x-axis is its *x co-ordinate,* and its position along the y-axis is its *y co-ordinate.*

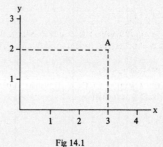

Fig 14.1

The x co-ordinate of a point is always given first. So at the point A(3,2), x = 3 and y = 2.

14.1.1 Examples for level 1

(1) (a) Give the co-ordinates of the point B as shown in fig 14.2.

  (b) Mark on the grid the point C(3,3).

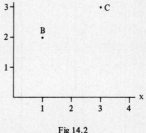

Fig 14.2

Solution (a) B is 1 along the x-axis, and 2 up the y-axis.

B is at (1,2)

  (b) Go 3 along the x-axis, and 3 up the y-axis. Mark in C as shown in fig 14.2.

(2) A temperature in Centigrade is converted to Fahrenheit by the formula $F = C \times ^9/_5 + 32$.

  Find the F values when C is 0°, 20°, 25°.

  Plot your results on a graph, using the Centigrade figures for the x values and the Fahrenheit figures for the y values.

Solution Apply the formula to the values.

  For C = 0, $F = 0 \times ^9/_5 + 32 = 32$.
  For C = 20, $F = 20 \times ^9/_5 + 32 = 36 + 32 = 68$
  For C = 25, $F = 25 \times ^9/_5 + 32 = 45 + 32 = 77$

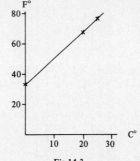

Fig 14.3

  Plot the points (0,32), (20,68), (25,77) on the graph of fig 14.3. Join them up by a straight line.

14.1.2 Exercises for level 1

(1) Fig 14.4 shows a map of a village. The church is in the square C5. Write down:

    (a) The square containing the station.

    (b) The squares through which Church Road passes.

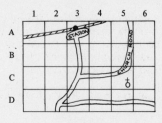

Fig 14.4

(2) Write down the co-ordinates of the points A, B, C, D of fig 14.5.

(3) Make a copy of the grid of fig 14.5. Mark on it the points X(1,1), Y(2,3), Z(4,1).

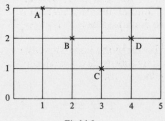

Fig 14.5

(4) On the graph paper of fig 14.6 mark in the points P(1,3), Q(5,3), R(5,1), S(1,1).

    Join up the sides to make a rectangle PQRS.

    What are the co-ordinates of the join of the diagonals PR and QS?

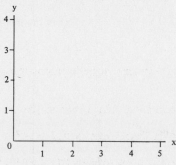

Fig 14.6

(5) On the graph of fig 14.7 mark the points A(1,3), B(1,2), C(4,2). Join the sides AB and BC.

    Find the point D on the graph so that ABCD is a rectangle. Write down its co-ordinates.

    Mark on the graph the midpoint of AC. What are its co-ordinates?

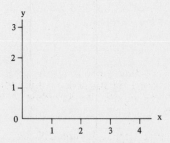

Fig 14.7

(6) Fig 14.8 represents a region of sea, in which the wind is coming from the North East. A sailing dinghy starts from (0,0), and has to tack against the wind. It sails 2 squares North, then 2 squares East, then 2 squares North, and so on.

Draw its route on the diagram, up to the point (6,6).

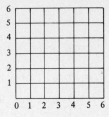

Fig 14.8

(7)   The cooking instructions for pork tell the cook to roast it for 30 minutes plus 20 minutes per pound. Calculate the roasting time for joints weighing 2 pounds, 3 pounds, 4 pounds.

Plot a graph showing this information, with the weight along the x-axis and the cooking time along the y-axis.

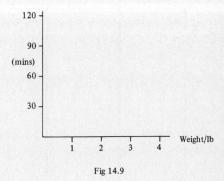

Fig 14.9

(8)   An electricity supply company offers its customers two schedules of payment for the power they use:

Schedule A: £20 per quarter plus 5 p. per unit.

Schedule B: £5 per quarter plus 8 p. per unit.

Complete the following table for the cost of electricity:

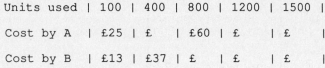

Units used	100	400	800	1200	1500
Cost by A	£25	£	£60	£	£
Cost by B	£13	£37	£	£	£

Plot these figures on the graph of fig 14.10. Join up the points for Schedule A and for Schedule B.

If you expect to use 200 units, which Schedule should you pay by?

When does it become cheaper to pay by Schedule A rather than by Schedule B?

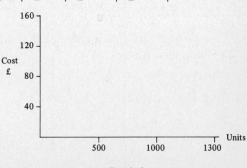

Fig 14.10

For level 2

The *gradient* of a line is the ratio of its y-change to its x-change.

A collection of points which are roughly but not exactly in a straight line is called a *scatter diagram*.

14.1.3 Examples for level 2

(1)   Find three pairs (x,y) such that y = 2x - 4. Plot these points on a graph, and join them up with a straight line.

Repeat this process with the equation 3x + 2y = 6, plotting the points on the same graph.

Use your graph to solve the simultaneous equations:

$$y - 2x = -4$$
$$3x + 2y = 6$$

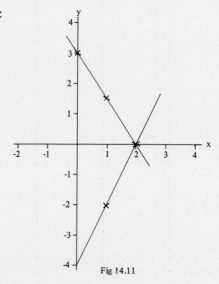

Fig 14.11

Solution   3 points which satisfy y = 2x - 4 are (0,-4), (1,-2), (2,0). They are plotted on the graph of fig 14.11.

3 points which obey 3x + 2y = 6 are (0,3), (2,0), (1,1¹/2). They are also plotted on fig 14.11.

The two equations are both true at the crossing point of the two lines. Read off the co-ordinates of this point:

x = 2 : y = 0

(2)   Different weights were put on the end of an elastic string, and the extension was measured. The weight in grams is given by x, and the extension in cm is given by y. The approximate results are given by the following table:

x	0	1	2	3	4	5
y	0.4	0.5	1.2	1.3	1.4	1.6

Plot these points on a scatter diagram. Draw a straight line through these points. Find the gradient of your line.

Solution   The points are shown plotted in fig 14.12. The line is shown.

The y change from end to end is 1.8 - 0.3 = 1.5. The x change is 5. Divide the y change by the x change.

The gradient is $^{1.5}/5 = 0.3$

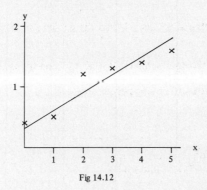

Fig 14.12

14.1.4 Exercises for level 2

(1) Find the gradient of the line segments in fig 14.13.

(2) Find three points of the form (x,y) which satisfy the equation:

$$y = 3x + 1.$$

Plot these points on the graph of fig 14.14. Find the gradient of the line.

(3) Repeat question 2, using the same graph paper, for the equation:

$$2y = x + 7.$$

Hence solve the simultaneous equations:

$$y = 3x + 1$$
$$2y = x + 7.$$

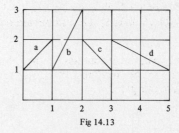

Fig 14.13

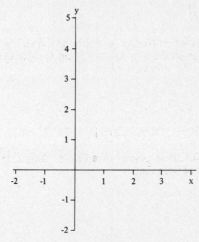

Fig 14.14

(4)   Use the method of question 2 to find the gradients of the lines given by the equations:

(a) $y = 4x - 3$ (b) $y = x + 1$ (c) $y = ^1/2x + 3$ (d) $4y + 3x = 2$.

143

(5)   Solve the following simultaneous equations, by means of drawing lines on graph paper and finding the point of intersection:

(a) y = x + 1   (b) y = 2x - 4   (c) 2y + 3x = 12   (d) y = $^1/2$x - 1
    y = 3x - 5   y = 11 - 3x   x + 3y = 11   4y = 3x - 8

(6)   Draw. on the same piece of graph paper, the lines corresponding to the equations y = x, y = x + 3, y = x - 1. What can you say about the lines?

(7)   For the following tables, plot the points as a scatter diagram on graph paper, and draw as close-fitting a straight line as you can. Find the gradient of your straight line.

(a)   x | 0 | 1 | 2 | 3 | 4 |

       y |-0.2 | 1.0 | 3.0 | 3.9 | 5.4 |

(b)   x | 2 | 3 | 4 | 5 | 6 |

       y | 2.2 | 2.8 | 4.0 | 4.7 | 5.0 |

(c)   x | 1 | 2 | 3 | 4 | 5 |

       y | 27 | 20 | 13 | 14 | 3 |

(8)   An income tax scheme is as follows.

No tax is paid on the first £3,000 of income.

The next £7,000 of income, i.e. up to £10,000, is taxed at 30%.

The next £10,000 of income, i.e. up to £20,000, is taxed at 40%.

All income over £20,000 is taxed at 50%.

Find how much tax is paid on incomes of (a) £2,500 (b) £7,500 (c) £15,000 (d) £25,000.

Plot a graph to show these figures, using 1 cm along the x-axis to represent £5,000 of income, and 1 cm along the y-axis to represent £1,000 of tax.

If a man pays £5,000 in tax, what is his income?

14.2 INTERPRETATION OF GRAPHS

14.2.1 Examples for level 1

(1)   Allen cycles to see his grandmother. On the way his bike gets a puncture. He repairs it and cycles on. Fig 14.15 shows a graph of the distance he has travelled in terms of the time.

(a) How long did he take in all?

(b) How far away is his grandmother?

(c) How long did he take to mend the puncture?

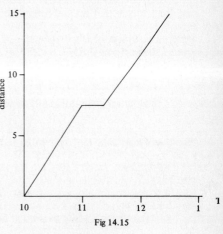

Fig 14.15

Solution   (a) The total time along the x-axis is $2^1/_2$ hours.

The journey took $2^1/_2$ hours.

(b) The distance along the y-axis is 15 miles.

His grandmother lives 15 miles away.

(c) He was delayed between 11 o'clock and 11.20.

He mended the puncture in 20 minutes

(2)   Jasmine has a fever. Her temperature, taken every 4 hours, was recorded as follows.

```
Time  |12.00|16.00|20.00|24.00|04.00|08.00|12.00|
Temp. | 98  | 102 | 104 | 105 | 105 | 102 | 99  |
```

Plot these figures on a graph of temperature against time. Estimate her highest temperature, and the length of time during which her temperature was over 100°.

Solution   The graph of temperature against time is shown in fig 14.16. Read off the highest value:

Her highest temperature was 105.5°

Her temperature was over 100° between 14.00 and 11.00.

Her temperature was over 100° for 21 hours.

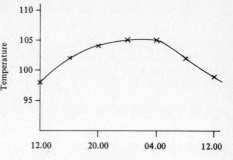

Fig 14.16

14.2.2 Exercises for level 1

(1) Every morning Karen walks to the bus-stop and waits for the bus which will take her to the gate of her school. The graph of her distance from home against time is shown in fig 14.17.

(a) How far is the bus-stop from her home? How long does it take to walk there?

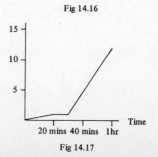

Fig 14.17

145

(b) How far is it from the bus-stop to school? How long does the bus take?

(c) How long does she have to wait at the bus-stop?

(2) George sets off for school one morning, but on his way he realises that he has forgotten his books. He returns home to collect them and sets off for school again. The graph of his distance from home against time is shown in fig 14.18.

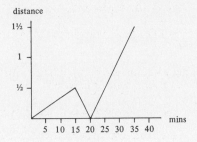

Fig 14.18

(a) How far had he gone when he turned back?

(b) How long did the total journey take?

(c) Why are the last parts of the journey steeper than the first part?

(3) The Jameson family sets off for their holiday by car. They drive from their home A to the motorway at B. Then they travel along the motorway to C, where they turn off and drive to the hotel at D. The graph of fig 14.19 shows the distance travelled against time.

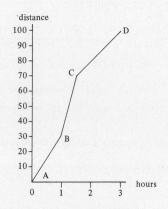

Fig 14.19

(a) Find the times taken for the three stages of the journey, AB, BC, CD.

(b) Find the distances of the three stages.

(c) Find the speeds of the three stages.

(d) Find the average speed for the whole journey.

(4) Dr Forster drives from her surgery to see a patient. On her way back she stops to visit another patient. The graph of her distance from the surgery against time is shown in fig 14.20.

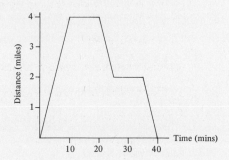

Fig 14.20

(a) How long did she spend with each patient?

(b) How far was the first patient from the surgery?

(c) What was her speed when she drove to the first patient?

(5) The speed of a tube train between stations is given in fig 14.21.

(a) How long did it take the train to travel between the two stations?

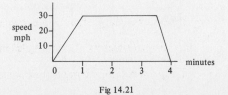

Fig 14.21

146

(b) What was the maximum speed?

(c) For how long did the train stay at a steady speed?

(6) Fig 14.22 shows the graph of the inflation rate in a country over a period of 6 years.

(a) For how long was inflation over 10%?

(b) When did inflation start to come down?

(c) If inflation continues down at a steady rate, when will it be zero?

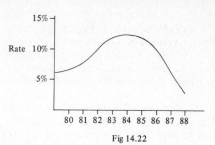

Fig 14.22

(7) Simon records the temperature every hour for 24 hours. His result are shown in the graph of fig 14.23.

(a) What was the greatest and the least temperatures?

(b) When was the temperature falling most rapidly?

(c) How long was the temperature below 5°?

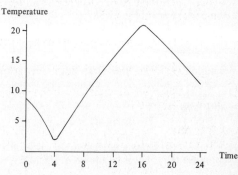

Fig 14.23

(8) A house agent keeps records of the average price of houses in his town. The results over 8 years are given by the following table:

```
Year (31 December) |1978|1979|1980|1981|1982|1983|1984|1985|
Price in £1,000's  | 17 | 19 | 20 | 18 | 21 | 26 | 30 | 34 |
```

Draw a graph to illustrate these figures, taking 1 cm along the x-axis to represent 1 year, and 1 cm up the y-axis to represent £5,000.

(a) When did prices fall?

(b) When were prices rising most steeply?

(c) When approximately was the average price equal to £25,000?

(9) A stone is thrown vertically up into the air. Its height afterwards is given by the following table:

```
Time in Secs. | 1 | 2 | 3 | 4 | 5 | 6 | 7 | 8 | 9 |
Height in m.  | 42| 74| 96|108|110|102| 84| 56| 18|
```

Plot these figures, taking 1 cm along the x-axis to represent 1 sec, and 1 cm up the y-axis to represent 10 m. Join the dots up as smoothly as you can.

(a) What was the greatest height reached? When was it reached?

(b) At what times was the stone 100 m above the ground?

(c) By continuing the graph until it reaches the x-axis, find how long the stone was in the air.

For level 2

The gradient of a distance time graph represents the *speed*.

The gradient of a curve is the gradient of its *tangent*.

14.2.3 Example for level 2

Maurice enters a marathon race. He starts off at a steady rate, then as he gets tired he slows down. The graph of distance against time is shown in fig 14.24. Find (a) his starting speed (b) his speed after 2 hours.

Solution (a) For the first hour the graph has a constant gradient. The y change is 6 miles, and the x change is 1 hour.

He starts off at 6 m.p.h.

(b) Draw a tangent to the curve at x = 2, and complete the triangle as shown in fig 14.25.

The y change of the tangent is 8 miles, and the x change is 2 hours.

After 2 hours he is running at $^8/_2 = 4$ m.p.h.

14.2.4 Exercises for level 2

(1) Michael sets off for school, then realises that he might be late and starts to run. The graph of his distance against time is shown in fig 14.26.

(a) What was his walking speed?

(b) What was his running speed?

(c) If he had continued to walk, how long would he have taken to get to school?

(2)   In question 1 of 14.2.2, what speed does Karen walk at? What speed is the bus?

(3)   In question 2 of 14.2.2, how fast did George walk in the three stages of his journey?

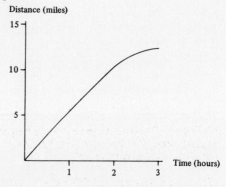

Fig 14.24

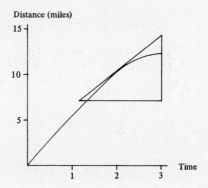

Fig 14.25

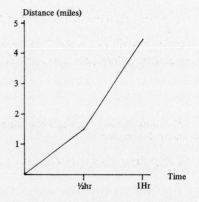

Fig 14.26

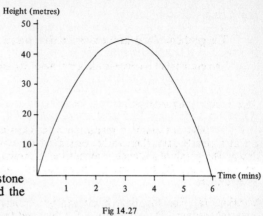

Fig 14.27

(4) Fig 14.27 shows the graph of the height of a stone against time. By drawing suitable tangents find the speeds at 1 sec and at 3 secs.

(5)   The height of a child is measured over the first 6 months of its life. The results are shown in the following table:

```
Age in months |  0  |  1  |  2  |  3  |  4  |  5  |  6  |
Height in in.  | 14  | 15  | 15.8| 16.5| 17.1| 17.6| 18.0|
```

Plot these values on a graph, taking 1 cm along the x-axis to represent 1 month and 1 cm up the y-axis to represent $\frac{1}{2}$ in. Join up the points as smoothly as you can. Draw tangents to the curve at 2 months and at 5 months. Find the gradients of these tangents, and hence find how fast the baby is growing at these times.

(6) The graph of fig 14.28 shows the amount of petrol in the storage tanks of a garage over a 24 hour period.

(a) What is the storage capacity of the tanks?

(b) When were the tanks refilled? How long did it take?

(c) When was the period of greatest demand?

(d) By drawing a tangent find out what was the demand, in gallons per hour, at 09.00.

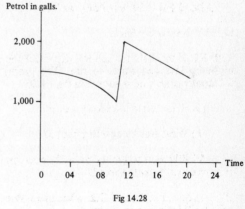

Fig 14.28

For level 3

The area under a speed-time curve corresponds to the *distance*. This area can often be approximated by *trapezia*.

The gradient of a speed-time curve corresponds to the *acceleration*.

14.2.5 Examples for level 3

(1)   The graph of fig 14.29 gives the speed of a sports-car as it accelerates from rest.

      (a) Find the acceleration at 3 secs.

      (b) By approximating the area by two trapezia and a triangle, find the distance gone in the first 6 seconds.

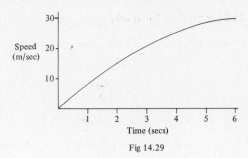

Fig 14.29

Solution   (a) Draw a tangent to the curve at x = 3, as in fig 14.30. Measure the gradient of this tangent.

$$\text{Acceleration} = \frac{20}{4} = 5 \text{ m/sec}^2$$

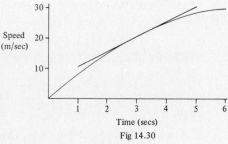

Fig 14.30

      (b) Draw vertical lines at x = 2 and x = 4, as in fig 14.31. The area under the curve is approximately the area in the two trapezia and the triangle.

$$\text{Distance} \simeq \tfrac{1}{2} \times 15 \times 2 + \tfrac{1}{2}(15+25) \times 2 + \tfrac{1}{2}(25+30) \times 2 = 110 \text{ m.}$$

(2)   Fred sets off for school, and walks the half mile to the bus-stop at a steady speed of 3 m.p.h. He waits 20 minutes for a bus, which then travels the 5 miles to school at 30 m.p.h. Draw a distance-time graph for his journey.

Solution   The first stage of the journey lasts for 10 minutes, and reaches $\frac{1}{2}$ mile.

      While he is waiting at the stop the graph is flat, for a period of 20 minutes.

      The final stage lasts for 10 mins and reaches $5\frac{1}{2}$ miles. The result is shown in fig 14.32.

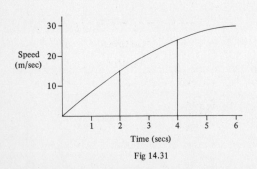

Fig 14.31

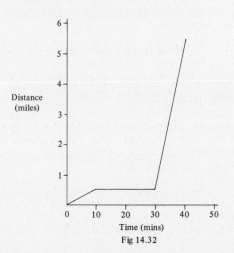

Fig 14.32

14.2.6 Exercises for level 3

(1) The speed of a tube train between successive stations is given by fig 14.33.

      (a) Find the acceleration when it starts and the deceleration when it stops.

      (b) Find the distance between the stations.

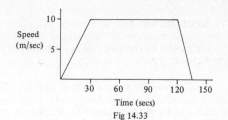

Fig 14.33

(2) The speed of a runner during a sprint race is shown in fig 14.34.

      (a) Find the acceleration of the runner after 1 second.

      (b) By splitting the region into 3 trapezia and a triangle estimate the total distance of the race.

(3) The speed of a lift is shown in fig 14.35.

      (a) What is the acceleration when it starts to rise?

      (b) What is the total height it rises?

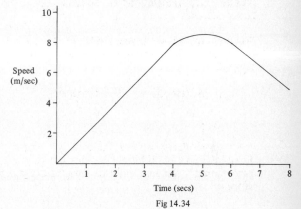

Fig 14.34

(4)   Jules lives 6 miles due North of Jim. They set out to visit each other at the same time; Jules walks South at $4\frac{1}{2}$ m.p.h. and Jim walks North at $3\frac{1}{2}$ m.p.h.

      On the diagram of fig 14.36 draw graphs showing their distances from Jim's house. When do they meet? How far North of Jim's house do they meet?

(5)   Laurie and Daphne are both driving to Bristol along the M4. Laurie's car cruises at 50 m.p.h., and he sets off from London at noon. Daphne's car cruises at 60 m.p.h., and she sets off 20 minutes later.

      On the diagram of fig 14.37 draw graphs showing their distances from London. When does she pass him? How far from London are they when she passes him?

(6)   Fred has two inaccurate clocks. At midday the first clock is 15 minutes fast, but it loses 5 minutes per hour. The second clock is 15 minutes slow, but it gains 15 minutes per hour.

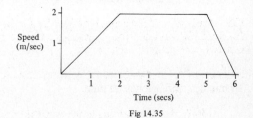

Fig 14.35

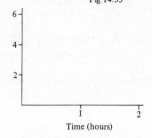

Fig. 14.36

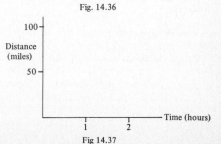

Fig 14.37

What times do the clocks show at 1 o'clock? On the graph of fig 14.38 draw lines representing the times shown by the two clocks. When do they show the same time?

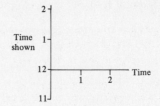

Fig 14.38

(7)   Val's bathtime is as follows. She fills the bath, then lies in it. After a while she adds more hot water, and lies in it for a few more minutes. She then pulls the plug.

Draw a graph, not to scale, showing the level of water in the bath against time.

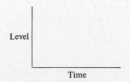

Fig 14.39

(8)   A tap can deliver 20 litres of water per minute. It is being used to fill a cylindrical tank which holds 80 litres. Half way up the tank is a hole, through which water pours out at 10 litres per minute.

Draw an accurate graph of the level of water against time.

(9)   Water is being poured at a steady rate into a conical container. Which of the following pictures represents the graph of the water level against time?

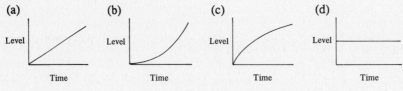

Fig 14.40

(10)   You read in a newspaper that the population of your town is still growing, but at a progressively slower rate. Draw a rough graph of population against time.

(11)   During a rainstorm an open cylinder was left out to catch the rain.

For the first 2 hours 6 cm fell at a steady rate. Then the rain slackened, and for the next hour $1^1/2$ cm fell. The rain then ceased, and 2 hours after that a plug was removed at the bottom of the cylinder, causing the water to drain away in 10 minutes.

Illustrate the information above, by means of an accurate graph of the water level in the cylinder against time.

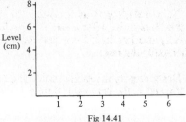

Fig 14.41

(12)   Dave is hitch-hiking. He gets a lift at 45 m.p.h. for 20 minutes. This takes him to the beginning of the motorway, where he waits for 30 minutes. Then he gets a lift at 60 m.p.h. for 120 miles.

Illustrate the above information, by means of a graph of distance against time.

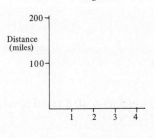

Fig 14.42

COMMON ERRORS

(1)   **Co-ordinates**
       (a) The x co-ordinate is always given first. The point at (3,2) is at x = 3 and y = 2. Be sure not to get these the wrong way round.

       (b) Make sure that the scale of your graph is uniform. It is misleading to write down a scale in which the x-values or y-values are not evenly spaced.

       (c) Make sure that you get the gradient the correct way up. It must be the y-change over the x-change, not the other way round.

       (d) The gradient is the *y-change* over the *x-change*. It is not the *y-value* over the *x-value*.

       (e) A line through a scatter diagram does not necessarily go through the first and last points. The gradient must be of the line you have drawn, not of the line through the first and last points.

       (f) A tangent should just *touch* a curve. It does not necessarily go through the origin (0,0).

       (g) The gradient of a curve at a point is the gradient of the tangent at that point. It is not the gradient of the line from the origin to that point.

(2)   **Interpretation**
       (a) Read the question very carefully, to make sure that you do not draw a distance-time graph instead of a speed-time graph, or the other way round.

       (b) To find the *average* speed, divide the total distance by the total time. To find the *instantaneous* speed, find the gradient of the tangent to a distance-time curve. Do not confuse these.

15 Functions and Flow Charts

15.1 GRAPHS OF FUNCTIONS. Levels 2 and 3

15.1.1 Examples for level 2

(1)   Complete the table for the function $y = (x + 1)^2$. Draw a graph of the function using a scale of 1 cm per unit. Find the gradient of the tangent at the point $x = 0$.

x	−3	−2	−1½	−1	−½	0	1
y	4			¼			

Solution   For $x = -2$, $y = (-2 + 1)^2 = (-1)^2 = 1$. Complete the rest of the table similarly. The final result is:

x	−3	−2	−1½	−1	−½	0	1
y	4	1	¼	0	¼	1	4

The graph is shown in fig 15.1.

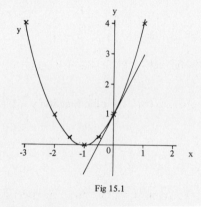

Fig 15.1

Draw a tangent at the point $x = 0$. The y change is 2 and the x change is 1.

The gradient of the tangent is 2

(2)   y is given in terms of x by the formula $y = 2 - 1/x$. Complete the following table.

x	¼	½	¾	1	1½	2	3
y	−2					1½	

Draw a graph of the function, using a scale of 1 cm per unit. Use your graph to solve the equation $2 - 1/x = 1/2$.

154

Solution For x = $^1/_2$, y = 2 - 1/($^1/_2$) = 2 - 2 = 0. The other values are found similarly. The completed table is:

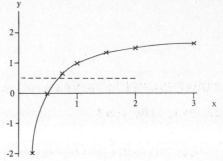

$$x| \ ^1/_4| \ ^1/_2| \ ^3/_4| \ 1|1^1/_2| \ \ 2 \ \ | \ 3 \ |$$

$$y|-2 \ \ | \ 0 \ \ |^2/_3 \ | \ 1|^4/_3 \ | \ 1^1/_2|^5/_3|$$

The graph is shown in fig 15.2.

Draw a line for y = $^1/_2$, and read off where it crosses the graph.

The solution of 2 - $^1/_x$ = $^1/_2$ is x = 0.7

Fig 15.2

15.1.2 Exercises for level 2

(1)   Complete the following table for the function y = (x - 1)^2.

$$x \ | \ \ -1 \ | \ \ 0 \ \ | \ \ ^1/_2 \ \ | \ \ 1 \ \ | \ \ 1^1/_2 \ \ | \ \ 2 \ \ | \ \ 3 \ \ |$$

$$y \ | \ \ 4 \ \ | \ \ \ \ \ | \ \ \ \ \ \ | \ \ \ \ \ | \ \ ^1/_4 \ \ | \ \ \ \ \ \ | \ \ \ \ \ |$$

Draw a graph of the function, using a scale of 1 cm per unit. Draw the tangent at x = -1 and find its gradient.

(2)   Complete the table for the function y = 1 + $^2/_x$.

$$x \ | \ \ ^1/_4 \ | \ \ ^1/_2 \ \ | \ \ ^3/_4 \ \ | \ \ 1 \ \ | \ \ 1^1/_2 \ \ | \ \ 2 \ \ | \ \ 3 \ \ |$$

$$y \ | \ \ 9 \ \ | \ \ \ \ \ \ | \ \ \ \ 3 \ | \ \ \ \ \ \ | \ \ \ \ \ \ | \ \ \ \ \ |$$

Draw a graph of the function. Use your graph to solve the equation:

$$1 + 2/x = 4.$$

(3)   Complete the following table for the function y = x^2 + 2x - 1.

$$x \ | \ \ -3 \ \ | \ \ -2 \ \ | \ \ -1^1/_2 \ | \ \ -1 \ \ | \ \ -^1/_2 \ \ | \ \ 0 \ \ | \ \ 1 \ \ |$$

$$y \ | \ \ \ \ | \ \ \ \ \ \ | \ \ \ \ \ \ \ | \ \ \ \ \ | \ \ \ \ \ \ | \ \ \ \ | \ \ \ \ |$$

Draw the graph of your function. Use your graph to find approximate solutions to the equation:

$$x^2 + 2x - 1 = 0$$

(4)   Complete the following table for the function $y = x + 1/x$.

x	$1/8$	$1/4$	$3/8$	$1/2$	$3/4$	1	$1^1/2$	2	3
y									

Draw a graph of the function. Draw tangents at $x = 1/2$ and at $x = 2$, and find their gradients.

(5)   Set up a table of values for the function $y = 3x - 2$, using the x-values 1, 2, 3, 4. Draw a graph for the function, using a scale of 1 cm per unit.

(6)   Set up a table of values for the function $y = x^2 - 3x + 1$, using the x-values 0, 1, $1^1/2$, 2, 3. Draw a graph of the function using a scale of 1 cm per unit.

Use your graph to solve the equation $x^2 - 3x + 1 = 0$.

(7)   Draw a graph of the function $y = 3 - x^2$, using the x-values -2, -1, $-1/2$, 0, $1/2$, 1, 2.

Use your graph to solve the equation $3 - x^2 = 1$.

Draw a tangent to the curve at $x = -1$, and find its gradient.

(8)   Draw a graph for the function $y = 3 - 1/x$, using the x-values $1/4$, $1/2$, 1, 2, 3, 4, 5.

Use your graph to solve the equation $3 - 1/x = 1 \ 1/2$.

Draw a tangent at $x = 2$ and find its gradient.

For level 3

The function $y = 3x + 2$ is sometimes written as:

$$f(x) = 3x + 2$$

$$\text{or as } f{:}x \rightarrow 3x + 2.$$

If g is given as $g{:}x \rightarrow 2x - 1$, then the *composition function* fg is defined by:

$$fg(x) = f(g(x)) = 3(2x - 1) + 2 = 6x - 1.$$

$$\text{So } fg{:} x \rightarrow 6x - 1.$$

15.1.3 Examples for level 3

(1)   A function f is given by $f(x) = x^3 - 2x + 1$. complete the following table:

x	-2	-1	$-1/2$	0	$1/2$	1	$1^1/2$	2
f(x)								

Draw a graph of the function, using a scale of 1 cm per unit. By drawing a suitable line find the solutions of the equation:

$$x^3 - 2x + 1 = x + 1$$

Solution   After it has been completed the table is as follows:

x	-2	-1	-^{1}/2	0	^{1}/2	1	1^1/2	2
f(x)	-3	2	1^7/8	1	^{1}/8	0	1^3/8	5

The graph is shown in fig 15.3.

Draw the line y = x + 1. Notice that it crosses the graph at 3 places.

The solutions are x = -1.7, 0, 1.7.

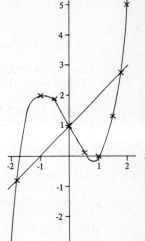

Fig 15.3

(2)   Let f be the function f:x → sin x. Fill up the following table of values:

x	0	45	90	135	180	225	270	315	360
f(x)									

Sketch the graph of this function, taking 1 cm to be 45° on the x-axis, and 2 cm to be 1 unit along the y-axis.

If g is defined by g:x → 2 - x, sketch the graph of gf.

Solution   Use your calculator to find the values of f. Complete the table:

x	0	45	90	135	180	225	270	315	360
f(x)	0	0.71	1	0.71	0	-0.71	-1	-0.71	0

The graph is shown as the filled in line of fig 15.4.

gf(x) = 2 - sin x. Subtract each of the f(x) values from 2 and plot the result. The graph is the dotted line of fig 15.4.

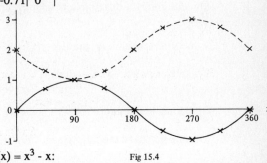

Fig 15.4

15.1.4 Exercises for level 3

(1)   Complete the table of values for the function f(x) = x^3 - x:

x	-2	-1	-^{1}/2	0	^{1}/2	1	1^1/2	2
f(x)								

Draw the graph of this function. By drawing an appropriate line solve the equation:

$$x^3 - x = {}^1/2x - 1.$$

(2)   Complete the table of values for the function $f(x) = x^3 - 2x^2$:

```
x     |-1 |-1/2 | 0 | 1/2 | 1 |11/2 | 2 |
f(x) |   |     |   |     |   |     |   |
```

Draw a graph of this function. By drawing an appropriate line solve the equation:

$$x^3 - 2x^2 = x - 2.$$

(3)   Complete the table of values for the function $y = x^2 - \frac{1}{x}$

```
x| 1/4 | 1/2 | 3/4 | 1 |11/2 | 2 |

y|     |     |     |   |     |   |
```

Draw the graph of this function. On the same paper draw the graph of the line $y = x - \frac{1}{2}$.

Show that where the graphs cross x obeys the equation

$$x^3 - x^2 + \frac{1}{2}x - 1 = 0.$$

Write down the solutions to this equation.

(4)   Complete the table of values for the function $f(x) = x + \frac{1}{x^2}$:

```
x     | 1/4 | 1/2 | 3/4 | 1 |11/2 | 2 | 3 |

f(x) |     |     |     |   |     |   |   |
```

Draw the graph of this function, taking 1 cm to be 1 unit along the x-axis and 1 cm to be 4 units up the y-axis.

On the same paper draw the line $g(x) = \frac{1}{2}x + 3$. Show that where the lines cross x obeys the equation:

$$x^3 - 6x^2 + 2 = 0.$$

Write down the solutions to this equation.

(5)   Draw the graphs of the following quadratics, indicating the axes of symmetry of the curves.

(a) $y = x^2 - 1$ (b) $y = x^2 - 2x + 3$

(c) $y = x^2 + 3x - 2$ (d) $y = 1 + x - x^2$

(6)   Draw the graph of $y = x - \frac{1}{2}x^3$, taking x-values of -2, -1, -$\frac{1}{2}$, 0, $\frac{1}{2}$, 1, 2.

Draw the tangent to the curve at $x = 1$ and find its gradient.

(7) Complete the table for the function c(x) = cos x:

x | 0 | 30| 60| 90|120|150|180|210|240|270|300|330|360|
c(x)| | | | | | | | | | | | | |

Draw the graph of c(x), taking 1 cm per 60° along the x-axis and 2 cm per unit up the y-axis.

Use your graph to solve the following equations:

(a) cos x = $^1/_2$ (b) cos x = $-^1/_2$ (c) cos x = 0.6 (d) cos x = -0.8

(8) Use the graph of question 7 to draw the graph of 2 + cos x.

(9) Use the graph of question 7 to draw the graph of 1 - 3 cos x.

(10) let f and g be functions defined by f:x $\rightarrow$ 3x - 1 and g:x $\rightarrow$ 3 - 2x.

Find (a) f(1)   (b) g(2)   (c) f(g(3))   (d) g(f(3)).

Express fg and gf in the form fg:x $\rightarrow$ .... and gf:x $\rightarrow$ ....

15.2 STRAIGHT LINE GRAPHS AND GRAPH PROBLEMS. Level 3

The graph of a straight line must be of the form y = mx + c, where m and c are constants.

The line has gradient m, and it crosses the y-axis at (0,c).

15.2.1 Examples for level 3

(1) Find the equation of the straight line which passes through (1,5) and (3,9)

Solution   The line goes through (1,5). Put these values into the equation y = mx + c:

$$[1] \quad 5 = m + c.$$

The line also goes through (3,9):

$$[2] \quad 9 = 3m + c.$$

Subtract [1] from [2].

$$4 = 2m.$$

Hence m = 2. Substitute to find that c = 3.

The equation is y = 2x + 3.

159

(2)   A farmer has 100 m. of fencing with which to enclose a rectangular field, one side of which is a stone wall.

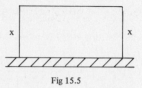

Fig 15.5

Let the two sides perpendicular to the wall be of length x. Show that the area enclosed is $100x - 2x^2$.

Draw the graph of this function, and find the greatest area that the farmer can enclose.

Solution   After x metres have been used for the two perpendicular sides, the amount remaining for the parallel side is 100 - 2x.

The area of the rectangle is the product of the length and the breadth:

$$x(100 - 2x) = 100x - 2x^2 \text{ m}^2$$

The graph of $y = 100x - 2x^2$ is in fig 15.6. The greatest area is the greatest height of the curve.

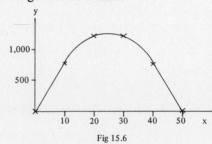

Fig 15.6

The greatest area is 1,250 m^2

15.2.2 Exercises for level 3

(1)   Find the equations of the straight lines which go through the following pairs of points:

   (a) (0,0) and (2,6)   (b) (1,2) and (4,5)   (c) (2,1) and (3,-2)

   (d) (3,2) and (5,3)   (e) (-1,-2) and (2,4)   (f) (-2,3) and (2,-5)

(2)   Find the equations of the straight lines shown in fig 15.7.

(a)
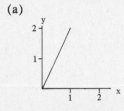

(b)
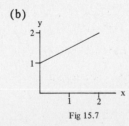

(c)
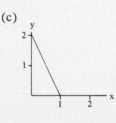

Fig 15.7

(3)   The profit of a firm in successive years after it has been started is shown by the following table, in which x is the number of years and P is its profit in £10,000's.

x	1	2	3	4	5	6	7
P	0.5	1.2	2.0	2.1	2.9	3.2	4.0

Plot these points on a scatter diagram. Draw a straight line to fit these points. Find the equation of your straight line.

From your equation estimate the profit after 8 years.

(4)   Find the equation of the line in example 2 of 14.1.3.

(5)   Find the equations of the lines in exercise 7 of 14.1.4.

(6)   A stone is thrown up in the air, so that x seconds later its distance y above the ground is 2 + 20x - 5x^2.

Plot the graph of this function, for x between 0 and 5. Use your graph to answer the following:

(a) What is its greatest height?

(b) When is it 10 m. high?

(c) For how long is it in the air?

(7)   The *Highway Code* states that the distance it takes for a car to stop is given by the formula D = s + s^2/20, where s is the speed in m.p.h. and D is the distance in feet.

Plot the graph of this function. Use your graph to solve the following:

(a) How long does a car travelling at 55 m.p.h. take to stop?

(b) If a car must be able to stop within 200 feet, what is the greatest speed it can do?

(8) In fig 15.8 ABCD is a square with side 10 cm. P, Q, R and S are points along the sides such that AP = BQ = CR = DS = x cm.

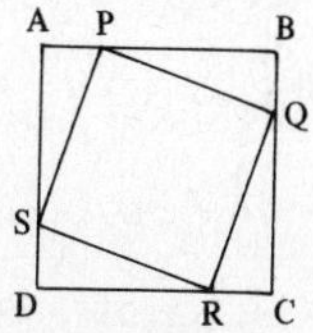

Show that the area of Δ APS is $^1/2$x(10 - x).

Show that the area y of PQRS is given by:

$$y = 100 - 2x(10 - x)$$

Fig 15.8

Plot the graph of y against x, taking values of x between 0 and 10. Hence find the value of x which makes y as small as possible.

(9) An open cardboard box is to be made by cutting off the corners from a 20 cm square sheet of cardboard and folding the sides over. Fig 15.9.

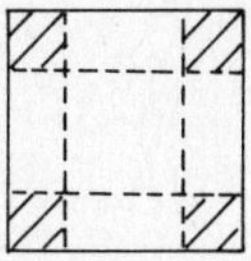

Fig 15.9

Let a square of side x cm be cut off from each corner. Show that the base of the box is a square of side 20 - 2x.

Show that the volume V of the box is x(20 - 2x)^2.

Draw a graph of V in terms of x, taking values of x from 0 to 10. Hence find the value of x which gives the largest volume.

(10)   An open rectangular box is to be made with a square base. The volume must be 4 in^3.

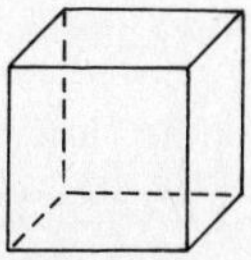

Let the height be h in. and the base have side x in. Show that hx^2 = 4.

Fig 15.10

Show that the surface area of the box is $A = x^2 + 4xh$.

Show that this can be re-written as $A = x^2 + {}^{16}/x$.

Draw a graph of A against x, taking values of x from $^1/2$ to 3. Hence find the least possible surface area of the box.

15.3 FLOW CHARTS

15.3.1 Example for level 1

The flow chart shown converts temperature in Fahrenheit to Centigrade. Use the flow chart to find the Centigrade equivalent of 32°, 50°, 131°.

Write a new flow chart which will convert Centigrade to Fahrenheit.

| Read F° | → | Subtract 32 | → | Divide by 9 | → | Multiply by 5 | → | Write C° |

Solution   Put in the values and apply the operations.

32°F	0	0	0	0°C
50°F	18	2	10	10°C
131°F	99	11	55	55°C

To convert Centigrade to Fahrenheit reverse the flow chart.

| Read C° | → | Divide by 5 | → | Multiply by 9 | → | Add 32 | → | Write F°. |

15.3.2 Exercises for level 1

(1)   The following flow-chart works out values of the expression $3(x + 5)$. Use it to evaluate the expression when x = 1, 3, 10.

| Read x | → | Add 5 | → | Multiply by 3 | → | Write y |

(2)   The standing charge for a telephone is £12.50, and the cost per unit used is 2 p. The following flow-chart finds the total bill £C in terms of the number x of units used.

| Read x | → | Multiply by 0.02 | → | Add £12.50 | → | Write £C |

Find the charge for bills which have used (a) 300 units (b) 500 units.

Write the flow chart which will convert the charge £C to the number x of units used.

(3)   Write down the flow-chart which will work out values of the expression $2(x - 7)$. Use your flow chart to evaluate the expression when x = 12, 20, 100.

(4)   Tracy writes a flow-chart to describe the making of toast. She puts the instructions in the wrong order. Re-arrange them in the right order.

| Put bread in | → | Cut slice | → | Turn toast | → | Take out toast | → | Light grill |

(5)   Re-order the following instructions in a flow-chart to describe the making of a telephone call.

| Lift receiver | → | Wait for answer | → | Dial number | → | Find Number |

(6)   Re-arrange the following instructions so that they form a flow-chart for washing hands.

| Pull plug out | → | Wait | → | Turn tap on | → | Turn tap off | → | Wash hands | → | Put plug in |

15.3.3 Example for level 3

The following flow-chart evaluates $\sqrt{2}$. Use it to fill in the table of values. How would you find $\sqrt{2}$ to a greater degree of accuracy?

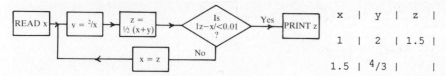

```
x  |  y  |  z  |

1  |  2  | 1.5 |

1.5 | 4/3 |     |
```

Solution   Fill in the table as shown:

```
  x    |   y   |   z    |
  1    |   2   | 1.5    |
 1.5   | 4/3   | 1.417  |
 1.417 | 1.412 | 1.414  |
```

The answer 1.414 will be printed.

To get a greater degree of accuracy change the entry in the diamond shaped box. Changing it to $|z - x| < 0.0001$ will make the answer accurate to 4 decimal places.

15.3.4 Exercises for level 3

(1)   The following flow chart evaluates $\sqrt{3}$. Use it to fill in the table, and hence to find $\sqrt{3}$ to 3 decimal places.

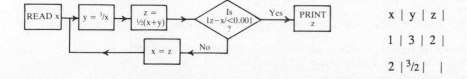

```
x | y | z |

1 | 3 | 2 |

2 | 3/2 |   |
```

(2) Write a flow-chart which will evaluate $\sqrt{10}$. Use it to find $\sqrt{10}$ to 3 decimal places.

(3) If n is a positive integer, n! is the product of all the numbers up to and including n. The following flow-chart evaluates n! Use it to find 4!, 6!.

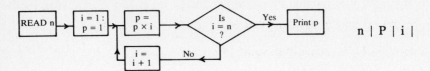

(4) The following flow-chart evaluates the sum of all the numbers up to and including n. Fill in the empty boxes.

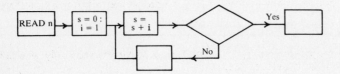

(5) Apply the following flow chart to the numbers 5, 8, 15, 17. Which sort of numbers end in List A and which in List B?

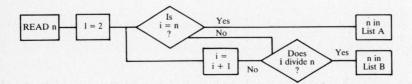

(6) When the score at tennis is "Deuce", the winner is the first player to be 2 clear points ahead. Fill in the boxes of the following flow-chart to describe the scoring at Deuce.

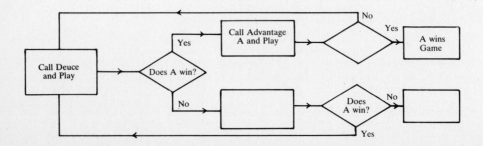

COMMON ERRORS

(1)   Graphs of functions
(a) Algebraic mistakes are very frequent when working out tables of values. Be very careful when negative numbers are involved. For example, it is easy to get the wrong sign when working out $1 - 2x^2$ for $x = -1$.

Be very careful also when dividing. When $x = {}^1/_2$, ${}^1/_x$ is equal to 2, not ${}^1/_2$.

Do not divide by 0. $1 + {}^1/_x$ has no meaning when $x = 0$.

(b) Once you have plotted the points of a graph, do not join them up by straight lines. Make as smooth a curve as you can to connect them.

(c) If your scale is not simple, then be very careful when plotting points. It is easy to make mistakes when the scale is, for example, 1 cm per 2 metres.

(d) There are various common errors to do with tangents and scatter diagrams, which are listed in the previous chapter.

(2)   Solving equations by graphs
(a) If you are solving an equation in x by the graph method, then you need only give the *x-value* of the crossing point. The y-value is unnecessary.

(b) If a line and a curve cross at more than one point, then give the x-values of *all* the points where they cross. They are all solutions of the equation.

(3)   Straight line graphs
If you know that the line $y = mx + c$ goes through (2,3) say, then that means that *x and y* are equal to 2 and 3 respectively. So the equation you obtain is:

$$3 = 2m + c$$

Do not put m = 2 and c = 3, to obtain $y = 2x + 3$

(4)   Flow-charts
(a) When a number is going through a flow-chart, its value may be changing. Be careful that you do not use the wrong value of the number.

(b) The convention for flow-charts is that *orders* are in rectangular boxes, and *questions* in diamond boxes. Make sure that you use the correct shapes.

(c) Do not be confused by the statement "S = S + 3". It can be read: "Change S to S + 3" or "Add 3 to S".

16 Constructions and Drawings

16.1 CONSTRUCTIONS AND MEASUREMENT

The main instruments used in geometrical drawings are the *ruler* (to measure distances), the *protractor* (to measure angles), and *compasses* (to draw circles).

16.1.1 Examples for level 1

(1)   Fig 16.1 shows a cylinder held on top of a ruler. What is the diameter of the cylinder?

Solution The left hand edge is at 5.6 cm. The right hand edge is at 8.3. Subtract one measurement from the other.

The diameter is 8.3 - 5.6 = 2.7 cm.

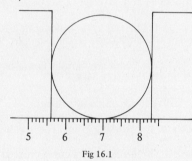

Fig 16.1

(2)   The triangle ABC has AB = 2 cm, BC = 3 cm, CA = 4 cm. Make an accurate drawing of the triangle, and measure the angle at A.

Solution   The steps are shown in fig 16.2.
        (a) Use a ruler to draw the line CA of length 4.

        (b) Separate compasses to a width of 3 cm, and with the point at C draw an arc.

Fig 16.2

        (c) Separate the compasses to 2 cm, and with the point at A draw an arc.

        (d) Where the arcs intersect is B. Join up the sides.

        Lay the protractor on AC so that the central point is at A. Notice that AB makes an angle of about 47°.

<BAC = 47°

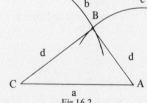

16.1.2 Exercises for level 1

(1) Fig 16.3 shows a ruler laid on top of a cube.
    Find the side of the cube.

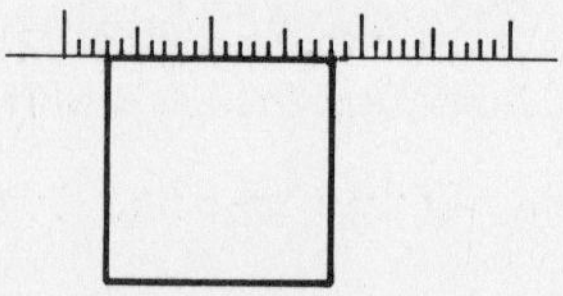

Fig 16.3

(2) Fig 16.4 shows a length of wood which is to be divided into 5 equal parts. Find the length of the wood. Divide that length by 5. Mark on the wood where it should be cut.

Fig 16.4

(3) Measure the angles x°, y°, z° of fig 16.5. Find x + y + z. What do you notice?

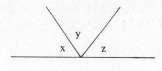

Fig 16.5

(4)   Draw a line AB of length 2 inches. Use your protractor to make an angle of 40° at A.

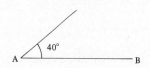

Fig 16.6

(5)   Draw a line AB of length 5 cm. Use your protractor and ruler to draw a line AC such that <BAC = 50° and AC is of length 7 cm. Find the length BC and the angle <ACB.

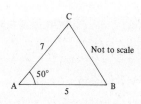

Fig 16.7

(6) The dots of fig 16.8 are 1 cm apart.

    (a) Measure AC.

    (b) Find the angle <ACB.

    (c) Separate your compasses to the length DC. Draw a circle with centre at D. How many dots does this circle go through?

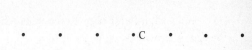
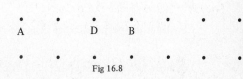

Fig 16.8

(7)   Draw a line PQ of length 6 cm. Use your compasses and ruler to find a point R such that PR = 7 cm and QR = $8^1/2$ cm. Find the angles of the triangle PQR.

(8)   Construct a triangle XYZ such that XY = 10 cm, YZ = 11 cm, XZ = 12 cm. Find the angle <XYZ.

(9)   Construct a triangle ABC such that AB = 5 cm, AC = 12 cm, BC = 13 cm. What is <A?

(10) Draw a line AB of length 6 cm. Use your protractor to draw a line through A which makes 55° with AB. Draw a line through B which makes 75° with BA.

    Let the two lines cross at C. Check that <ACB = 50°. Measure AC and CB.

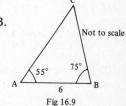

Fig 16.9

(11)   Construct a triangle PQR such that PQ = 7 cm, <RPQ = 82° and <RQP = 72°. What are RQ and RP?

(12)   Construct a triangle XYZ such that XY = 4 in., <ZXY = 35°, <ZYX = 47°. Find ZX and ZY.

(13)   Draw a line AB of length 7 cm. Draw a line through A which makes 46° with AB. Let C be on this line, with AC = 5¹/2 cm.

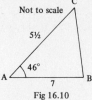

Fig 16.10

    (a) Find BC.

    (b) Find <ACB.

(14)   Construct a triangle PQR with PQ = 7 cm, <RPQ = 65°, RP = 6¹/4 cm.

    (a) Find RQ

    (b) Find <RQP.

(15)   Construct a triangle XYZ with XY = XZ = 11 cm and <YXZ = 50°. Find YZ.

16.1.3 Examples for level 2

(1)   Make an accurate copy of the lines AB and AC of fig 16.11.

    Draw the bisector of the angle <BAC.

    Construct the perpendicular bisector of the line AB.

    Let these two new lines meet at D. Measure AD.

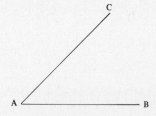

Fig 16.11

Solution   The bisector of <BAC is constructed as follows.

    (a) Put the point of your compasses on A, and draw an arc cutting AB and AC at E and F.

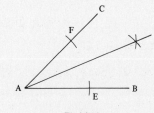

Fig 16.12

168

(b) Draw further arcs with the compass point at E and at F. Join up A with the meet of these further arcs.

The perpendicular bisector of AB is constructed as follows.

(c) Draw equal arcs with the point of your compasses at A and at B, meeting AB at G and H.

Draw further arcs from G and H, and join up their meeting points.

Finally measure AD.

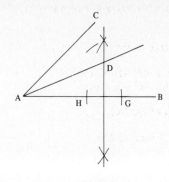

Fig 16.13

AD = 2.4 cm.

(2)   A ship is sailing due East at 10 m.p.h. An enemy gun site with a range of 3 miles is 5 miles East and 2 miles South.

Draw a scale diagram, using 1 cm to represent 1 mile, of the ship's course and the region which is within the range of the gun. For how long is the ship in danger?

Solution   The positions of ship and gun-site are shown in fig 16.14.

The ship's course is the horizontal line SD.

The region within range of the gun is the circle of radius 3 cm. with centre at G.

The ship is in danger while it is within the circle. The length of the chord CD is 4.5 cm, which represents 4.5 miles.

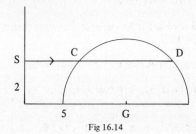

Fig 16.14

The ship is in danger for $^{4.5}/10 = 0.45$ hours

16.1.4 Exercises for level 2

(1)   Construct, using compasses and ruler, the bisectors of the angles below.

(a)   (b)   (c)

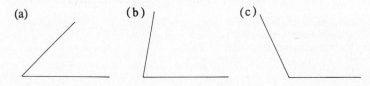

Fig 16.15

(2)   Without measuring the lengths of the lines below construct their perpendicular bisectors.

   (a)   (b)   (c)

Fig 16.16

(3) Make a copy of the triangle ABC of fig 16.17. Construct as accurately as you can the perpendicular bisectors of the three sides. The three bisectors should all go through the same point G.

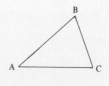

Fig 16.17

      Put the point of your compasses on G and draw a circle radius AG. What do you notice?

(4)   Make another copy of the triangle ABC of fig 16.17. Construct as accurately as you can the bisectors of the three angles. What do you notice about the intersection points?
      Put your compass point on the intersection point and draw a circle which just touches AB. What do you notice?

(5)   Construct a triangle ABC with sides 5 cm, 6 cm, 7 cm. By the method of question 3 construct the circle which goes through the points A, B, C. What is the radius of this circle?

(6)   Construct a triangle PQR with sides 7 cm, 9 cm, 13 cm. By the method of question 4 construct the circle which touches the three sides of the triangle. What is the radius of this circle?

(7)   Construct a quadrilateral ABCD with AB = 5 cm, <A = 100°, DA = 7 cm, <B = 120°, BC = 12 cm. Measure DC and <C.

(8) ABCD is a parallelogram with AB = 10 cm, AD = 8 cm, and <A = 75°. Construct an accurate diagram of ABCD.

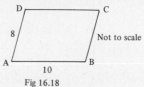

Fig 16.18

(9) The parallelogram ABCD has AB = 5 cm, and the diagonals AC and BD are 6 cm and 11 cm respectively. Use the fact that the diagonals of a parallelogram bisect each other to construct ABCD. What is AD?

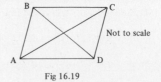

Fig 16.19

(10) The diagonals of a rhombus are 10 cm and 14 cm. Use the fact that the diagonals of a rhombus bisect each other perpendicularly to construct the rhombus. What are the lengths of the sides?

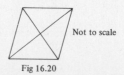

Fig 16.20

Fig 16.21

(11) Draw a circle with radius 5 cm. With your compasses at the same separation mark out points A, B, C, D, E, F on the circumference, as in fig 16.21. What is the figure ABCDEF? What is the figure ACF? Measure AC.

(12) Draw any triangle ABC. Construct the midpoints L, M, N of BC, CA, AB respectively. Join AL, BM, CN. What do you notice?

(13) Fig 16.22 shows a line AB and a point P not on the line.

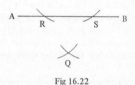

Fig 16.22

Put the point of your compasses on P and draw an arc to cut AB at two points R and S.

Put the point of your compasses on R and draw an arc below AB. Do the same with S.

Let the two arcs meet at Q. Join PQ. What can you say about the lines AB and PQ?

(14) Fig 16.23 shows a triangle ABC which has been rotated through 90° to triangle A'B'C'.

The *centre of rotation* must be the same distance from A as from A'.

Construct the perpendicular bisectors of AA' and of BB'. Let them cross at G.

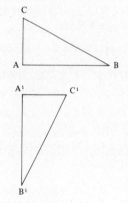

Measure the distances GC and GC'. What do you notice?

Fig 16.23

(15) Draw a square ABCD with side 8 cm.

(a) Shade the region of those points which are less than 3 cm from the side AB.

(b) Shade the region of those points which are less than 2 cm from the point A.

(c) Mark the point X which is 4 cm from AD and 3 cm from AB.

(d) Mark the point Y which is 2 cm from BC and 3 cm from B.

171

(16)   Draw a rectangle PQRS with PQ = 10 cm and PS = 8 cm.

(a) Shade the region of those points which are less than 2 cm from P.

(b) Mark the point Z which is 7 cm from P and 5 cm from Q.

(c) Mark the points which are within 2 cm of the diagonal PR and within 4 cm from R.

(17) Fig 16.24 shows a scale drawing of a room, in which 1 cm represents 0.5 m. X marks the only power point. A lamp has a cable of 1.5 m. Shade the region in the diagram corresponding to the places where the lamp could be put.

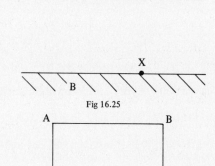

Fig 16.24

(18) Fig 16.25 shows a map of two countries, A and B, with sea in between. The scale is 1 cm to 50 km.
   (a) Country A does not allow boats from B to fish within 20 km of its coast. Shade the region on the map where B's boats may not fish.

(b) A fishing smack can only travel 100 km from its port at X. Shade the region within which it can fish.

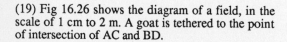

Fig 16.25

(19) Fig 16.26 shows the diagram of a field, in the scale of 1 cm to 2 m. A goat is tethered to the point of intersection of AC and BD.

(a) Mark the point where the goat is tethered.

(b) The goat is on a 3 m. lead. Shade the region within which it can graze.

Fig 16.26

(c) A fierce dog is kennelled at A, on a leash of 2 m. Shade the region within which the goat is in danger from the dog.

(20) An old manuscript tells of buried treasure near the walls of an abbey.

The directions are: *the treasure is buried 3 feet from the South wall, and is 5 feet from the oak tree.*

Mark on the map the places where the treasure might be.

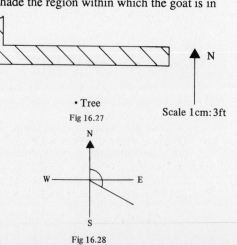

Fig 16.27

Scale 1cm: 3ft

16.2 SCALE DIAGRAMS AND MAPS

Directions are often given in the form of a *bearing*. This is the angle of the direction, measured from North in a clockwise direction.

Fig 16.28

16.2.1 Examples for level 1

(1)   Fig 16.29 shows a map of Scotland. The scale is 1 cm for 50 miles.

   (a) How far is it from Glasgow to Aberdeen?

   (b) What is the bearing of Perth from Stranraer?

Solution   (a) Measure the distance on the map. It comes to 2.2 cm. Multiply this by 50.

The distance from Glasgow to Aberdeen is 110 miles.

   (b) The line from Stranraer to Perth makes 28° with the North direction.

The bearing of Perth from Stranraer is 028°.

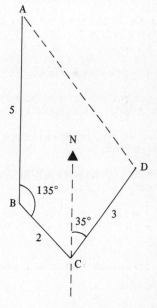

Fig 16.29

(2)   Explorers in the Sahara set off from base camp. They travel South for 50 km, then SE for 20 km, then for 30 km on a bearing of 035°.

   Use a scale of 1 cm to 10 km to make an accurate drawing of the journey. How far are they from the base camp and what is its bearing?

Solution   The first leg of the journey is the 5 cm line AB.

   The second stage BC makes 135° with this line.

   The third stage CD makes 35° with the North South line.

Measure DA:

They are 50 km from base

Measure the angle DA makes with North South.

<DAB = 38°.

Subtract from 360°:

The bearing is 322°

Fig 16.30

16.2.2 Exercises for level 1

(1)   What bearings are the following equivalent to?

   (a) East (b) South (c) North (d) North East (e) South West.

(2)   Express the following bearings in terms of compass directions:

   (a) 090° (b) 270° (c) 135° (d) 315°.

(3) Fig 16.31 shows part of a map of Kent. The scale is 1 cm per 5 km.

    (a) Find the distance from Tonbridge to Goudhurst.

    (b) What is the bearing of Paddock Wood from Tunbridge Wells?

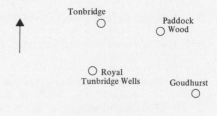

Fig 16.31

(4) The scale of fig 16.32 is 1 cm per 10 km.

    (a) What is the distance of Blantyre from Zomba?

    (b) What is the bearing of Blantyre from Zomba?

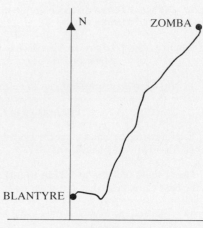

Fig 16.32

(5) Fig 16.33 shows a map of Southern England. The scale is 1 cm to 50 km.

    (a) What is the distance and the bearing of London from Southampton?

    (b) Which town lies on a bearing of 270° from London?

    (c) Which town is 50 km from Ipswich?

Fig 16.33

(6)   A ship leaves harbour and sails North for 20 miles. It then turns and sails East for 10 miles. Using a scale of 1 cm per 5 miles draw a map of the ship's voyage. How far is the ship from harbour? What is the bearing of the ship from the harbour?

(7) Fig 16.34 shows part of a map of the Lleyn peninsula. The scale is 1 inch per mile. Gareth starts from Botwnnog, marked A. What is the bearing of the peak at point B?

He walks 2 miles on a bearing of 340°. Mark his journey on the map. What is the distance and bearing of the peak at B?

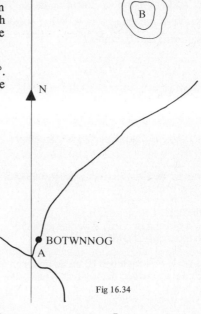

Fig 16.34

(8) From point X the bearing of a target Z is 028°. From point Y, 100 m East of X, the bearing is 340°. Use a scale of 1 cm to 10 m to draw an accurate diagram of the triangle XYZ. What is the distance of Z from X?

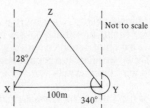

Fig 16.35

(9)   At point A I see a church spire on a bearing of 350°. I walk due North for 200 m to point B and the spire is now on a bearing of 260°.

Using a scale of 1 cm per 50 m. make an accurate diagram of my journey. How far is the spire from point B?

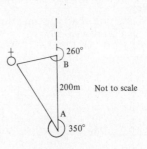

Fig 16.36

175

(10) Aytown is 10 km North of Beetown. The bearing of Ceetown from Beetown is 030° and its bearing from Aytown is 100°.

Using a scale of 1 cm for 2 km make an accurate diagram of the three towns. What is the distance of Ceetown from Beetown?

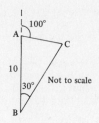

Fig 16.37

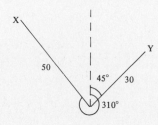

Fig 16.38

(11) From the top of a mountain Eckston is 50 miles away on a bearing of 310°. Wyechester is 30 miles away on a bearing of 045°.

Using a scale of 1 inch per 10 miles make an accurate map of the mountain and the two towns. What is the distance between the towns? What is the bearing of Wyechester from Eckston?

COMMON ERRORS

(1) **Use of instruments**

(a) Make sure that you get the best out of your instruments. You cannot make accurate diagrams with a blunt pencil or with a wobbly compass.

(b) After making a construction, do not rub anything out. The lines and arcs are part of your answer, and the examiner will want to see them.

(c) Do not give your answers to too high a degree of accuracy. With a ruler, you can measure distances to about 0.5 mm. It is misleading to give your answers to a greater degree of accuracy than that.

(2) **Bearings**

Make sure that you measure bearings *clockwise* round from North. Due West, for example, is on a bearing of 270°.

17 Areas and Volumes

17.1 LENGTHS AND AREAS

The *area* of a rectangle is the product of the length and the breadth.

$$A = l \times b$$

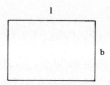

The *perimeter* of a rectangle is the sum of twice the length and twice the breadth.

$$P = 2l + 2b$$

The area of a triangle is half the product of the base and the height.

$$A = \tfrac{1}{2} b \times h$$

The perimeter or *circumference* of a circle is 2π times the radius.

$$C = 2\pi \times r$$

Fig 17.1

17.1.1 Examples for level 1

(1) A rectangular lawn measures 5 metres by 16 metres. What is its area and its perimeter?

Solution   For the area multiply the length by the breadth.

$$A = 5 \times 16 = 80 \text{ m}^2.$$

Fig 17.2

For the perimeter add twice the length to twice the breadth.

$$L = 2 \times 5 + 2 \times 16 = 42 \text{ m.}$$

(2)   The dots of fig 17.3 are 1 cm apart. Find the area of the triangle shown.

Solution   The base of the triangle is 3 cm. Its height is 2 cm.

$$A = \tfrac{1}{2} \times 3 \times 2 = 3 \text{ cm}^2$$

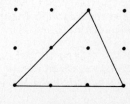

Fig 17.3

(3)   A string of length 15 inches just goes round a circular post. Find the radius of the post. (Take π = 3.142).

Solution   The circumference of the post is 15 inches. This is 2π times the radius. Divide 15 by 2π:

$$\textbf{Radius} = 15 \div 2\pi = \textbf{2.387 inches.}$$

17.1.2 Exercises for level 1

(1)   Find the areas of the rectangles below.

(a)   (b)   (c)   (d)

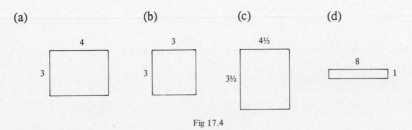

Fig 17.4

(2)   Find the perimeters of the rectangles in question 1.

(3)   A lawn measures 30 feet by 63 feet.

   (a) What is its area and its perimeter?

   (b) What would it cost to spread it with fertiliser at $^1/2$ p. per square foot?

   (c) What would it cost to surround it with a fence costing 35 p. per foot?

(4)   A kitchen floor measures $3^1/2$ metres by 2 metres. What would it cost to cover it with linoleum at £2.30 per square metre?

(5)   A bathroom wall is 2 m. high and 3 m. long. What is its area? What would it cost to tile it with 10 by 10 cm tiles costing 16 p each?

(6)   One side of a rectangle is 20 cm, and its area is 600 cm^2. What is the length of the other side?

(7)   The perimeter of a square is 100 m. What is its area?

(8)   Find the areas of the triangles shown.

(a)   (b)   (c)   (d)

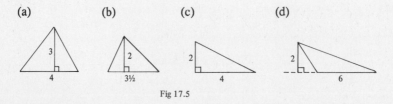

Fig 17.5

(9)   The dots of fig 17.6 are 1 cm. apart. Find the areas of the triangles and rectangles.

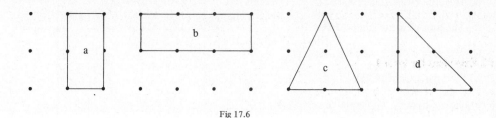

Fig 17.6

(10)   The base of a triangle is 3 ft., and its area is 18 ft^2. What is its height?

(11)   The height of a triangle is 5 cm., and its area is 25 cm^2. What is its base?

(12)   A rectangle on graph paper has its vertices at (1,1), (3,1), (3,4), (1,4). What is its area and its perimeter?

(13)   A triangle on graph paper has its vertices at (1,1), (5,1), (2,4). What is its area?

(14)   A triangle on graph paper has its vertices at (2,1), (6,3), (2,4). What is its area?

(15)   The radius of a running track is 50 m. What is its circumference?

(16)   The circumference of a race track is 1,200 m. What is its radius?

(17)   The radius of the Earth is 6,400 km. What is its circumference?

(18)   A cotton reel is 2.6 cm in diameter. Cotton is wound round it 200 times. What is the length of cotton?

(19)   A square carpet of side 2 metres is on the floor of a room 2^1/2 metres by 4 metres. What is the area of the uncarpeted floor?

(20)   A garden is 40 ft by 20 ft. The flower beds are 2 ft wide on all sides. What area of the garden is covered by lawn?

Fig 17.7

(21)   A picture which is 25 cm by 15 cm is framed within a border 4 cm wide. What is the total area of picture and border?

(22)   Fig 17.8 shows a semicircle and a quarter circle. What are their perimeters?

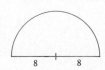

Fig 17.8

(23) The dots of fig 17.9 are 1 cm apart. Find:

    (a) The area of the square ABCD.

    (b) The area of the triangle AEF.

    (c) The area of the square EFGH.

(24) (a) Find the area of PQRS.

    (b) Draw a triangle with equal area to PQRS.

    (c) Draw a rectangle with twice the area of PQRS.

    (d) Draw a rectangle, with one side of length 1 cm, with the same perimeter as PQRS.

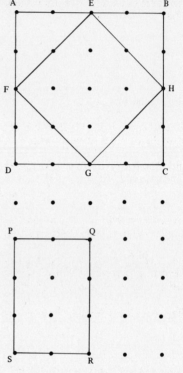

Fig 17.9

For level 2

    The area of a parallelogram is the product of the base and the height.

$$A = b \times h$$

    The area of a trapezium is the product of the height and the average of the parallel sides.

$$A = h \times (d + e)/2$$

180

The area of a circle is π times the square of the radius.

$$A = \pi r^2$$

Fig 17.10

17.1.3 Examples for level 2

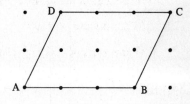

Fig 17.11

(1)   The dots of fig 17.11 are 1 cm apart. Find the area of the parallelogram shown.

Solution   The base AB is 3 cm. long, and the height is 2 cm.

The area is 3 x 2 = 6 cm^2.

(2)   A garden shed is 2 m. from back to front: the back wall is $1^1/2$ m. high and the roof slopes up uniformly to the front wall which is $2^1/2$ metres high. Find the area of a side wall.

Fig 17.12

Solution   Fig 17.12 shows a diagram of the shed. The side wall is a trapezium, in which the parallel sides are of length $1^1/2$ and $2^1/2$. The height is 2.

The area is 2 x ($1^1/2 + 2^1/2$)/2 = 4 m^2.

(3)   A rectangular lawn which is 8 m. by 14 m. contains a circular flower-bed of radius 2 m. Find the area of grass.

Solution   Fig 17.13 shows a diagram of the lawn.

The area of grass is found by subtracting the area of the flower-bed from the whole area.

Fig 17.13

Area = 8 x 14 - $\pi 2^2$ = 99 m^2.

17.1.4 Exercises for level 2

(1)   Find the area of the parallelograms shown.

(a)

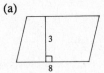

(b)

(c)

Fig 17.14

181

(2)   Find the areas of the trapezia shown.

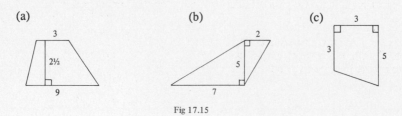

Fig 17.15

(3)   The dots of fig 17.16 are 1 cm apart. Find the areas of the parallelograms and trapezia in the figure.

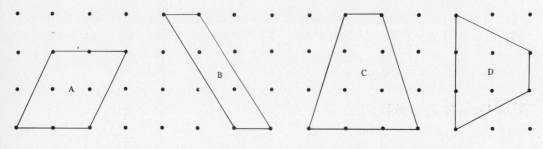

Fig 17.16

(4)   The parallelogram ABCD has its vertices at (1,1), (2,2), (5,2), (4,1). Find its area.

(5)   The trapezium ABCD has its vertices at (4,2), (4,7), (3,6), (3,4). Find its area.

(6)   The kite of fig 17.17 has diagonals of length 4 cm. and 7 cm. Find its area.

Fig 17.17

(7)   Find the areas of the circles with:

(a) Radius 3 cm. (b) Diameter 4 ft. (c) Radius 12 m. (d) Diameter $1/2$ in.

(8)   Find the radii of the circles with:

(a) Diameter 6 cm. (b) Circumference 12 in. (c) Area 9 m^2. (d) Area 5 ft^2.

(9)   A long-playing record has diameter 12 inches, and the playing surface ends 2 inches from the centre. What is the area of the playing surface?

(10) Julian is helping to make mince pies. He cuts circles of radius 3 cm, from a strip of pastry 6 cm. wide, as shown in fig 17.18. What proportion of pastry is not used?

Fig 17.18

(11) Fig 17.19 shows a square of side 4 cm, in which arcs of radius 2 cm. have been drawn from centres at the four corners of the square. Find the shaded area.

Fig 17.19

17.2 VOLUMES AND SURFACE AREAS

The *volume* of a cuboid is the product of its length, breadth and height.

$$V = l \times b \times h$$

The density of a solid is its mass divided by its volume.

Fig 17.20

17.2.1 Examples for level 1

(1) Find the volume of the cuboid of fig 17.21.

Solution The height is 6 cm., the width 5 cm. and the length 4 cm.

The volume is 4 x 5 x 6 = 120 cm².

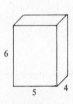

Fig 17.21

(2) The surface of a pond is a rectangle 50 cm. by 30 cm.. During a cold spell it froze to a depth of 2 cm. Find the mass of ice, given that the density of ice is 0.9 g/cm³.

Solution The ice may be thought of as a cuboid, 50 cm long, 30 cm wide and 2 cm deep. Its volume in cm³ is:

$$50 \times 30 \times 2 = 3,000 \text{ cm}^3.$$

Its mass is:

$$\textbf{0.9 x 3,000 = 2,700 g. = 2.7 kg.}$$

17.2.2 Exercises for level 1

(1) Find the volumes of the cuboids in fig 17.22.

(a)   (b)   (c)

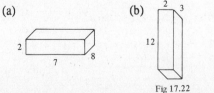

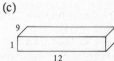

Fig 17.22

(2)   Squash balls are sold in cubes of side 4 cm. How many of these cubes can be packed in a carton which is 20 cm. by 24 cm. by 16 cm.?

(3)   A field is in the shape of a rectangle 50 m. by 40 m. Snow falls on the field to a depth of $1^{1}/2$ m. What is the weight of snow when it melts, given that the density of snow is 50 kg/m^3.?

(4)   A garden is a rectangle 6 m wide by 13 m long. 0.04 m of rain falls: what is the volume of water?

(5)   A cuboid has a square base of side 3 cm. What is its height, given that its volume is 36 cm^3.?

(6)   A water tank is a cuboid, with a base of 1.2 m. by 0.8 m. How deep is the water when the tank contains 0.384 m^3 of water?

(7)   A classroom is 5 m. by 6 m. by 3 m. Health regulations require that each pupil must have at least 5 m^3 of air. How many pupils can the classroom legally hold?

(8)   Sugar cubes are cubes of side $1^{1}/2$ cm. How many fit in a box which is 24 cm. by 6 cm. by 12 cm.? If the density of sugar is 1.5 g/cm^3, what is the mass of one sugar cube?

(9)   The dots of Fig 17.23 are 1 cm apart. (a) is the net of a cube and (b) is the net of a cuboid. Find their areas and their volumes.

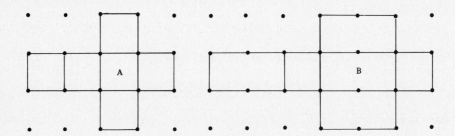

Fig 17.23

For level 2

The volume of a cylinder is the product of the height and the area of the base circle.

$$V = \pi r^2 h$$

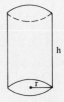

Fig 17.24

The volume of a prism is the product of the height and the area of the cross-section.

$$V = A \times h$$

Fig 17.25

17.2.3 Examples for level 2

(1)   Find the volume of a soup can which is 20 cm high and which has base radius 3 cm.

Solution   The can is a cylinder, with h = 20 and r = 3.

The volume is $\pi 3^2$ x 20 = 565 cm³.

(2)   Fig 17.25 shows a wedge of cheese, with the dimensions as shown. Find its price, given that the density of the cheese is 1.5 g/cm³ and that it costs £2.25 per kilogram.

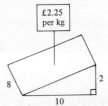

Solution   The wedge is a triangular prism. The cross-sectional area is a right-angled triangle, with area $^1/_2$ x 2 x 10 = 10 cm².

Volume = 8 x 10 = 80 cm³.

Weight = 1.5 x 80 = 120 grams = 0.12 kg.

Price = 0.12 x 2.25 = £0.27 = 27 p.

(3)   A room is 2 m. high, 4 m. long and 3 m. wide. Find the surface area of the walls and ceiling.

Solution   The ceiling is 4 by 3. There are two walls which are 2 by 4, and two walls which are 2 by 3. The total area is therefore:

3x4 + 2x2x4 + 2x2x3 = 40 m²

17.2.4 Exercises for level 2

(1)   Find the volumes of the cylinders of fig 17.26, giving your answers to 3 significant figures.

(a)   (b)   (c)

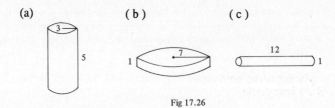

Fig 17.26

(2)   Find the volumes of the prisms of fig 17.27.

(a)   (b)

Fig 17.27

(3)   A tin of shoe-polish is 2 cm. high, and the diameter of its base is 7 cm. Find its volume.

(4)   What is the volume of 1,000 cm. of wire, if the cross-section is a circle with diameter 0.05 cm.?

(5)   What is the area of the walls and ceiling of a room which is $2^1/2$ m. high, 4 m. long and $3^1/2$ m. wide? If a coat of paint is 0.01 cm thick, what is the volume of paint needed to provide one coat of paint? (Give your answer in cubic centimetres.)

(6)   What is the surface area of a cube of side 3 cm.?

(7)   A tank is 1 m. high and its base is a rectangle which is 0.8 by 0.6 m. What is the surface area of the sides and base of the tank?

(8) Fig 17.28 shows the cross section of a copper pipe. The inside radius is 2 cm and the outside radius is $2^1/2$ cm.

   (a) Find the shaded area.

   (b) Find the volume of copper in 40 m. of the pipe.

Fig 17.28

(9)   An electrical cable consists of a copper cylinder surrounded by rubber. The diameter of the copper is $1^1/2$ cm, and the rubber is $1/4$ cm. thick. The cable is 100 m. long.

   (a) Find the volume of the copper.

   (b) Find the total volume of the cable.

   (c) Find the volume of the rubber.

(10)   The base radius of a cylinder is 2 cm. If its volume is 25 cm^3, what is its height?

(11)   What length of wire with radius of cross-section 0.5 mm. can be made from 20 cm^3 of copper?

(12)   A cylinder which holds 1 litre has height 20 cm. What is the area of its cross-section?

(13)   A measuring jar is a cylinder with radius of cross-section 5 cm. If 500 c.c. of water is poured into it, how deep will the water be?

(14) Fig 17.29 shows a shed which is 3 m. long, $1^1/2$ m. deep, and 2 m. high at the back and $1^3/4$ m. high at the front.

   (a) Find the area of a side wall.

   (b) Find the volume of the shed.

Fig 17.29

(15)   Wire is made with a cross-sectional area of 0.1 cm^2. What length of this wire can be made from 100 cm^3 of metal? How many times will the wire wrap round a cylinder of radius 4 cm?

For level 3

The volume V and the surface area A of a sphere with radius r are given by:

$$V = (4\pi r^3)/3$$

$$A = 4\pi r^2.$$

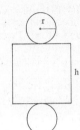

The surface area of a cylinder, with height h and base radius r, is the sum of the areas of the two circles at the ends and the curved surface.

$$A = 2\pi r^2 + 2\pi rh$$

The volume V of a cone with base radius r and height H is given by:

$$V = (\pi r^2 h)/3$$

The volume V of a pyramid with base area A and height h is given by:

$$V = \frac{Ah}{3}$$

Fig 17.30

17.2.5 Examples for level 3

(1)   A hollow rubber ball has outside radius 6 cm. and inside radius $5^{1}/_{2}$ cm. Find the volume of rubber.

Solution   The volume is found by subtracting the volume of the hollow sphere inside from the volume of the whole ball.

$$\textbf{Volume} = (4\pi 6^3)/3 - 4\pi(5^1/_2)^3/3 = 208 \text{ cm}^3$$

(2)   A cone with height 6 cm. and base radius 3 cm. is immersed in a cylinder of water with base radius 7 cm. By how much does the water rise?

Solution See fig 17.31. The extra volume introduced is the volume of the cone:

$$\pi(3^2 \times 6)/3 = 18\pi \text{ cm}^3$$

Fig 17.31

This extra volume adds an extra cylinder of water. The base radius of the extra cylinder is still 7 cm. The height of the extra cylinder is found by dividing the extra volume by the base area.

$$\textbf{Rise} = 18\pi \div (\pi 7^2) = 18 \div 49 = 0.367 \text{ cm.}$$

(3)   A pyramid is of height 12 in. and its base is a square of side 9 in. It is cut by a plane parallel to the base and 4 inches from the peak. What is the volume of the lower part?

Solution   Fig 17.32 shows the solid. The top part is similar to the original pyramid. It is a third of the height of the original. The side of its base must be a third of the original side.

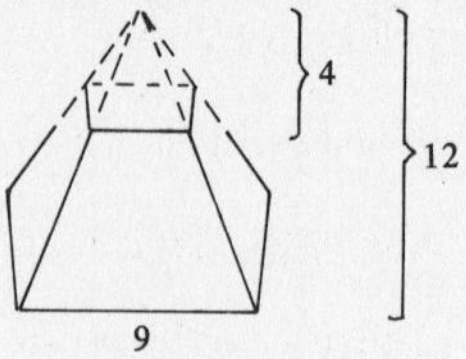

Fig 17.32

Subtract the volume of the small pyramid from the volume of the original pyramid.

Volume of lower part = $(12 \times 9^2)/3 - (4 \times 3^2)/3 = 312$ in^3

17.2.6 Exercises for level 3

(1)   Find the volumes of the spheres with radii:

(a) 5 cm. (b) 2 m. (c) $7^1/2$ in.

(2)   Find the surface areas of the spheres in question 1.

(3)   Find the surface areas of the cylinders in question 1 of 17.2.4.

(4)   A cube has volume 24 cm^3. What is its side?

(5)   A cylinder of height 5 in. has volume 64 cm^3. What is the radius of its base?

(6)   Find the radii of the spheres with volumes:

(a) 46 cm^3 (b) 12 in^3 (c) 5.56 mm^3.

(7)   A sphere has volume 3.44 cm^3. What is its surface area?

(8)   A sphere has surface area 66 mm^2. What is its volume?

(9)   The great pyramid of Gizah has a square base of side 230 m. and is 148 m. high. What is its volume?

(10)   A cone has base radius 6 cm. and is 7 cm. high. What is its volume?

(11)   A sphere of radius 20 cm. is painted. A coat of paint is 0.01 cm thick. Evaluate the volume of paint by the following two methods:

(a) Multiply the surface area of the sphere by the thickness of the coat of paint.

(b) Subtract the volume of the shell before painting from the volume of the shell after painting.

(c) Compare your answers to (a) and (b). Which is more accurate?

(12)   A cylindrical jar with a base radius of 5 cm contains water to a depth of 10 cm. Find the rise in water level if the following are immersed:

(a) A sphere of radius 2 cm.

(b) A cube of side $1^1/2$ cm.

(c) A cone with base radius 3 cm and height $4^1/2$ cm.

A tall cylindrical rod with base radius 2 cm is placed upright in the jar, so that the rod is *not* immersed. By how much does the water rise?

(13)   A hemispherical bowl of radius 7 cm. is full of water. It is then emptied into a cylindrical jar of base radius 5 cm. How deep is the water in the cylinder?

(14)   Paper which is 0.03 cm thick is rolled round a cylindrical core of radius 8 cm. The completed roll of paper is a cylinder with radius 20 cm. Find the length of the paper.

(15)   A certain sort of ball is a sphere with radius $2^1/2$ cm. It is made by winding rubber string round a spherical core of radius 1 cm.

(a) Find the volume of rubber.

(b) If the rubber string has area of cross-section 0.02 cm^2 find the length of rubber.

(16)   A cylindrical pipe delivers 2 litres of water per second. If the radius of cross-section is $1^1/2$ cm find the speed of water in the pipe.

(17)   A hose-pipe delivers 3 litres of water per second, at a speed of 17 m/sec. What is the radius of cross-section of the pipe?

(18)   Fig 17.33 shows a cone which has been cut in two by a plane parallel to its base, halfway from the base. The original height was 24 cm and the base radius is 4 cm.

Fig 17.33

(a) Find the volume of the original cone.

(b) Find the radius of the cut circle.

(c) Find the volume of the cone which has been removed.

(d) Find the volume of the remaining solid.

Show that the volume of the remaining solid is $^7/8$ of the original volume.

(19)   A cone is cut in two by a plane parallel to its base and $^1/4$ of the height down from the apex.

(a) Show that the upper cone is similar to the original cone.

(b) Show that the upper cone has volume $^1/64$ of the original volume.

(c) What is the ratio of the volumes of the two parts?

(20) A yoghourt carton consists of the lower half of a cone after the top has been cut off. The height of the carton is 10 cm, and the radii at top and bottom are 2 cm and 4 cm respectively.

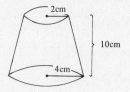

Fig 17.34

(a) Find the height of the original cone.

(b) Find the volume of the carton.

(21)   If a circle is rotated about a diameter, the solid region it passes through is a sphere.

(a) If an isosceles triangle is rotated about its axis of symmetry, what solid region does it pass through?

(b) What plane figure would you rotate in order to pass through a cylindrical region of space?

17.3 COMBINATIONS OF FIGURES. Levels 2 and 3

17.3.1 Examples for level 2

(1)   Find the area of the shape in fig 17.35.

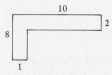

Fig 17.35

Solution   Divide the shape into two rectangles, one 10 by 2, the other 6 by 1.

The area is 10x2 + 6x1 = 26

(2)   A barn is 12 m. long, 3 m. deep and $2^1/2$ m. high. The roof consists of half a cylinder. Find the total volume.

Solution   The shape is shown in fig 17.36. The side walls are in the shape of a semi-circle of diameter 3 m. over a 3 by $2^1/2$ rectangle.

Fig 17.36

Surface area of end wall $=(\pi(1^1/2)^2)/2 + 3 \times 2^1/2 = 11$ m^2.

Volume = length x surface area of end = 12 x 11 = 132 m^3

17.3.2 Exercises for level 2

(1)   Find the areas of the shapes in fig 17.37.

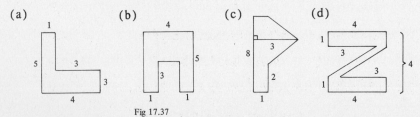

Fig 17.37

190

(2)   The dots of fig 17.38 are 1 cm apart. Find the areas of the shapes shown.

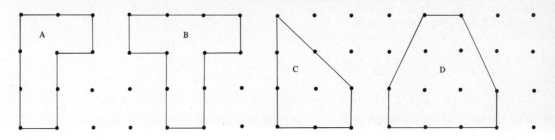

Fig 17.38

(3) A running track consists of two parallel straight sections of length 50 m., which are 20 m. apart, and semi-circular ends. Find the total length and the area enclosed by the track.

Fig 17.39

(4) A steel girder has the cross-section shown in fig 17.40. Find the area of cross section and the volume of 6 m. of the girder.

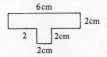

Fig 17.40

(5) The end wall of a hut is shown in fig 17.41. Find the area of the wall. If the hut is 2 $^1/_2$ m. long find its volume.

Fig 17.41

(6)   A window consist of a semi-circle on top of a rectangle: the rectangle is 60 cm high and 40 cm wide. Find the area of the window.

(7)   A box is made out of wood 1 cm thick, so that its interior dimensions are 20 by 15 by 12, and its exterior dimensions are 22 by 17 by 14. By subtracting the interior volume from the exterior volume find the volume of the wood.

17.3.3 Examples for level 3

(1)   Fig 17.42 shows a sector of a circle with radius 2 cm. Find the area of the sector ABC and of the shaded segment.

Solution   The area of the sector is a quarter of the area of the circle.

$$\text{Area} = {}^1/_4\pi2^2 = 3.14 \text{ cm}^2$$

Fig 17.42

The segment is the difference between the sector ABC and the triangle ABC.

$$\text{Area} = {}^1/_4\pi2^2 - {}^1/_2x2x2 = 1.14 \text{ cm}^2$$

191

(2)   A house is built on a rectangular base which is 6 m. by 8 m. The walls are 10 m. high, and the roof tapers to a point 12 m. above the ground. Find the total volume of the house.

Solution   The house is shown in fig 17.43. It is a cuboid with a pyramid on top. The total volume is:

6x8x10 + 6x8x²/3 = 512 m³

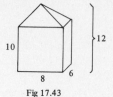

Fig 17.43

(3)   A cone is 24 cm high and its base radius is 7 cm. Find the area of the curved side.

Solution   The slant height l of the cone, from the apex to the rim of the base, is found by Pythagoras' Theorem:

$$l = \sqrt{7^2 + 24^2} = 25.$$

Fig 17.44

If the cone is cut along the dotted line and laid flat, the shape is a sector of a circle with radius 25. Fig 17.45.

The arc length AB is the perimeter of the base circle, 2π7.

Reduce the area of the full circle in the ratio 2π7:2π25

Area = π25² x (2π7/2π25) = π25x7 = 550 cm²

Fig 17.45

17.3.4 Exercises for level 3

(1)   Find the areas of the sectors in fig 17.46.

(a)   (b)   (c)

        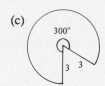

Fig 17.46

(2)   Find the arc lengths of the sectors in question 1.

(3)   A square is drawn inside a circle of radius 5 cm. Find the area of the square, and hence find the area between the circle and the square.

Fig 17.47

(4) Four cog wheels of radius 2 cm. have their centres at the corners of a square of side 5 cm. A chain passes tightly round the wheels: how long is the chain?

Fig 17.48

(5) In fig 17.49 part of a circle of radius 1 has been drawn from the centre (1,1), so that it touches the x-axis and the y-axis. The shape between the axes and the circle is shaded. Find the area and the perimeter of the shaded region.

Fig 17.49

(6) In fig 17.50 a square of side 8 cm. has had one corner cut by a circle radius 4 cm. The centre of the circle is at the centre of the square. Find the area and the perimeter of the shape.

Fig 17.50

(7)   A test-tube consists of a cylinder with a hemisphere at one end. The length of the cylinder is 20 cm, and its radius of cross-section is 1 cm. Find the volume of the test-tube and its surface area.

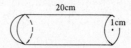

Fig 17.51

(8)   Before it is sharpened, a pencil is a cylinder 15 cm long with diameter 1 cm. Find its volume.

   It is now sharpened, so that the last 2 cm of the pencil forms a cone. How much volume has been lost?

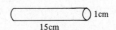

Fig 17.52

(9) An cornet ice-cream is a cone surmounted by a hemisphere. The total height is 5 inches and the radius of the hemisphere is 1 inch. Find the total volume.

Fig 17.53

(10)   A peg-top consists of a hemisphere on top of a cone. The hemisphere has radius 2 cm. How high should the cone be if it is to have the same volume as the hemisphere?

Fig 17.54

(11)   A boiler consist of a cylinder, with base radius 20 cm and length 50 cm, with hemispheres at both ends. Find its volume and its surface area.

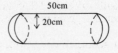

Fig 17.55

(12)   A sphere of radius 3 cm fits exactly inside a cylinder of base radius 3 cm and height 6 cm. Show that their surface areas are equal. Find the ratio of their volumes.

Fig 17.56

(13) A child makes a model of a "Dalek" as follows. He takes a cone with base radius 5 cm and height 15 cm and cuts off the top 9 cm. The top now has radius 3 cm.

The cone he has cut off is replaced by a hemisphere with the same radius. Find the volume of the "Dalek".

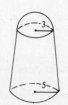

Fig 17.57

(14)   Find the curved surface area of a cone with height 4 cm and base radius 3 cm.

(15)   Find the angle of the sector obtained when the surface of question 14 is laid flat.

(16)   A tent is cone shaped, with height 8 ft and base radius 6 feet. Find its volume, and the area of canvas of its sides.

(17) Fig 17.58 shows a sector of a circle with angle 60° and radius 8 cm. It is folded and the two radii are joined to form a cone.

(a) What is the arc length AB?

(b) What is the base radius of the cone?

(c) What is the height of the cone?

(d) What is the volume of the cone?

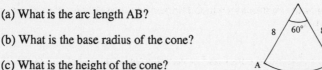

Fig 17.58

(18)   One sphere has twice the radius of another. What is the ratio of their surface areas? What is the ratio of their volumes?

(19)   A cylinder has its base radius halved and its height tripled. What has the volume been multiplied by?

(20)   A cone has its height halved and its base radius multiplied by 4. What has the volume been multiplied by?

(21)   Two golf balls differ by 1% in their radii. Find the ratio of their radii. What is the ratio of their areas? By what percentage do they differ in area? By what percentage do they differ in volume?

(22)   A sphere of radius 10 cm is hollow, the cavity being a sphere of radius 5 cm.

(a) What is the ratio of the volume of the whole sphere to the volume of the cavity?

(b) What proportion of the whole sphere is solid matter?

(23)   A cone is 24 cm high, and its base radius is 7 cm. The top 4 cm. is cut away.

(a) What is the volume and curved surface area of the original cone?

(b) What is the ratio of the heights of the cut away cone and the original cone?

(c) What is the ratio of the surface areas of the cut away cone and the original cone?

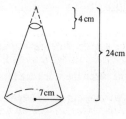

Fig 17.59

(d) What is the curved surface area of the remaining solid?

COMMON ERRORS

(1)   **Areas**
(a) Be careful that you do not confuse the radius and the diameter of a circle. Read the question carefully to see which is being used.

(b) The area of a triangle or a parallelogram is found by the product of the base and the *height*. It is not found by the product of adjacent sides, unless the triangle or parallelogram happens to be right-angled.

(c) Mistakes are common with questions in which the units are mixed. It is safest to express all the measurements in the same units.

(2)   **Volumes**
The formulas for the volumes of cones and pyramid refer to the height of the solid. This means the *vertical* height, not the slant height.

(3)   **Combinations**
If you have two similar solids, then the ratio of their lengths is not the same as the ratio of their areas or of their volumes.

If the length ratio is a:b, then the area ratio is $a^2:b^2$ and the volume ratio is $a^3:b^3$.

18 Trigonometry

The equations of *trigonometry* connect together the sides and angles of a triangle.

18.1 RIGHT-ANGLED TRIANGLES. Level 2

The right-angled triangle ABC of fig 18.1 is labelled as shown.

The ratios between the sides are functions of the angle P°.

sin P° = OPP/HYP   (The sine of P°)

cos P° = ADJ/HYP   (The cosine of P°)

tan P° = OPP/ADJ   (The tangent of P°)

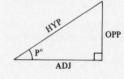

Fig 18.1

These can be remembered by the word:

OHSAHCOAT

(Opposite over Hypoteneuse equals Sine, Adjacent over Hypoteneuse equals Cosine, Opposite over Adjacent equals Tangent)

or SOHCAHTOA

(Sine is Opp over Hyp, Cos is Adj over Hyp, Tan is Opp over Adj).

These equations give the ratios in terms of the angle. The *inverse functions* give the angle in terms of the ratios.

$$P° = sin^{-1}OPP/HYP = cos^{-1}ADJ/HYP = tan^{-1}OPP/ADJ.$$

When using a calculator to find ratios, the angle is pressed first. To find sin 43°, press the following buttons:

| 4 | 3 | SIN |

The answer 0.682 will appear.

When using a calculator to find angles, the ratio is pressed first. To find cos^{-1}0.3, press the following buttons:

| . | 3 | INV | COS |

The answer 72.5° will appear.

18.1.1 Examples for level 2

(1)   In the triangle ABC of fig 18.2, <A = 90°, <B = 53°, and BC = 7 cm. Find the other two sides.

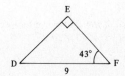

Fig 18.2

Solution   Comparing with fig 18.1, the HYP side is BC, OPP is AC, ADJ is AB. Using the equations above:

$$\sin 53° = {}^{OPP}/_7$$

Use tables or a calculator to find that:

$$\sin 53° = 0.7986$$

This gives:

$$0.7986 = {}^{OPP}/_7$$

$$AC = 7 \times 0.7986 = 5.59$$

Similarly $\cos 53° = {}^{ADJ}/_7$

$$AB = 7 \times \cos 53° = 7 \times 0.6018 = 4.21$$

(2)   In the triangle DEF of fig 18.3 <E = 90°, <F = 43°, DF = 9 metres. Find the other sides.

Solution   In this figure the triangle is not in the same position as in fig 18.1. The sides are labelled as follows:

DF = HYP
(Because it is farthest from the right-angle)

ED = OPP
(Because it is farthest from the angle of 43°)

EF = ADJ
(Because it is next to the angle of 43°)

$$EF = 9 \times \cos 43° = 9 \times 0.7314 = 6.58 \text{ m}$$

$$ED = 9 \times \sin 43° = 6.14 \text{ m}$$

Fig 18.3

(3)   In triangle GHI, <H = 90°, HG = 3, HI = 5. find the angle <I.

Solution   Here OPP = GH = 3, and ADJ = HI = 5. This gives the equation:

$$\tan I° = {}^{OPP}/_{ADJ} = {}^3/_5.$$

Fig 18.4

197

Use the inverse tan button:

$$I° = \tan^{-1} 0.6 = 31°$$

18.1.2 Exercises for level 2

(1)   Use your calculator to find the following, giving the answers to 3 decimal places:

(a) sin 23°   (b) cos 47°   (c) tan 75°   (d) sin 64.34°   (e) cos 0.934°

(2)   Use your calculator to find the following, giving the answers to the nearest tenth of a degree:

(a) $\sin^{-1}0.8$ (b) $\cos^{-1}0.24$ (d) $\tan^{-1}0.63$ (e) $\tan^{-1}4.9$

(3)   In the triangles shown, find the unknown sides.

Fig 18.5

(4)   In the triangles shown, find the unknown sides.

  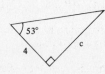

Fig 18.6

(5)   In the triangles shown, find the angles.

Fig 18.7

(6)   In the triangles shown, find the angles.

Fig 18.8

198

(7)   In the triangle ABC, <A = 90°, <B = 27° and BC = 8 cm. Find AB and CA.

(8)   In the triangle PQR, <P = 90°, QR = 57 metres and QP = 23 metres. Find the angles at Q and at R.

(9)   ABC is a triangle in which <A = 90°, AB = 17 cm and AC = 12 cm. Find the other angles.

(10)   In the triangle LMN, <L = 90°, <M = 43 ° and MN = 83 m. Find LM and LN.

18.1.3 Examples for level 3

(1)   In the triangle ABC of fig 18.9, <B = 90°, <A = 66°, BC = 6 cm. Find BA.

Solution   Apply the basic formula:

$$\tan 66° = {}^6/\text{BA}$$

Multiply both sides by BA:

$$\text{BA} \times \tan 66° = 6$$

Now *divide* both sides by tan 66°:

$$\text{BA} = {}^6/\tan 66°$$

$$\textbf{BA} = {}^6/\textbf{2.2460} = \textbf{2.67}$$

Fig 18.9

(2)   In the triangle ABC of fig 18.10, AB = AC = 8cm, BC = 4 cm. Find the angles of the triangle.

Solution The trigonometric ratios are only defined for a right-angled triangle.

        Convert the isosceles triangle into two right-angled triangles by dropping a perpendicular. Fig 18.11.

        Now the equations of trig can be used.

$$\sin \text{BAX} = {}^2/8$$

$$\text{BAX} = \sin^{-1} {}^1/4 = 14.5°.$$

$$\textbf{A} = \textbf{2} \times \textbf{14.5} = \textbf{29°ʼ and B = C = 75.5°}$$

Fig 18.10

Fig 18.11

18.1.4 Exercises for level 3

(1)   Find the sides of the triangles shown.

 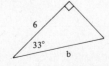

Fig 18.12

(2)   Find the angles of the triangles shown.

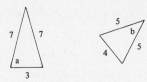

Fig 18.13

(3)   In the quadrilateral ABCD, <ACB = <DAC = 90°, AB = 10 cm, <ABC = 50° and <DCA = 40°. Find AC and DC.

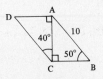

Fig 18.14

(4)   For the triangle ABC, BC = 3 and <B = <C = 57°. Find AB.

(5)   For the triangle ABC, AB = AC = 7, BC = 4. Find <A.

(6)   The diagonals of a rhombus are 8 cm and 4 cm in length. Find the angles of the rhombus.

(7)   A pair of compasses is 3 inches in length, and can be separated to 54°. Find the area of the largest circle which can be drawn with it.

(8)   When the neck of a giraffe is at 20° to the vertical, it can reach leaves which are $4\frac{1}{2}$ metres above the ground. Assuming that the shoulders of the giraffe are 2 metres high, how long is its neck?

(9) Pincers whose blades are 20 cm long can just grasp a pipe of radius 5 cm. Find the angle of separation of the blades.

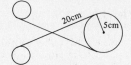

Fig 18.15

(10)   A line joins the point (1,5) and (6,6). Find the angle this line makes with the x-axis.

(11)   Find the angle that the line y = 2x + 3 makes with the x-axis. Find the angle between this line and y = $\frac{1}{2}$x - 4.

(12)   In a circle of radius 12 cm a chord AB of length 7 cm is drawn. Find the angle this chord subtends at the centre of the circle.

(13) A circle of radius 6 cm has a regular pentagon ABCDE inscribed in it.

    (a) Find the angle subtended at the centre by AB.

    (b) Find the length of the perpendicular from the centre to AB.

    (c) Find the length of AB.

    (d) Find the area of the pentagon.

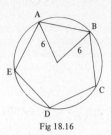

Fig 18.16

18.2 ELEVATION, DEPRESSION, BEARINGS. Level 2

When we look up at something, the angle our line of vision makes with the horizontal is called the *angle of elevation.*

When we look down on something, the angle between the horizontal and our line of vision is called the *angle of depression.*

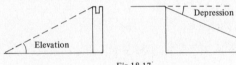

Fig 18.17

A direction can be described by the angle its path makes with due North, measured clockwise. This angle is called its *bearing.* Fig 18.18.

18.2.1 Examples for level 2

(1)   From the top of a cliff, the angle of depression of a boat is 5°. If the boat is 2000 m. out to sea, how high is the cliff?

Solution   Let d be the height of the cliff. From the picture:

$$\tan 5° = {}^d/_{2000}$$

d = 2000 x tan 5° = 175 m.

(2)   A kite on 75 m of string is 55 m. high. What angle does it make with the horizontal?

Solution   Let P° be the angle. From the picture:

$$\sin P° = {}^{55}/_{75}$$

$$\mathbf{P° = \sin^{-1} {}^{55}/_{75} = 47°}$$

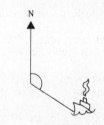

Fig 18.18

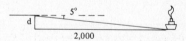

Fig 18.19

Fig 18.20

(3)   A ship sails 50 km on a bearing of 070°. How far North and how far East has it gone?

Solution   In the triangle as shown, the actual path of the ship is the Hypoteneuse of the triangle. The North distance N is the Adjacent side and the Easterly distance E is the Opposite side.

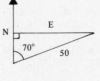

$$\cos 70° = {}^N/_{50}$$

$$\sin 70° = {}^E/_{50}$$

Fig 18.21

N = 17.1 km and E = 47 km.

18.2.2 Exercises for level 2

(1) 200 metres from the base of a tower the angle of elevation of the top is 24°. Find the height of the tower.

Fig 18.22

(2) A cliff is 75 metres high, and a boat is 3000 metres out to sea. Find the angle of depression of the boat from the cliff.

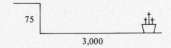

Fig 18.23

(3) A ladder of length 2 metres leans against a wall at an angle of 35° to the vertical. Find how far up the wall it reaches.

Fig 18.24

(4) A flagpole 12 feet high throws a shadow of length 20 feet. What is the angle of elevation of the sun?

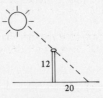

Fig 18.25

(5)   A girl's eyes are 5 feet above the ground. When she is 30 feet from the base of a tree, the angle of elevation of the top is 35°. Find the height of the tree.

(6)   Two minarets of a mosque are 40 m. and 55 m. in height. Their bases are 30 m. from each other. What is the angle of depression of the lower minaret from the higher?

(7)   A terrace of seats at a football ground is 30 metres long and rises 4 metres from top to bottom. Find the angle of slope of the terrace.

(8)   A crane is 15 metres long. Its base is on the ground. When the end of the crane is 7 metres above the ground, what angle does it make with the vertical?

(9)   A rocket is fired at an elevation of 50°. If it travels at 40 metres per second, how high is it after 5 seconds?

(10)   I stand on the bank of a river, and see a tree directly opposite me on the other side. I now walk 20 metres upstream, and the line from me to the tree makes 83° with the bank. How wide is the river?

(11)   100 m. from a church the base of the spire has angle of elevation 12°, and the top of the spire has elevation 23°. What is the height of the spire?

(12)   The gradient of a hill is 1 in 10. (This means that we go 1 unit vertically for every 10 units horizontally.) Find (a) the angle of slope (b) How far we have climbed after walking 100 metres.

(13)   A plane flies at 6,000 metres above the ground. I spot it directly above me, and 10 seconds later its angle of elevation is 72°. How fast is the plane flying?

(14)   A line is drawn from the origin to the point (5,2). What angle does this line make with the x-axis?

(15)   A plane flies 400 km at 057°. How far North and how far East is it from the starting point?

(16)   I walk 5 miles North and then 3 miles West. What is the bearing of my starting point?

(17)   A ship is 30 km due South of a port, and is sailing on a bearing of 048°. It sails until it is due East of the port. How far East will it then be?

18.2.3 Exercises for level 3

(1)   A spire is 50 metres high, and the angle of elevation is 15°. How far away am I from the base? If I now walk 150 metres towards the tower, what is the new angle of elevation?

(2)   A ship travels 50 km on a bearing of 050°, then for 60 km on a bearing of 135°. How far away is it from its original starting point? What is the bearing of the ship from its starting point?

(3)   A gun is due South of its target. From an observation post 50 m. due West of the gun, the target is on a bearing of 005°. Find the distance of the target from the gun.

(4)   Two men stand on opposite sides of a flagpole, which is 12 metres taller than they are. The angles of elevation of the top of the flagpole from the two men are 10° and 15°. How far apart are the men?

(5)   From the top of a 50 m cliff, the angle of depression of a ship is 3°. How far away is the ship from the top of the cliff?

(6)   At midday in the tropics the sun is vertically overhead. What is the angle of elevation of the sun at 4 o'clock in the afternoon?

What is the length of the shadow of a 6 foot man at this time?

18.3 OBTUSE ANGLES. Level 3

The trigonometric ratios of sin, cos and tan can be defined for obtuse angles, (angles between 90° and 180°), as follows:

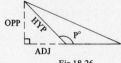

Fig 18.26

$$[s] \quad \sin P = {}^{OPP}/_{HYP} = \sin (180° - P)$$

$$[c] \quad \cos P = {}^{ADJ}/_{HYP} = - \cos (180° - P)$$

$$[t] \quad \tan P = {}^{OPP}/_{ADJ} = - \tan (180° - P)$$

18.3.1 Examples for level 3

(1)   Express in terms of the ratios of acute angles:

(a) sin 124° (b) cos 163° (c) tan 98°

Solution   Put P = 124° into the formula [s] above.

$$\textbf{sin 124°} = \textbf{sin (180° - 124°)} = \textbf{sin 56°}$$

Similarly, for P = 163° and P = 98°:

$$\textbf{cos 163°} = \textbf{- cos (180° - 163°)} = \textbf{- cos 17°}$$

$$\textbf{tan 98°} = \textbf{- tan (180° - 98°)} = \textbf{- tan 82°}$$

(2)   Find obtuse angles for which:

(a) sin P = 0.75 (b) cos Q = - 0.8

Solution   The acute angle whose sine is 0.75 is:

$$\sin^{-1}0.75 = 48.6°.$$

Use the formula [s] for the sine of obtuse angles:

$$\sin 48.6° = \sin (180° - 48.6°) = \sin 131.4°.$$

$$\textbf{P = 131.4°}$$

The acute angle with the cosine of + 0.8 is:

$$\cos^{-1}0.8 = 36.9°$$

Use the formula [c] for the cosine of obtuse angles:

$$- \cos 36.9° = \cos (180° - 36.9°) = \cos 143.1°$$

$$\textbf{Q = 143.1°}$$

18.3.2 Exercises for level 3

(1)   Express in terms of the ratios of acute angles:

(a) sin 138° (b) cos 99° (c) tan 176° (d) sin 118° (e) cos 103°

(2)   Find obtuse angles which satisfy the following:

(a) sin x = 0.5   (b) cos y = - 0.76   (c) tan z = - 0.44   (d) sin w = 0.2
 (e) cos p = - 0.68   (f) tan q = - 2.7.

(3)   Find the range of values of x if sin x > 0.5

(4)   Find the range of values of y if cos y ⩽ - 0.1

(5)   Find the range of values of z if tan z ⩾ 10

18.4 THE SINE AND COSINE RULES. Level 3

The *sine rule* and the *cos rule* extend trigonometry for triangles which are not necessarily right-angled.

When labelling the sides and angles of the triangle ABC, the convention is that each side is labelled with the same letter as the opposite angle. So side c is opposite <C etc.

Fig 18.27

The *sine* rule is:

$$\frac{\sin <A}{a} = \frac{\sin <B}{b} = \frac{\sin <C}{c}$$

The rule can be written the other way up:

$$\frac{a}{\sin <A} = \frac{b}{\sin <B} = \frac{c}{\sin <C}$$

The *cosine* rule is:

$$c^2 = a^2 + b^2 - 2bc \cos <C.$$

This rule gives a side in terms of the other sides and the opposite angle. The rule can be re-arranged to give the angles in terms of the sides.

$$\cos <C = \frac{a^2 + b^2 - c^2}{2\,a\,b}$$

18.4.1 Examples for level 3

(1)   In fig 18.28, <Q = 68°, PQ = 4 and PR = 5. Find <R.

Solution   The first version of the sine rule is used.

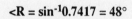

$$\frac{\sin <R}{4} = \frac{\sin 68°}{5}$$

Multiply across to obtain:

$\sin <R = \sin 68 \times {}^4/_5$

<R = sin^{-1}0.7417 = 48°

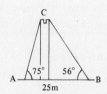

Fig 18.28

(2)   Two men are 25 m. apart. One is North of a tower, and one is South. They measure the angles of elevation of the top of the tower as 75° and 56°. Find the height of the tower.

Solution   Let C be the top of the tower, and A and B the two men. AB = 25, <C = 180° - 56° - 75° = 49°.

Use of the sine rule gives:

$AC = \sin 56° \times {}^{25}/_{\sin 49°} = 27.46.$

Trigonometry now gives:

Height = AC x sin 75° = 27 metres.

Fig 18.29

(3)   In the triangle PQR, PQ = 5, QR = 6.3, RP = 7.4. Find <P.

Solution   The cosine rule gives, on putting <C = <P, c = 6.3, a = 5, b = 7.4:

$$\cos <P = \frac{5^2 + 7.4^2 - 6.3^2}{2 \times 5 \times 7.4}$$

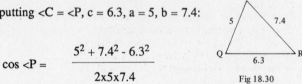

Fig 18.30

$\cos <P = 0.5415$

<P = 57.2°

(4)   The hour hand of a clock is 15 cm long, and the minute hand is 20 cm. long. How far apart are the tips of the hands at five o'clock?

Solution   At five o'clock the angle between the hands is $360 \times {}^5/_{12} = 150°$.

Let d be the distance between the tips. Use the cosine rule:

$$d^2 = 15^2 + 20^2 - 2 \times 15 \times 20 \times \cos 150°$$

d = 33.8 cm.

Fig 18.31

18.4.2 Exercises for level 3

Sine rule exercises

(1)   Find the unknown sides in the following triangles:

Fig 18.32

(2)   Find the unknown angles in the following triangles:

Fig 18.33

(3)   In the triangle ABC, <A = 23°, <B = 68°, AB = 5.6. Find the sides AC and CB.

(4)   In the triangle DEF, DE = 10, EF = 13, <D = 62°. Find the angles <F and <E.

(5)   I measure the angle of elevation of a tree as 15°. I then walk 10 metres towards the tower, and the angle of elevation is now 23°. Find my distance from the top of the tree, and hence find the height of the tree.

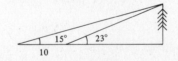

Fig 18.34

(6)   From a gun emplacement, the bearing of its target is 053°. From an observation point 30 metres west of the gun, the bearing of the target is 054°. What is the distance from the gun to its target?

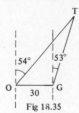

Fig 18.35

(7)   A wall leans at 15° to the vertical. A ladder of length 2 metres is placed so that one end of the ladder is 1 metre from the base of the wall. What angle is the ladder leaning at? How high up the wall does the ladder reach?

Fig 18.36

(8)   Ship A leaves from port C at a bearing of 123°. Ship B leaves from port D, 200 miles south of C, on a bearing of 046°. How far from C do their paths cross?

Cosine rule exercises

(9)   Find the unknown sides in the following triangles.

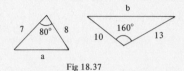

Fig 18.37

(10) Find the unknown angles in the following triangles.

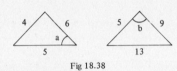

Fig 18.38

(11) In the triangle ABC, AB = 17, BC = 23, <ABC = 43°. Find AC.

(12) In the triangle ABC, AB = 45, BC = 52, AC = 27. Find <B.

(13) A pilot flies his plane for 50 miles, then turns through 20° and flies a further 73 miles. How far is he from home?

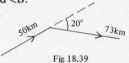

Fig 18.39

(14) The lower jaw of a crocodile is 58 cm, and the upper jaw is 52 cm. It can open its jaws to an angle of 43°. What is the greatest width of object that it can grasp in its jaws?

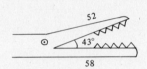

Fig 18.40

(15) A parallelogram has sides of length 12 and 7 cm, and the longer diagonal is 15 cm. Find the angles of the parallelogram. Find the length of the shorter diagonal.

Fig 18.41

(16) Aytown is 50 miles due north of Beetown; Ceetown is 27 miles from Aytown and 32 miles from Beetown, and is west of them. Find the bearings of Ceetown from Aytown and from Beetown.

Fig 18.42

(17) Show that if we put <C = 90° in the cosine rule, we obtain Pythagoras' theorem.

(18) A flagpole of length 5 metres leans at 17° to the vertical, and the length of its shadow is 6 metres. How far is it from the end of the flagpole to the end of the shadow?

208

Miscellaneous exercises

(19)   Find the unknown sides of the following triangles.

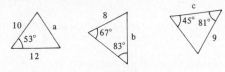

Fig 18.43

(20)   Find the unknown angles of the following triangles.

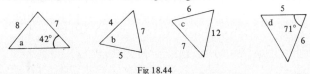

Fig 18.44

(21)   A quadrilateral ABCD has sides given by AB = 12, BC = 14, CD = 13.5, DA = 11.2. <A = 58°. Find BD and <C.

(22)   In triangle ABC, AB = 4.5, BC = 5.3, <B = 39°. Find <A.

COMMON ERRORS

(1)   **Right-angled triangles**
          Be careful that you do not apply trigonometry to the wrong sort of triangle. The ratios are only defined for those triangles which are *right-angled*. If there is not any right-angle in the triangle, then more work has to be done so that there is one. This was done in the case of an isosceles triangle (Example 2 of 18.1.3)

(2)   **The three ratios**
          Make sure that you do not use the wrong ratio for trig. Usually this error occurs because the triangle has been wrongly labelled.

          It does not matter which way up the triangle is, the longest side, (the side farthest from the right-angle) is always the HYP. The side which is next to the angle we are dealing with is the ADJ, and the side opposite the angle we are dealing with is the OPP.

          If you have any difficulty with these, make sure that the triangle is labelled with HYP, ADJ and OPP before you start working out the ratios.

(3)   **Depression**
          The angle of depression is the angle which the line of vision makes with the *horizontal*, not with the vertical.

(4)   **Algebra**
          Mistakes in algebra are easy to make. When the unknown term is on the bottom of the ratio instead of the top, it is very common to get the wrong answer. For example, from the equation $\sin 67° = {}^6/h$, the answer is h = 6/sin67° = 6.518.

          If you obtain h = 5.523, it means that you must have multiplied by sin 67° instead of dividing.

There is always a quick check to make sure that you have done the correct thing. The hypoteneuse is always the largest side of the triangle. So if you find that your value for HYP is smaller than your value for OPP or ADJ, then you must have multiplied instead of divided.

(5)   Use of calculator
(a) Your calculator works "backwards" for the sin, cos and tan ratios. Do not press the sin button before the angle button.

(b) Do not press the = button after the function button.

(c) Be careful that you give the answer to the right degree of accuracy. If you are asked to give an answer to 3 decimal places, then it is a good idea to work to at least 4 decimal places.

For example, to find 53 x sin 26°:

53 x sin 26° = 53 x 0.43837 = 23.23361 = 23.234 (to 3 decimal places).

(6)   Sine and Cosine rules
(a) When using either the sine rule or the cosine rule, be sure that you have labelled the sides and angles correctly. Side a is opposite angle <A, b is opposite <B, c is opposite <C.

(b) Calculator errors are very common, especially when using the cosine rule for finding an angle from the sides. The formula is:

$$\cos <A = \frac{b^2 + c^2 - a^2}{2xbxc}$$

After working out the top line, be sure to press the = button. Otherwise you will only divide the a^2 term by 2bc. Then, you must *divide* the top line by all the terms 2, b, c.

For example, if you have the formula:

$$\cos <C = \frac{6^2 + 9^2 - 7^2}{2x6x9}$$

The answer you should obtain is cos <C = 0.6296

If you obtain cos <C = 1836 it means that you have multiplied by 6 and by 9 instead of dividing by them.

19 Pythagoras' Theorem

19.1 PYTHAGORAS IN TWO DIMENSIONS. Level 2

Pythagoras' Theorem gives a relationship between the sides of a right-angled triangle.

The square on the hypoteneuse is equal to the sum of the squares on the other two sides.

In fig 19.1, the side farthest from the right-angle is the hypoteneuse. The theorem says:

$$c^2 = a^2 + b^2.$$

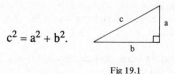

Fig 19.1

19.1.1 Examples for level 2

(1)   In the triangle of fig 19.2, $<C = 90°$, $CB = 3$ and $AC = 5$. Find BA.

Fig 19.2

Solution   Comparing with fig 19.1, $BA = c$, $AC = b$, $BC = a$. The theorem now gives:

$$c^2 = 3^2 + 5^2$$

$$c^2 = 9 + 25 = 34$$

$$c = \sqrt{34} = 5.831$$

(2)   A ladder of length $2^1/2$ metres leans against a wall, so that the foot of the ladder is 1 metre from the base of the wall. How far up the wall does the ladder reach?

Solution   Here the length of the ladder is c, and b is the distance to the base of the wall. The height up the wall is a. The theorem gives:

$$2^1/2^2 = a^2 + 1^2$$

$$a^2 = 6^1/4 - 1 = 5^1/4$$

$$a = \sqrt{5^1/4} = 2.291 \text{ m}$$

Fig 19.3

19.1.2 Exercises for level 2

(1)   For the following triangles, find the unknown sides.

Fig 19.4

(2) I travel 3 miles North and 5 miles East. How far am I from my starting point?

Fig 19.5

(3) A rectangle is 7 cm by 8 cm. How long are the diagonals?

Fig 19.6

(4) A kite is flying so that it is 55 feet high, and is above a point 75 from the flyer. How long is the string of the kite?

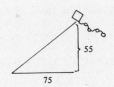

Fig 19.7

(5) The diagonals of a rhombus are 10 cm and 13 cm. What is the length of the side of the rhombus?

Fig 19.8

(6) P is a point 6 cm from the centre of a circle of radius 2 cm. A tangent is drawn from P, touching the circle at T. How long is the tangent PT?

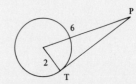

Fig 19.9

(7) A chord of length 5 is drawn in a circle of radius 20. What is the shortest distance from the centre of the circle to the chord?

Fig 19.10

(8) A cylindrical horizontal drain has radius 4 inches. There is water in it to a depth of 1 inch. How far below the centre of the drain is the water? How wide is the water in the drain?

Fig 19.11

(9)   A road has a gradient of 1 in 20. (Which means that it rises 1 metre for every 20 metres horizontally.) If I walk so that I have risen $2^1/2$ metres, how far have I travelled horizontally? How far have I walked along the road?

(10)   A stick of length 5 feet is held at an angle, so that when the sun is vertically above the stick the shadow is 3 feet long. How high is the end of the stick above the ground?

(11) A pendulum 2 metres long is pulled aside so that the bob has moved 10 cm to the left. How far is the bob below the support? By how much has the bob risen?

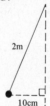

Fig 19.12

(12)   The sides of a triangle are 7, 11 and 12. Is this a right-angled triangle?

(13)   The sides of a triangle are 15, 8 and 17. Is this a right-angled triangle?

(14)   An isosceles triangle ABC has sides AB = AC = 7cm and BC = 6 cm. Find the length of the perpendicular from A to BC.

(15)   PQR is an equilateral triangle of side 4 cm. Find the length of the perpendicular from P to QR. Find the area of the triangle.

(16)   The dots of the figure below are 1 cm. apart. Find the lengths of: (a) AB (b) AC (c) CD.

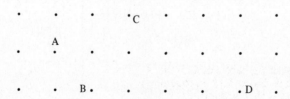

Fig 19.13

213

(17)   A graph is drawn to a scale of 1 cm per unit. Find the following distances:

(a) (0,0) to (1,2)   (b) (0,0) to (5,-3)   (c) (1,1) to (5,7)   (d) (1,3) to (6,1).

19.2 PYTHAGORAS AND TRIGONOMETRY IN THREE DIMENSIONS. Level 3

19.2.1 Examples for level 3

(1)   A room ABCDEFGH is a cuboid, $2^1/2$ metres high, 4 metres long, 3 metres wide. What is the length of the "space diagonal" AG? What angle does this line make with the horizontal?

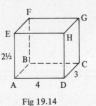

Fig 19.14

Solution   The diagonal of the floor, AC, has length given by:

$$AC^2 = 3^2 + 4^2$$

$$AC = \sqrt{9 + 16} = \sqrt{25} = 5.$$

In the triangle AGC, <ACG is a right angle. This then gives:

$$AG^2 = AC^2 + CG^2$$

$$AG = \sqrt{25 + 6^1/4} = \sqrt{31.25} = 5.59 \text{ metres}$$

For the angle, trigonometry gives:

$$<GAC = \tan^{-1} (2^1/2)/5 = 26.6°.$$

AG makes 26.6° with the horizontal

(2)   A section of a hillside is shown in fig 19.15.

ABFE is a horizontal plane; AB, EF and CD are horizontal lines of length 50 metres. BF is 12 metres and FD and CE are vertical lines of length 5 metres. Find:

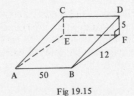

Fig 19.15

(a) BD (b) CB (c) The slope of DB (d) The slope of CB.

Solution   BDF is a right-angled triangle. This then gives:

$$BD^2 = FD^2 + BF^2$$

$$BD = \sqrt{169} = 13$$

BCD is a right-angled triangle. This gives:

$$BC^2 = CD^2 + DB^2$$

$$BC = \sqrt{2500 + 169} = \sqrt{2669} = 52 \text{ metres}$$

Using trigonometry in triangle BFD:

tan <FBD = $^5/_{12}$

BD makes tan^{-1}($^5/_{12}$)= 22.6° with the horizontal

Using trigonometry in triangle BFE:

sin <CBE = $^5/_{CB}$

CB makes sin^{-1}($^5/_{52}$)= 6° with the horizontal

19.2.2 Exercises for level 3

(1)   A cuboid ABCDEFGH is shown in fig 19.16.

Find the diagonal FD and the angles <FDB, <EHB, <FDC.

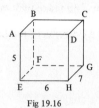

Fig 19.16

(2) A wedge ABCDEF is shown in fig 19.17. Find the lengths AF, AC, AD and the angles <ACE, <BCA.

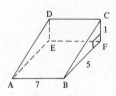

Fig 19.17

(3) In fig 19.18 the pyramid VABCD has a square base ABCD of side 4 cm. VA = VB = VC = VD = 6 cm. Find AC, and the height of V above the plane ABCD. Find the angles <VAC and <VAD.

Fig 19.18

(4) A prism is of height 10 cm and its cross-section is an equilateral triangle of side 3 cm. (fig 19.19). Find the length AE. Find the angle <BAE.

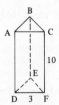

Fig 19.19

(5)   A cardboard carton is a cuboid with sides 2 ft, $1^1/_2$ ft, $2^1/_2$ ft. Find the length of the longest stick which can be put in the carton. Find the angles that the stick now makes with the sides of the carton.

(6) In fig 19.20 the pyramid VABCD is on a square base ABCD of side 100 metres, and V is 50 metres above the ground. Find:

(a) The diagonal AC of the base.

(b) The length of VA.

(c) The distance from V to the midpoint X of AB.

(d) The angle <VAC.

(e) The inclination of VX to the horizontal.

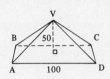

Fig 19.20

(7) In fig 19.21 ABCD is a regular tetrahedron of side 8 cm. It is placed with BCD horizontal, so that A is vertically above the centre O of triangle BCD. X is the midpoint of BD. Find:

(a) AX and XC.

(b) The angle <AXC.

(c) XO and OC.

(d) The angle <ACO.

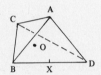

Fig 19.21

19.3 PROBLEMS OF AREAS AND VOLUMES. Level 3

A useful formula for the area R of the triangle ABC is:

$$R = \tfrac{1}{2}AB \times AC \times \sin <A$$

19.3.1 Examples for level 3

(1) A regular octagon is inscribed inside a circle of radius 10 cm. Find the area of the octagon.

Solution Join up the points of the octagon to the centre of the circle as shown in fig 19.22.

Fig 19.22

The angle at the centre of each of the triangles is $^{360}/_8 = 45°$. Using the formula, the area of each little triangle is:

$$^1/_2 \times 10 \times 10 \times \sin 45°.$$

There are eight little triangles, so the total area is:

$$8 \times {}^1/_2 \times 10 \times 10 \times \sin 45° = 283 \text{ cm}^2$$

216

(2)   A stick of chocolate is shaped as a triangular prism 20 cm long. Each end is an equilateral triangle of side 2 cm. Find its volume and its surface area.

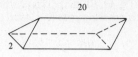

Fig 19.23

Solution   Use the formula to find the area of each triangle.

$$^1/_2 \text{ x } 2 \text{ x } 2 \text{ x sin } 60°.$$

The volume is:

$$20 \text{ x } ^1/_2 \text{ x } 2 \text{ x } 2 \text{ x sin } 60° = 34.6 \text{ cm}^3.$$

The surface area consists of the two ends and the three sides.

$$2\text{x}^1/_2\text{x}2\text{x}2\text{xsin } 60° + 3\text{x}20\text{x}2 = 123.5 \text{ cm}^2.$$

19.3.2 Exercises for level 3

(1)   Find the areas of the triangles shown below:

(a)   (b)   (c)

 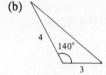

Fig 19.24

(2)   For the triangle LMN, ML = 3.4, NL = 5.6, <L = 66°. Find the area of the triangle.

(3)   A regular hexagon is inscribed inside a circle of radius 8 cm. Find the area of the hexagon.

(4) A regular pentagon of side 12 cm is inscribed inside a circle. Find the radius of the circle. Find the area of the pentagon.

Fig 19.25

(5) A circle of radius 13 cm is inscribed inside a regular 10 sided figure. Find the side of the figure. Find the area of the figure.

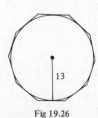

Fig 19.26

(6)   The dots of the figure below are 1 cm apart. Find the perimeter of the triangle ABC. Mark the points D, E such that BCDE is a square, and find its area.

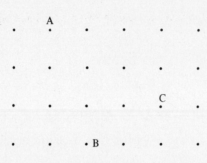

Fig 19.27

(7) The prism of fig 19.28 is 12 cm long, and its cross section is a triangle ABC with AB = 4 cm, BC = 5 cm, <ABC = 55°. Find the area of ΔABC and hence find the volume of the prism.

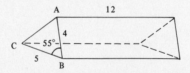

Fig 19.28

(8)   A prism is 10 cm long, and its cross section is an isosceles triangle with sides of length 7 cm, 7 cm, 8 cm. Find the area of the triangle, and hence find the volume of the prism.

COMMON ERRORS

(1)   Right-angled triangles

(a) Do not use Pythagoras for a triangle which is not right- angled. The theorem only holds if the triangle does have a right-angle.

(b) Do not use the wrong sides for the theorem. The longest side is always the hypoteneuse. Make sure that you have labelled the sides correctly before using the equation $c^2 = a^2 + b^2$.

(2)   Errors in algebra

When using the theorem, the a^2 and b^2 terms must be added before the square root is taken. Do not take the square root first.

$$\sqrt{a^2 + b^2} \neq a + b.$$

(3)   Three-dimensional problems

It is very easy to get the wrong triangle, or to label the sides incorrectly. Any three-dimensional diagram is in perspective, so a right-angled triangle may not look like one on the diagram.

It is often a good idea to make a separate two-dimensional picture of the triangle you are using.

20 Statistical Diagrams

A *statistical diagram* makes sense of a whole mass of data by summarizing the important results in a picture.

20.1 TABLES, BAR CHARTS, PIE CHARTS, PICTOGRAMS

A set of statistical figures can often be divided into several groups. There are several ways of displaying these groups.

A *table* gives the numbers of figures in each of the different groups.

A *bar-chart* shows the sizes of each group by bars, whose lengths are proportional to the sizes.

A *pie-chart* compares the sizes by representing them as the slices of a pie.

A *pictogram* gives a picture, from which the numbers in each group can be found.

20.1.1 Examples for level 1

(1)   A group of schoolchildren were asked how they had come to school that morning. The following table gives their answers.

Type of transport	Number of children
Car	7
Bus	12
Train	9
Cycle	5
Walking	17

(a) How many walked to school?
(b) How many children were there in all?

Solution   (a) Read from the table, to find that:

17 walked to school

(b) Add up all the figures, to find that:

7+12+9+5+17 = 50 children were asked

(2)   Erica's record collection has pop, jazz, classical and country records. She shows how many she has of each by a table and by a bar chart.

Classical   6
Country   3
Pop   12
Jazz   ?

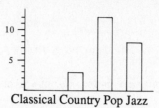

Fig 20.1

(a) How many jazz records does she have?

(b) Fill in the bar corresponding to her Classical records.

(c) How many records does she have in all?

Solution   (a) From the bar chart, the bar for jazz is 8 high.

She has 8 jazz records

Fig 20.2

(b) The bar corresponding to Classical records is 6 high. Fill in the bar chart as shown.

(c) Add together all the figures in the table.

She has 12+8+6+3 = 29 records in all

(3)   All the children in a class of 24 were asked what their favourite sport was. The answers were illustrated in the pie chart shown.

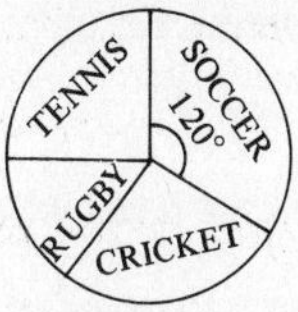

(a) How many preferred Soccer?

Fig 20.3

(b) If 4 children liked rugger best, what is the size of the angle in the rugger slice?

Solution   (a) There are 120° in the soccer slice, out of a total of 360°. 120° is a third of 360°. Hence the number of children who liked soccer best is one third of the total:

1/3x24 = 8 children liked soccer best

(b) 4 children out of 24 prefer rugger. 1 in 6 of the children prefer rugger. The slice is one sixth of the whole pie.

The angle is $360°/6 = 60°$

(4)   The numbers of children in two schools are shown by the following pictogram, in which each figure represents 100 children.

School 1

School 2

Fig 20.4

(a) How many children are there in each school?

(b) Draw an extra row of figures, to show the number of children in a third school with 550 pupils.

Solution   (a) There are 8 figures in the first row, and $7^1/2$ in the second. Each figure represents 100 children, so there are:

800 children in the first school, 750 in the second.

(b) Here there must be $550 ÷ 100 = 5\,^1/2$ figures.

Fig 20.5

20.1.2 Exercises for level 1

(1)   A class of children were asked how many brothers or sisters they had. The answers were displayed in the following table:

Number of brothers or sisters	0	1	2	3	4	5	6+
Frequency	5	9	8	4	2	1	0

(a) What was the most common number of brothers and sisters?
(b) How many only children were there?
(c) How many children were there in the class?

(2)   A survey of the colour of cars was carried out in a car-park. The results were as follows:

Colour	Red	Black	White	Blue	Yellow	Green	Other
Number	4	7	6	8	3	4	8

(a) How many red cars were there?
(b) How many cars were blue or green?
(c) How many cars were not black?

(3)   The results in an exam were as follows:

Grade	A	B	C	D	E	F	U
Numbers	23	47	124	98	65	43	21

(a) How many candidates got grade A?
(b) How many got below grade D?
(c) How many took the exam in all?
(d) If grades A, B and C represent a pass, how many passed the exam?

(4)   The cinemas in a town have shown 60 films throughout the year. The category of film is shown by the pie chart in fig 20.6.

(a) How many PG films were there?
(b) What is the angle for the U sector of the pie?

Fig 20.6

(5)   A boy makes a pie-chart showing how he spends each day. It is given in fig 20.7.

(a) For how many hours is he at school?
(b) What is the angle for the meals sector of the chart?
(c) For how many hours is he awake?

Fig 20.7

(6)   A girl receives some money for her birthday. She decides to spend it as shown in the pie-chart of fig 20.8.

(a) How much money did she receive in all?
(b) How much did she spend on magazines?

Fig 20.8

(7)   The inflation rate for 5 countries is shown by the bar-chart in fig 20.9. Fill in the table below to show the same information.

Country	UK	USA	France	Japan	Germany
Inflation					

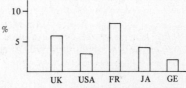

Fig 20.9

(8)   A survey of 60 people gave the following results about where they had gone for their holidays.

Holiday	Britain	Spain	Italy	Greece	Other
Numbers	23	12	6	5	14

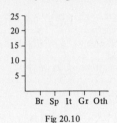

Fig 20.10

Draw a bar chart on fig 20.10 to show the same information.

(9) The bar-chart in fig 20.11 shows the numbers of different types of ice-cream sold on a certain afternoon.

(a) How many vanilla flavoured ones were sold?
(b) What was the most popular flavour?
(c) How many ice-creams were sold in all?

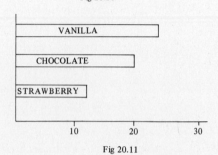

Fig 20.11

(10) Out of 120 cars which were taken to a garage for repair, 30 had defective brakes, 45 were put in for service, 25 had to have body work done, and the remaining 20 were put in for other reasons. Complete the bar-chart of fig 20.12 to illustrate this.

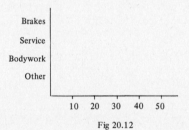

Fig 20.12

(11)   The strengths of two opposing armies is shown by the following pictogram, in which each figure stands for 10,000 soldiers.

Blue Army   Green Army

Fig 20.13

(a)   What are the strengths of the two armies?

(b)   The Blue Airforce contains 500 planes, and the Green Airforce contains 650 planes. Illustrate the relative strength of the two airforces, using the symbol to represent 100 planes.

224

20.2 CONSTRUCTION OF PIE-CHARTS. FREQUENCY TABLES. HISTOGRAMS. Level 2

A *frequency* table for different groups shows the numbers in each group.

A *histogram* is a special sort of bar-chart. The successive groups must be linked in a definite numerical order.

20.2.1 Examples for level 2

(1)   A girl did a statistical survey of the traffic passing along the road outside her house. She made up a tally sheet as shown. Convert it to a frequency table.

	Cars	Lorries	Buses	Motorcycles
Tally	ꜰꜰꜰ ꜰꜰꜰ	ꜰꜰꜰ	ꜰꜰꜰ	ꜰꜰꜰ ///
	ꜰꜰꜰ ꜰꜰꜰ	ꜰꜰꜰ	/	ꜰꜰꜰ
	ꜰꜰꜰ ꜰꜰꜰ	//		ꜰꜰꜰ
	ꜰꜰꜰ /			ꜰꜰꜰ

Solution   Each little bundle of figures ꜰꜰꜰ corresponds to 5 units. Add up all the figures in each group, to obtain:

Cars	Lorries	Buses	Motorcycles
36	12	6	23

(2)   A survey into the tobacco habits of 400 people found how many people consumed the different sorts of tobacco. The results are given by the following frequency table:

Cigarettes	Pipes	Cigars	Snuff	Non-user
100	60	40	10	190

Construct a pie-chart to illustrate these figures.

Solution   For each group we construct a sector, whose angle is proportional to the number of people in that group. The cigarette smokers are a quarter of the whole, so their sector has angle $360°/4 = 90°$. The other figures are:

Pipe-smokers: $60/400 \times 360 = 54°$ Cigar-smokers: $40/400 \times 360 = 36°$
Snuff-takers: $10/400 \times 360 = 9°$ Non-Users: $190/400 \times 360 = 171°$

Now construct the pie-chart, dividing the circle into sectors corresponding to each group.

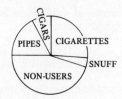

Fig 20.14

(3)   The ages of 100 members of a football club were given by the following frequency table:

Ages	15-20	20-25	25-30	30-35
Frequency	26	37	23	14

Construct a histogram to show these figures.

Solution   The construction of a histogram is the same as for a bar-chart.

There is a numerical relationship between the groups. Along the x-axis measure the ages of the players. The result is shown in fig 20.15.

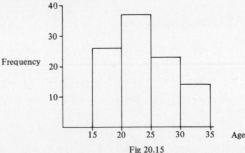

Fig 20.15

20.2.2 Exercises for level 2

(1)   A paragraph of text was analysed for the length of words occuring in it. The following tally chart was made:

Number of letters 1 2 3 4 5 6 7 8 9+

Tally

Draw up a frequency table for the numbers of letters in a word.

(2)   A die is rolled 60 times. The following tally chart shows how often each of the numbers came up:

Score 1 2 3 4 5 6

Tally

Draw up a frequency table for this information.

(3)   Throw a die 30 times, and draw up a table showing the frequency of each number.

(4)   Ask the members of your class how they travelled to school this morning. Record the information in a frequency table.

(5)   60 people were asked what they drank first in the morning. 23 drank tea, 15 coffee, 12 chocolate and 10 other drinks. Construct a pie-chart to show this information.

(6)   An education authority spends its money as shown in the following table:

Type of Education	Expenditure in £millions
Primary	120
Secondary	135
Further and Higher	120
Other	165

Construct a pie-chart to show this information.

(7)   A woman calculates that of her income of £200 per week, £40 goes in tax, £50 on rent, £30 on food, £30 on entertainment, and £50 for other items. Draw up a pie-chart to show this.

(8)   The survey of a county showed that 20% of it was built-up, 50% was farmland, and 30% formed part of a National Park. Display this information on a pie-chart.

(9)   The ages of a school orchestra were found to be as given in the following frequency table:

Age	11-12	12-13	13-14	14-15	15-16
Frequency	2	7	11	15	12

Construct a histogram to show this information.

(10)   The marks for 120 candidates were as follows:

Marks out of 40	0-9	10-19	20-29	30-40
Frequency	20	35	43	22

Construct a histogram to display these marks.

(11)   The manager of a restaurant did a survey on how much his customers spent per meal. His results were as follows:

Price in £s	3.00-4.99	5.00-6.99	7.00-8.99	9.00-10.99
Frequency	10	39	27	14

Draw a histogram to show these results.

(12)   In an effort to cut down on its telephone bill, a company did a survey of the length of time of calls from its office. The results were shown on the following table:

Length in Minutes	0-3	3-6	6-9	9-12	12-15
Frequency in 10's of calls	53	25	15	7	2

Construct a histogram to illustrate these figures.

20.3 HISTOGRAMS WITH UNEQUAL INTERVALS. Level 3

The height of a bar on a histogram represents the frequency *per unit interval*. Hence if an interval is twice the normal width, then its height must be halved. If an interval is one third the normal width, then its height must be tripled.

20.3.1 Example for level 3

The marks of 120 candidates in an exam were given by the following table:

% Marks	0-29	30-39	40-49	50-59	60-64	65-69	70-79	80-100
Frequency	12	10	16	24	17	14	19	8

(a) Construct a histogram to show this information.

(b) What is the modal class?

(c) Estimate the proportion of candidates who got at least 55%.

Solution   Notice that the intervals are not even. There must be a corresponding adjustment to the height of the bars.

The most common interval is of 10 marks.

The intervals at the end and the beginning are of 20 and 30 marks respectively. Divide each of their frequencies by 2 and 3 respectively.

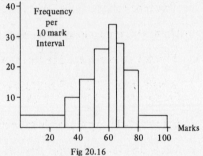

Fig 20.16

The intervals in the 60's are of 5 marks each. Double the frequencies in those intervals.

(a) The histogram is in fig 20.16.

(b) The modal class is the one in which the population is densest. This is equivalent to the highest bar on the histogram.

The modal group is 60-64

(c) Assume that half the candidates in the 50-59 range got at least 55%. This gives 12+17+14+19+8 = 70 candidates in all.

The proportion of candidates with at least 55% is $^{70}/_{120} = ^{7}/_{12}$

20.3.2 Exercises for level 3

(1)   The ages of a group of 500 people at a cinema were as follows:

Age range | 10-20 | 20-25 | 25-30 | 30-35 | 35-40 | 40-70 |
Frequency | 74 | 73 | 104 | 116 | 67 | 66 |

(a) Construct a histogram, using a span of 5 years as the basic interval.

(b) What is the modal class?

(c) Estimate from your graph the proportion of the audience who were over $32^{1}/_{2}$.

(2)   The averages of 80 cricketers were compared. The figures are shown in the following table:

Average | 0-20 | 20-30 | 30-40 | 40-50 | 50-80 |
Frequency | 22 | 18 | 16 | 12 | 12 |

(a) Draw a histogram to illustrate this data.

(b) Find the modal class.

(c) Estimate the percentage of cricketers whose average was less than 35.

(3)   The salaries of 100 employees of a firm were analysed, and the following table shows the distribution:

Salary in £1,000's | 7-9 | 9-10 | 10-11 | 11-12 | 12-13 | 13-16|
Frequency | 24 | 15 | 17 | 19 | 10 | 15 |

Draw a histogram to show the distribution.

(4)   60 cats were weighed. The results were as follows:

Weight in kg. | 1-3 | 3-3.5 | 3.5-4 | 4-4.5 | 4.5-6 |
Frequency | 20 | 9 | 15 | 10 | 6 |

Construct a histogram to illustrate these figures.

(5)   50 children ran 100 metres. Their times are given in the following table:

Time in secs.| 12-14 | 14-15 | 15-15.5 | 15.5-16 | 16-17 | 17-20 |
Frequency | 4 | 13 | 9 | 11 | 4 | 9 |

Construct a histogram to illustrate these times.

(6)   Out of 160 families, 33 had 0 children, 26 had one child, 45 had 2 children, 26 had 3 children, and 30 had 4, 5 or 6 children. Construct a histogram to show these figures.

(7)   A bridge player always counts the number of hearts in his hand. After 130 games, he finds that he has the following figures:

Number of hearts | 0 or 1 | 2 | 3 | 4 | 5 | 6 | 7-13 |
Frequency | 23 | 19| 25| 27| 13| 9 | 14 |

Show this information on a histogram.

COMMON ERRORS

(1)   **Tables.**
         (a) When measuring the frequencies of numerical data, be careful not to confuse a number with the frequency of that number. Suppose, for example, a die is thrown several times, and the number 5 comes up 20 times. Be sure that 20, not 5, goes into the frequency box.

         (b) When constructing a frequency table from a tally chart, be sure that each ⫻ symbol counts for 5, not 4. The cross stroke represents an entry as well as the vertical strokes.

(2)   **Bar-charts**
         If you are drawing a bar-chart, and not a histogram, draw the bars separately and do not join up them up.

         Suppose you are drawing a bar-chart representing car-production in different countries. Draw the bars for France and Japan apart from each other. It makes no sense to join up the bars, as there is no country "half-way" between France and Japan.

Fig 20.17

(3) **Histograms**

(a) If a histogram is measuring continuous data, such as time or height or weight, then the bars should be touching each other.

Suppose your histogram compares the weights of certain items. The bar corresponding to the range 30-40 grams should touch the bar corresponding to 40-50 grams.

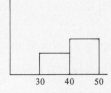

Fig 20.18

(b) If your histogram is measuring continuous data, then do not worry that the same figure is the right end of one interval and the left end of another, as in the example just above. It is almost impossible that an item weighs *exactly* 40 grams.

It would definitely be wrong to have intervals of 30-39 grams and of 40-50 grams. Where would you put an item of 39.5 grams?

(4) **Histograms with unequal intervals**

(a) Be sure that you adjust the height of the bars to compensate for the different widths of the intervals. If you do not do so, the histogram will be misleading. The example of 20.3.1, without such compensation, would be as shown.

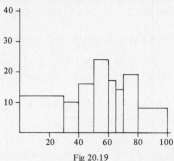

Fig 20.19

We can see that too much importance is given to the end intervals, and too little importance to the middle intervals.

(b) Be sure that you label the vertical axis clearly. It is not enough to label it "Frequency", you must also explain what the basic interval is. Label it "Frequency per 5 marks", or "Frequency per 10 cm." etc.

(c) The modal group is not necessarily the group with the greatest number in it. It is the group with the greatest density, i.e. it is the group with the tallest bar.

21 Averages and Cumulative Frequency

21.1 MEANS, MODES, MEDIANS

An *average* summarizes a collection of numbers.

The *mean* of a set of numbers is what is normally meant by the word average. The mean age of ten children is the sum of all their ages divided by 10.

The *median* of a set of numbers is the middle number. Half of the numbers are less than it and half of the numbers are greater than it.

The *mode* of a set of numbers is the most commonly occuring number.

The *range* of a set of numbers is the difference between the highest and the lowest numbers.

21.1.1 Examples for level 1

(1)   12 children took a test, and their marks out of 10 were as follows:

8, 8, 7, 7, 6, 8, 9, 5, 6, 7, 7, 6.

Find the mean, the mode and the range.

Solution   For the mean, find the sum of all the marks.

$8 + 8 + 7 + 7 + 6 + 8 + 9 + 5 + 6 + 7 + 7 + 6 = 84$.

Divide by 12 to obtain:

The mean mark is $^{84}/_{12} = 7$

The most common mark is 7, which occurs 4 times.

The mode is 7

The highest score is 9, and the lowest is 5.

The range is 4

(2)   In 8 successive innings, a batsman scored the following:

45, 65, 4, 0, 76, 12, 8, 30.

Find his median score.

Solution   Re-arrange the scores in increasing order:

0, 4, 8, 12, 30, 45, 65, 76.

The median score is between his fourth best and his fifth best. Take halfway between 12 and 30:

The median is 21

21.1.2 Exercises for level 1

(1)   Find the mean, mode and range of the following numbers:

5, 6, 4, 5, 10, 3, 9, 4, 9, 5.

(2)   Find the mean, median and range of the following numbers:

23, 38, 20, 30, 31, 19, 27, 24, 32, 26.

(3)   6 people were asked to estimate the distance of a church. They gave the answers:

100 m, 120 m, 90 m, 100 m, 150 m, 150 m.

Find the mean and the median of these distances.

(4)   8 children were asked to measure the width of their right hands. The results were (in cm):

6.3, 7.1, 5.8, 6.0, 5.5, 5.8, 6.2, 5.3.

Find the mean and the median of these widths.

(5)   A golfer had the following scores in 9 rounds:

103, 96, 110, 99, 100, 92, 105, 109, 113.

Find the mean, median and range of these scores.

(6)   A football team scored the following number of goals in 12 matches:

0, 3, 1, 0, 2, 3, 6, 3, 4, 0, 2, 0.

Find the mean and mode of these scores.

(7)   Richard bought 10 matchboxes. He counted the number of matches in each. The results were as follows:

48, 50, 48, 51, 47, 45, 43, 47, 49, 52.

What is the mean number of matches?

(8)   The weights of 5 apples were as follows:

43 g, 50 g, 46 g, 50 g, 47 g.

Find the mean weight.

(9)   The salaries of 8 employees of a firm were:

£8,000, £7,500, £6,500, £10,000, £6,000, £7,400, £5,000, £22,000.

Find the mean salary and the median salary.

(10) A boy surveyed the cars coming through the school gates, and noted down the number of people (including the driver) in each car. The results for the first 10 cars were:

1, 3, 1, 1, 2, 3, 4, 4, 1, 1.

Find the mean and mode of these figures.

(11) A shopper compared the prices of a jar of honey in 8 different shops. The prices were:

80p, 75p, 88p, 83p, 81p, 77p, 90p, 86p.

Find the mean price and the median price.

(12) 8 packets of butter were weighed and found to be:
252 g, 247 g, 253 g, 250 g, 250 g, 248 g, 247 g, 246 g.

Find the mean weight and the median weight.

(13) For the seven days of his holiday George measured the hours of sunshine. His results were:

8.3, 7.3, 2.1, 9.0, 6.5, 7.3, 8.5.

Find the mean and median of these figures. Find the range of values.

(14) On 5 successive days a television company estimated the number of people who had watched its news programme. The results, in millions of viewers, were as follows:

4.3, 5.3, 4.2, 4.6, 3.8.

Find the mean and the median number of viewers.

(15) 9 batteries were used until they faded. The lifetimes in hours were:

53, 47, 38, 56, 22, 50, 45, 39, 35.

Find the mean lifetime and the median lifetime. Find the range.

(16) For her first 7 litters, a cat has the following numbers of kittens:

1, 2, 3, 3, 5, 4, 3.

Find the mean number of kittens per litter and the modal number.

(17) The 8 classes of a year at a school contain the following number of pupils:

27, 28, 24, 25, 26, 25, 26, 25.

Find the mean and median number of pupils per class.

(18) The 24 children in a class were asked how they had come to school. Their answers were as follows:

Type of transport | Bus | Bike | Train | Walk | Car |
Frequency | 6 | 3 | 8 | 5 | 2 |

What is the modal type of transport? Why does it not make sense to ask for the mean or the median type of transport?

(19) Over a period of 3 months Bill takes 15 books out of the library. 6 are Science-fiction, 5 are Crime books, and 4 are Thrillers. What is the modal group? Can one refer to the mean book or the median book?

21.2 AVERAGES FROM FREQUENCY TABLES. Level 2

21.2.1 Examples for level 2

(1) A survey was made of 100 families, to investigate the number of children in each family. The results were given in the following frequency table:

Number of children | 0 | 1 | 2 | 3 | 4 | 5 | 6 | 7 | 8+ |
Number of families | 11| 24| 35| 16| 9 | 3 | 1 | 1 | 0 |

Find the mean number of children per family and the median number.

Solution   The total number of children is found by multiplying the numbers of children by the frequencies.

$$0x11 + 1x24 + 2x35 + 3x16 + 4x9 + 5x3 + 6x1 + 7x1 + 8x0 = 206$$

There are 100 families in all, which gives:

The mean number is $^{206}/_{100} = 2.06$

The median number is the number of children such that half the families have more than that number, and half the families have less. Only a very imperfect answer is possible here: 35 families have less than 2 children and 30 have more than 2. Hence:

The median number is 2

(2) The Post Office conducted a survey of the weights of 500 letters in the lowest weight range. The results of the survey were as follows:

Weight in grams | 0-10 | 10-20 | 20-30 | 30-40 | 40-50 | 50-60 |
Frequency | 53 | 176 | 113 | 64 | 53 | 41 |

Find the modal class and estimate the mean weight of the letters.

Solution   The modal class is the class with the largest number of letters in it.

The modal class is 10-20 grams

The table does not give the exact weight of all the letters. Without any more information the best that can be done is to assume that they are evenly spread along each interval.

Assume that the letters in the 0-10 range average at 5 grams each. Make a similar assumption in the other ranges. Work out the total weight:

$$5 \times 53 + 15 \times 176 + 25 \times 113 + 35 \times 64 + 45 \times 53 + 55 \times 41 = 12{,}610 \text{ grams.}$$

There are 500 letters in all.

Mean weight = $^{12{,}610}/_{500}$ = 25.22 grams.

21.2.2 Exercises for level 2

(1)   The first 500 words of a book were examined to see how many letters they contained. The results are shown in the following table:

Number of letters | 1 | 2 | 3 | 4 | 5 | 6 | 7 | 8 | 9 | 10 | 11+|
Frequency | 15| 34| 45| 76|106| 94| 79| 41| 6 | 3 | 1 |

Find the mean, the mode and the median number of letters per word.

(2)   A class of 24 children were asked how many books they had read in the previous week. 6 had read none, 8 had read 1, 6 had read 2, 3 had read 3 and one had read 4. Find the mean number of books read.

(3)   The manager of a hotel investigated how long his guests stayed. The results were:

Number of nights | 1 | 2 | 3 | 4 | 5 | 6 | 7 | 8 | 9 | 10+ |
Frequency | 48| 24| 19| 6 | 5 | 2 | 12| 3 | 1 | 0 |

Find the mean and median number of nights stayed at the hotel.

(4)   The Ministry of Transport conducts a survey in a certain area to see how many driving tests people have before they pass. The results are as follows:

Number of tests | 1 | 2 | 3 | 4 | 5 | 6 | 7 | 8+|
Frequency |131| 52| 25| 19| 11| 7 | 4 | 1 |

Find the mean number of tests needed to pass.

(5)   A traffic survey counted the number of people (including the driver) in cars during the rush hour period. The results were:

Number of people | 1 | 2 | 3 | 4 | 5 |
Frequency |217|103| 47| 30| 3 |

Find the modal number, and find the mean number.

(6)   A supermarket sells oranges in packs of 5. There are complaints about the number of over-ripe fruit. 150 packs are opened, and the number of rotten oranges in each pack are counted. The results are:

Number of bad oranges | 0 | 1 | 2 | 3 | 4 | 5 |
Frequency | 65 | 52 | 27 | 5 | 1 | 0 |

Find the mean and median number of bad oranges in a pack.

(7)   20 children were asked how many foreign countries they had visited. The answers are given in the table:

Number of countries | 0 | 1 | 2 | 3 | 4 |
Frequency | 6 | 7 | 5 | 0 | 2 |

Find the mode and the mean number of countries visited.

(8)   Of 400 pupils entering a university, 50 had 2 A-levels, 297 had 3 A-levels, 42 had 4 and 11 had 5. Find the mean number of A-levels per pupil.

(9)   The 50 people at a party had ages given by the following frequency table:

Age range | 15-20 | 20-25 | 25-30 | 30-35 | 35-40 |
Frequency | 12 | 23 | 6 | 8 | 1 |

Find the modal class and estimate the mean age.

(10)   The heights of 100 boys were as follows:

Height in cm | 50-60 | 60-70 | 70-80 | 80-90 |
Frequency | 27 | 34 | 29 | 10 |

Find the modal class and estimate the mean height.

(11)   The salaries of 50 employees of a firm were as follows:

Salary in £1,000's | 6-8 | 8-10 | 10-12 | 12-14 | 14-16 | 16-18 |
Frequency | 8 | 13 | 17 | 7 | 3 | 2 |

Estimate the mean salary.

(12)   30 children were asked how much television they watched each week. The answers were as follows:

Number of hours | 0-2 | 2-4 | 4-6 | 6-8 | 8-10 | 10-12 | 12-14 |
Frequency | 6 | 11 | 5 | 3 | 2 | 1 | 2 |

What is the modal class? What is the mean number of hours watched?

(13)   British Telecom surveys the length of telephone calls in a particular area. The lengths of calls, measured to the nearest minute, are shown by the following table:

Length of call	0	1	2	3	4	5	6	7	8	9	10+
Frequency	52	197	110	85	42	36	23	19	15	8	13

Find the modal class and the mean length of a telephone call.

(14)   A gun was fired 80 times, and the range was measured to the nearest 100 m. The results were:

Range in 100 m's	23	24	25	26	27	28
Frequency	5	17	29	24	4	1

Estimate the mean range.

(15)   120 candidates sat an exam, and the distribution of the marks is given by the following table:

% Marks	0-29	30-49	50-59	60-69	70-100
Frequency	16	33	31	26	14

Estimate the mean mark.

21.3 CUMULATIVE FREQUENCY. Level 3

The *cumulative frequency* at a value is the running total of all the frequencies up to that value.

A *cumulative frequency* table gives the cumulative frequencies at each value.

A *cumulative frequency* chart gives a graph of the cumulative frequency.

The cumulative frequency up to the median is $1/2$.

The cumulative frequency up to the *lower quartile* is $1/4$.

The cumulative frequency up to the *upper quartile* is $3/4$.

The *inter-quartile range* is the difference between the quartiles.

21.3.1 Example for level 3

A man did a survey of the prices of second-hand cars advertised in his local newspaper. The results were as follows:

Price Range	0-500	500-1000	1000-2000	2000-3000	3000-4000
Frequency	7	10	27	23	13

(a) Fill up a cumulative frequency table.

(b) Draw a cumulative frequency curve.

(c) Find the median, the quartiles and the inter-quartile range.

(d) What proportion of cars cost less than £1,500?

Solution   (a) Fill in the third row of the table, adding up the frequencies as you go along:

Price range	0-500	500-1000	1000-2000	2000-3000	3000-4000
Frequency	7	10	27	23	13
Cum. Frequ.	7	17	44	67	80

(b) The horizontal axis represents price. Label it from £0 to £4,000. The vertical axis represents cumulative frequency. Label it from 0 to 80. Now draw the graph.

The points on the graph correspond to the *endpoints* of the intervals. 44 cars cost less than £2,000, so the point (2000,44) must lie on the graph. The graph is shown in fig 21.1

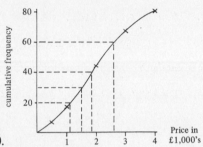

Fig 21.1

(c) From the graph, 40 cars cost less than £1800.

The median is £1,800

20 cars cost less than £1,100, and 60 cars less than £2,600.

The lower quartile is £1,100. The upper quartile is £2,600

The interquartile range is £1,500

(d) Take £1,500 on the horizontal axis. This corresponds to a cumulative frequency of 30.

The proportion of cars costing less than £1,500 is $^{30}/_{80} = ^3/_8$

21.3.2 Exercises for level 3

(1)   The heights of 60 boys were measured. The results are shown on the following table:

Height in cm	150-160	160-170	170-180	180-190
Frequency	10	23	18	9

(a) Fill up a cumulative frequency table.

(b) Draw a cumulative frequency curve.

(c) Find from your curve the median, the quartiles and the inter-quartile range.

(d) What proportion of boys were more than 165 cm in height?

(2)   The weights of 80 eggs were found to be as follows:

Weight in grams	50-60	60-70	70-80	80-100	100-120
Frequency	7	15	28	19	11

(a) Construct a cumulative frequency curve.

(b) Draw a cumulative frequency graph.

(c) From your graph find the median, the quartiles and the inter-quartile range.

(d) What proportion of eggs lie between 60 and 90 grams?

(3) to (7). For each of the situations in questions (1) to (5) of exercise 20.3.2 do the following:

(a) Find the cumulative frequencies.

(b) Draw a cumulative frequency chart.

(c) Find the median, the quartiles and the inter-quartile range.

(8)   The same exam was taken by class A and class B. The results were as follows:

% Mark	0-29	30-39	40-49	50-59	60-69	70-100
Class A frequency	1	3	10	14	2	0
Class B frequency	2	5	7	9	4	3

Construct cumulative frequency tables for both the classes, and draw the cumulative frequency curves on the same sheet of graph paper. What is the difference between the curves?

Find the inter-quartile ranges for the two classes. What does the difference between the values tell you about the two classes?

(9)   Two golfers Lucy and Liz regularly play together over the same course. After they have each been round 40 times their scores are given by the following table:

Score	70-74	75-79	80-84	85-89	90-99	100-109
Lucy's frequency	2	3	23	7	5	0
Liz's frequency	5	8	14	5	4	4

Construct cumulative frequency tables for both players, and draw the cumulative frequency graphs on the same sheet of graph paper. Comment on the difference.

Evaluate the inter-quartile ranges for the two women. Explain the difference. Who do you think is the better player?

COMMON ERRORS

(1)   Be sure that you know the difference between the three averages, mean, median, mode. They are often confused.

(2)   **Mean**
(a) If the mean of a set of values is not an integer, it is wrong to round it to the nearest whole number. Though a family must have a whole number of children, it still makes sense to say that:

"The mean family size is 2.4 children."

(b) When finding a mean from a frequency table, be sure that you are finding the mean of the *figures*, not of the frequencies.

(c) When finding a mean from a frequency table, be sure that you divide by the number of figures, not by the number of classes. If your frequency table is:

Number of children in family	0	1	2	3	4	5
Frequency	10	27	34	22	15	12

Be sure that you divide by 120 (the number of families) not by 6 (the number of classes). If you do the wrong thing you will get an absurdly large answer.

(d) When the figures in a frequency table are given by a range of values, make sure that you take the middle of the range for working out the mean. It may not be perfect, but it is the best that can be done in the circumstances.

(3)   **Median**
When working out the median, be sure that you take the middle value. The median of 17 numbers is not 9, it is the *9th number*.

(4)   **Cumulative Frequency**
(a) The points on a cumulative frequency curve must be at the *right endpoints* of the intervals, not at the middles.

If your graph is wrong then all your subsequent calculations of the median and quartiles will be wrong also.

(b) The median and quartiles must refer to the measurements, not to the frequencies. Suppose you are measuring the heights of 80 people. The lower quartile is not the number 20, it is the height of the *20th person*.

22 Probability

Probability is measured on a scale between 0 and 1.

If an event is certain, then it has a probability of 1. If it is impossible, then it has a probability of 0.

If an event may or may not happen, then its probability is the proportion of times in which it does happen.

The probability of an event A is written P(A).

22.1 PROBABILITY OF A SINGLE EVENT

An experiment or game may have several different outcomes. If a card is drawn from a pack, then it may be any one of the 52 cards. When a six-sided die is thrown, any one of the six faces may be uppermost.

If there are n equally likely outcomes of an experiment or game, then the probability of each outcome is $1/n$.

22.1.1 Examples for level 1

(1)   A fair six-sided die is rolled. What is the probability that it shows a 5?

Solution   Because the die is fair, each of the 6 numbers is equally likely to come up. The probability that the number is 5 is:

$$1/6$$

(2)   In her cupboard a woman has 4 cans of tomato soup and 6 cans of celery soup. She shuts her eyes and picks a can at random: what is the probability that it is a can of tomato soup?

Solution   She is equally likely to draw out any of the ten cans. The probability that she selects a tomato can is:

$$4/10 = 2/5$$

(3)   A card is drawn at random from a well-shuffled pack. What is the probability that it is a club?

Solution   Because the pack is well-shuffled, each of the 52 cards is equally likely to be drawn. The probability that one of the 13 clubs is drawn is therefore:

$$13/52 = 1/4$$

22.1.2 Exercises for level 1

(1)   A fair six-sided die is rolled. What is the probability that a 3 is uppermost?

(2)   A fair coin is tossed. What is the probability that it comes up heads?

(3)   A solid in the shape of a regular tetrahedron (four-sides) has the numbers 1 to 4 on its faces. It is thrown. What is the probability that it lands on the face labelled 2?

(4)   A bag contains 8 red and 17 blue marbles. A child draws one at random. What is the probability that the marble is blue?

(5)   In an election 4,000 people voted Labour, 4,500 voted Conservative, and 2,500 voted Liberal. If a voter is picked at random, what is the probability that she voted Labour?

(6)   A multiple choice question has 5 possible answers. If a candidate picks the answer at random, what is the probability that it is correct?

(7)   Out of 100 batteries made by a certain firm, 7 are faulty. If I buy one battery, what is the probability that it works?

(8)   A roulette wheel has slots numbered 1 to 36, and a zero slot. Assuming that it is fair, what is the probability that the ball lands in the slot numbered 19?

(9)   A box of sweets contains 10 toffees, 8 liquorices and 7 chocolates. If a sweet is drawn at random what is the probability that it is a toffee?

(10)   A bingo caller has balls labelled 1 to 99. If he draws one at random, what is the probability that it is 11?

(11)   A card is drawn from a well-shuffled pack. What is the probability that it is an Ace?

(12)   The letters of the word MATHEMATICS are each written on squares of cardboard, which are then shuffled. If one is drawn, what are the probabilities that the letter is (a) S (b) T?

(13) In the fairground game shown in fig 22.1 the pointer is spun until it comes to rest over one of the sectors. Find the probabilities that it points at:

(a) 7   (b) An odd number.

Fig 22.1

(14)   A 4 sided die and a 6 sided die are thrown together. The following table gives the possible total score:

		Score on 6 sided die					
		1	2	3	4	5	6
	1	2	3	4	5	6	7
Score on 4 sided die	2	3	4	5	6	7	8
	3	4	5	6	7	8	9
	4	5	6	7	8	9	10

Find the probabilities that:

(a) The total is 10   (b) The total is 7   (c) The total is less than 4.

(15) Two fair six sided dice are thrown. Complete the following table of the total score:

```
                        Score  on  first  die
                    1     2     3     4     5     6
                1 |  2 |  3 |  4 |  5 |  6 |  7 |
Score on second die 2 |  3 |    |    |    |    |  8 |
                3 |    |    |    |  7 |    |    |
                4 |    |    |    |    |  9 | 10 |
                5 |    |    |    |    |    |    |
                6 |    |    |    |    | 11 |    |
```

Find the probabilities that the total is (a) 2   (b) 7.

22.2 COMBINATIONS OF PROBABILITIES. Level 2

If the probability of an event is p, then the probability that the event does *not* happen is 1 - p.

If two unconnected events have probabilities p and q, then the probability of them *both* happening is pxq.

22.2.1 Examples for level 2

(1)   A card is drawn from a well-shuffled pack. What is the probability that it is not an ace?

Solution   The probability that the card is an ace is $^1/_{13}$. The probability that it is not an ace is therefore:

$$1 - {}^1/_{13} = {}^{12}/_{13}$$

(2)   Two fair six-sided dice are thrown. What is the probability that the result is a double six?

Solution   For each single die, the probability of a six is $^1/_6$. The probability that both dice give sixes is therefore:

$$^1/_6 \times {}^1/_6 = {}^1/_{36}$$

(3)   A class contains 6 boys and 10 girls. Two are chosen at random. What is the probability that they are both boys?

Solution   The probability that the first is a boy is $^6/_{16}$.

There are now 5 boys out of 15 children.

The probability that the second is a boy is $^5/_{15}$.

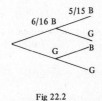

These probabilities are shown on a tree diagram. The probability that both are boys is found by multiplying the probabilities along the top branch.

Fig 22.2

$$\textbf{P(Both are boys)} = {}^6/_{16} \times {}^5/_{15} = {}^{30}/_{240} = 1/8$$

22.2.2 Exercises for level 2

(1)   A fair six-sided die is rolled. What is the probability that the result is not a 6?

(2)   A roulette wheel has the numbers 1 to 36 and a zero. Find the probability that the zero does not come up.

(3)   A card is drawn from a well-shuffled pack of cards. What is the probability that it is not a heart?

(4)   In a multiple choice exam, each question has 5 possible answers. A candidate answers two questions at random: find the probabilities that (a) both questions are right (b) both questions are wrong.

(5)   A fair coin is tossed twice. Find the probabilities of (a) two heads (b) two tails (c) a head and a tail, in any order.

(6)   A coin is biased so that it is twice as likely to come up heads as tails. What is the probability that it comes up heads?

(7)   A roulette wheel has the numbers 1 to 36. For the first spin I bet that an even number will come up, and for the second spin that a number divisible by 3 will come up. What is the probability that I win both my bets?

(8)   A football team has probabilities $1/2$, $1/3$, $1/6$ of winning, losing, drawing respectively. It the team plays two matches, find the probabilities that:

        (a) Both matches are drawn.
        (b) The first match is won and the second lost.
        (c) The team did not lose both matches.
        (d) Exactly one match was drawn.

(9)   Two fair dice are thrown and the sum is recorded. Find the probabilities that:

(a) Both are sixes (b) Both show the same number (c) The first number is less than the second.

(10)   Two letters are chosen from the word SCROOGE. What is the probability that both O's are chosen?

(11)   Two people are chosen from 5 women and 4 men. What is the probability that they are both men?

(12)   A box contains 7 red and 12 yellow counters. Two are chosen. What is the probability that they are both yellow?

22.3 EXCLUSIVE AND INDEPENDENT EVENTS. Level 3

Two events are *exclusive* if they cannot happen simultaneously.

If two event A and B are exclusive, then the probability of one or other of them occurring is the sum of their probabilities.

$$P(A \text{ or } B) = P(A) + P(B)$$

Two events are *independent* if one of them does not make the other either more or less likely.

If two events A and B are independent, then the probability of them both happening is the product of their probabilities.

$$P(A \text{ and } B) = P(A) \times P(B)$$

22.3.1 Examples for level 3

(1)   A bag contains 5 red and 7 blue marbles. Two are drawn in succession. Find the probabilities that:

    (a) The first is red.
    (b) Both are red.
    (c) They are both the same colour.
    (d) The second is red.

Solution   Draw a tree-diagram as shown. Mark in the probabilities at each fork.

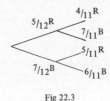

Fig 22.3

(a) The probability that the first is red is $^5/_{12}$

(b) The probability that both are red is the probability of the top branch.

$$\textbf{P(Both are red)} = {}^5/_{12} \times {}^4/_{11} = {}^{20}/_{132} = {}^5/_{33}$$

(c) The probability that they are of the same colour is the sum of the probabilities of the first and last branches.

$$\textbf{P(Both same colour)} = {}^5/_{12} \times {}^4/_{11} + {}^7/_{12} \times {}^6/_{11} = {}^{62}/_{132} = {}^{31}/_{66}$$

(d) The second marble is red in the first and third branches.

$$\textbf{P(Second red)} = {}^5/_{12} \times {}^4/_{11} + {}^7/_{12} \times {}^5/_{11} = {}^{55}/_{132} = {}^5/_{12}$$

(2)   In a class of 20 boys, 10 learn French, 12 learn German, and 7 learn neither. If a boy is selected at random, find the probabilities that:
            (a) He studies both languages.
            (b) He studies French but not German.

Solution   The two events F (= he studies French) and G (= he studies German) may not be independent. Hence the two probabilities cannot be multiplied together.

12 + 10 + 7 = 29. 9 boys must have been counted twice. These boys must be the ones who study both languages.

$$P(F \text{ and } G) = {}^9\!/_{20}$$

Of the 10 who study French, 9 also study German. Hence only 1 studies French and not German.

P(French but not German) = $^1/_{20}$

22.3.2 Exercises for level 3

(1) One in ten flash cubes is faulty. I buy 2. Complete the tree diagram.

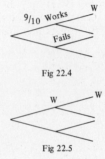

Find the probabilities that   (a) Both work (b) exactly one works.

Fig 22.4

(2) 3 white and 4 black balls are in a bag. 2 are drawn at random. Complete the tree diagram:

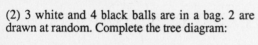

Find the probabilities that   (a) the second is black   (b) both are the same colour.

Fig 22.5

(3)   In his drawer a man has 8 left shoes and 11 right shoes. He picks out 2 at random. Find the probabilities that:

      (a) Both are left shoes.
      (b) Both are left or both are right.
      (c) He draws a left and a right shoe.

(4)   Three quarters of a batch of tulip bulbs give flowers. If I plant 2, find the probabilities that:

(a) Both flower.   (b) At least one flowers.

(5)   Two fair six-sided dice are rolled. Find the probabilities that:

      (a) Both are sixes.
      (b) Neither is a six.
      (c) At least one is a six.

(6)   A box of chocolates contains 12 soft-centered and 14 hard-centered chocolates. Two are selected; find the probabilities that:

      (a) Both are soft-centred.
      (b) One is soft-centred and the other hard-centred.

(7)   A woman goes to work by bus, by car or on foot with probabilities $^1/_2$, $^1/_3$, $^1/_6$ respectively. For each type of transport, the probabilities that she will be late are $^1/_{10}$, $^1/_5$, $^1/_{50}$ respectively. Find the probabilities that:

        (a) She will go by car and be late.
        (b) She will go by bus and be on time.
        (c) She will go by bus or by car and be late.
        (d) She will be late.

(8) The diagram shows a street map of a village. A motorist enters the village at A, and at each crossroads he drives straight ahead with probability $^1/_2$, and turns left or right with equal probability $^1/_4$. Find the probabilities that:

        (a) He goes to C.
        (b) He goes to D.
        (c) He goes to F.

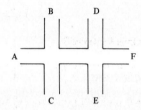

Fig 22.6

(9)   Two different letters are chosen from the word COMPUTER. Find the probability that at least one is a vowel.

(10)   If it is fine today, the probability that it will rain tomorrow is $^1/_5$. If it is rainy today, the probability that it will rain tomorrow is $^2/_3$. The probability that today is fine is $^1/_2$. Find the probabilities that:

        (a) It will be fine on both days.
        (b) There will be different weather today and tomorrow.

(11)   Two cards are drawn from a well-shuffled pack. Find the probabilities that:

        (a) Both are hearts.
        (b) They are of the same suit.
        (c) The first is a heart and the second is a King.

(12)   In order to start at a certain game each player must throw a six with a fair die. Find the probabilities that a player starts after:

(a) One throw (b) Two throws (c) Three throws.

(13)   A number is chosen at random from 1 to 20 inclusive. Find the probabilities that:

        (a) It is even.
        (b) It is greater than 17.
        (c) It is odd and greater than seventeen.
        (d) It is a prime greater that 10.

(14)   A poker hand consists of 5 cards. The first four cards I receive are the 6, 7, 8, 9 of hearts. If the fifth card is another heart, then I have a flush. If the fifth card is a 5 or a 10, then I have a straight. Find the probabilities that I get:

        (a) A flush.
        (b) A straight.
        (c) A straight flush.
        (d) Either a flush or a straight.

(15)   Out of 30 boys, 18 play soccer and 12 play rugger. 8 play both. if a boy is picked at random, find the probabilities that:

(a) He plays soccer but not rugger   (b) He does not play either sport.

(16)   The probability that chips are on the menu for dinner is $2/3$, and the probability of beans is $1/4$. The probability of both is $1/5$. Find the probabilities of:

        (a) Chips but not beans.
        (b) Neither chips nor beans.

COMMON ERRORS

(1)   Simple probability

If there are n outcomes to an experiment, then the probability of each individual outcome is $1/n$ if all the outcomes are *equally likely*. Do not forget the "equally likely" condition.

If two coins are tossed, then there are either 0, 1 or 2 heads. But it does not follows that P(1 head) = $1/3$. This outcome is more likely than the others.

(2)   Addition of probabilities

Probabilities are only added together if the events cannot happen together. If events A and B can happen together, then it is not true that P(A or B) = P(A) + P(B).

If your answer gives a probability of more than 1, then you must have made a mistake of this sort.

(3)   Multiplication of probabilities

Probabilities are only multiplied together if the events do not affect each other.

If event A makes event B either more likely or less likely, then it is not true that P(A & B) = P(A) x P(B).

(4)   Exclusive and independent events

Do not confuse these terms. They refer to *pairs* of events, not to single events.

They are not opposites - if a pair of events is not exclusive then it does not follow that they are independent.

(5)   Independence of successive events

Read a question carefully to make sure that you do not confuse the following sorts of probability:

Rolling a die twice. If the first roll is a six then the second roll is just as likely to be a six.

Drawing two cards. If the first is an Ace then the second is less likely to be an Ace.

23 Transformations

A *transformation* changes the positions of points and shapes in the plane. The standard transformations are as follows:

A *translation* shifts all points in a fixed direction.

A *rotation* turns all points through a fixed angle.

A *reflection* moves each point to its mirror image on the other side of a fixed line.

An *enlargement* increases the size of shapes by a constant factor.

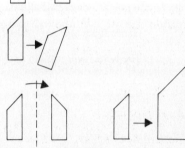

Fig 23.1

If a shape is unchanged after reflection in a line, then it has *symmetry* about that line.

If a shape is unchanged after rotation round a point, then it has *symmetry* about that point. The *order of symmetry* is the number of turns before it returns to the original position.

23.1 SINGLE TRANSFORMATIONS

23.1.1 Examples for level 1

1/   The letter F is shown on the grid of fig 23.2. Show its position after it has been:

(a) Reflected in the dotted line.
(b) Translated 1 up and 2 to the right.
(c) Rotated through 90° clockwise.
(d) Enlarged by a factor of 2.

Fig 23.2

Solution   The effect of these operations is shown in fig 23.3.

(a)   (b)   (c)   (d)

Fig 23.3

249

2/ Describe the symmetries of the letter H and the three-spoked wheel of fig 23.4.

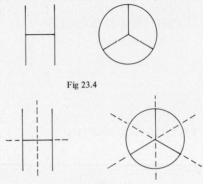

Fig 23.4

Solution   The letter H will remain unchanged if it is reflected about either of the dotted lines shown in fig 23.5. It will also be unchanged if it is rotated about the central point through half a turn.

Fig 23.5

Hence there are two lines of symmetry, and the letter has rotational symmetry of order 2.

The wheel will remain unchanged if it is reflected about any of the spokes. It will also remain unchanged if it is rotated through a third of a circle about the centre.

Hence there are three lines of symmetry, and the symbol has rotational symmetry of order 3.

23.1.2 Exercises for level 1

(1)   Reflect each of the shapes below in the horizontal dotted line.

Fig 23.6

(2)   Reflect each of the shapes shown in the vertical dotted line.

Fig 23.7

(3)   Rotate the picture of fig 23.8 through 90°, 180°, 270°.

(4)   Make an enlargement of the picture in fig 23.8, by a scale factor of 2.

Fig 23.8

(5) Complete each of the shapes in fig 23.9 by reflecting them in the dotted line.

Fig 23.9

(6) Complete each of the shapes in fig 23.10 by rotating them through 180°, about the point X.

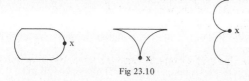

Fig 23.10

(7) Complete each of the shapes below by rotating them through 90°, 180° and 270°, about the point X.

Fig 23.11

(8) Make copies of the figures below, enlarged by a scale factor of 3.

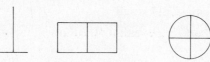

Fig 23.12

(9) Draw dotted lines to show the lines of symmetry of the following figures:

Fig 23.13

(10) Do the following figures have any rotational symmetry? If they do, give the order of symmetry.

Fig 23.14

(11)   The year 1881 was the last year which was symmetrical about a vertical line through the centre. When will be the next such year?

(12)   The year 1961 was the last year which had rotational symmetry. When will be the next such year?

(13)   Fig 23.15 shows a 4 by 6 grid, lettered A to X.

Fig 23.15

   (a) What translation will take G to P?

   (b) What translation will take N to E?

   (c) A translation takes M to J. Where will this translation take G? Where will it take N?

   (d) Describe the transformation which takes the shape made from the blocks AGMN to the blocks EKQP.

   (e) Describe the transformation which takes the AGMN shape to QKED.

   (f) Describe the transformation which takes the AGMN shape to LKJP.

   (g) Describe the transformation which takes the MST block to AGMSTUV.

(14)   A wallpaper pattern is obtained by reflection, rotation and translation of a simple shape. Complete the following patterns:

(a)   (b)

Fig 23.16

(15) The triangles A, B, C, D, E of fig 23.17 are all congruent. Describe the transformations which take:

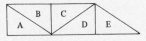

Fig 23.17

(a) A to E.   (b) B to C.   (c) C to A.

23.1.3 Examples for level 2

(1)   A triangle is drawn on the grid of fig 23.18 by joining up the points A(1,3), B(1,4), C(3,3). Draw this triangle after reflection in the mirror line y = x.

Fig 23.18

Solution   The point C is on the mirror line, so it remains where it is. A moves to A'(3,1), and B to B'(4,1). The new triangle is shown in fig 23.19.

Fig 23.19

(2)   The triangle ABC of question 1 is rotated through 90° clockwise about A. Draw the new triangle.

Solution Because A is the centre of rotation, it remains fixed. The vertical line AB moves to a horizontal line AB', where B' is at (2,3). The horizontal line AC moves to a vertical line AC', where C' is at (1,1). The new triangle is shown in fig 23.20.

Fig 23.20

(3)   The triangle ABC is gived a translation (1,-2). Find the new co-ordinates, and draw the new triangle.

Solution A translation (1,-2) means that 1 is added to the x co-ordinate and 2 is subtracted from the y co-ordinate. The new positions are A'(2,1), B'(2,2), C'(4,1). The triangle is shown in fig 23.21.

23.1.4 Exercises for level 2

Fig 23.21

(1)   Draw axes on a sheet of graph paper. Mark the axes from -5 to 10, allowing 1 cm per unit. On the paper draw the triangle with vertices A(1,1), B(3,1), C(2,5).

Draw the triangles obtained by applying the following transformations to $\triangle$ ABC. Write down the co-ordinates of the vertices:

(a) A reflection in the x-axis.

(b) A reflection in the y-axis.

(c) A reflection in the line y = x.

(d) A reflection in the line y = -x.

(2)   Draw axes as in question 1. Draw the triangle with vertices D(1,2), E(1,3), F(5,2). Apply the following transformations to $\triangle$ DEF: draw the transformed triangles and write down their vertices.

(a) A rotation of 180° about D.

(b) A rotation of 180° about E.

(c) A rotation of 180° about the origin (0,0).

(d) A rotation of 90° clockwise about F.

(e) A rotation of 90° anti-clockwise about (1,1).

(3)   The triangle with vertices A(1,1), B(1,3), C(4,3) is given a transformation so that it moves to A'(-1,1), B'(-1,3), C'(-4,3). Draw both triangles on graph paper, and describe the transformation.

(4)   The triangle ABC of question 3 is transformed to A"(1,1), B"(3,1), C"(3,4). Draw the new triangle on graph paper, and describe the transformation.

(5) The rectangle with vertices A(1,1), B(1,3), C(2,3), D(2,1) is transformed to the rectangle with vertices A'(1,-1), B'(3,-1), C'(3,-2), D'(1,-2). Draw both rectangles on graph paper and describe the transformation.

(6) The rectangle ABCD of question 5 is transformed to A"(-1,-1), B"(-1,-3), C"(-2,-3), D"(-2,-1). Draw this rectangle and describe the transformation.

(7) A translation of (1,-2) is applied to the triangle ABC of question 3. Find the new co-ordinates, and draw the translated triangle.

(8) A translation takes the point (3,4) to (6,2). Where would the same translation take the triangle with vertices at A(1,1), B(2,3), C(0,4)?

(9) The triangle ABC of question 3 is transformed to A'''(1,1), B'''(1,5), C'''(7,5). Describe the transformation.

23.1.5 Examples for level 3

(1) The rectangle with vertices at A(1,1), B(1,3), C(2,3), D(2,1) is transformed to the rectangle with vertices at A'(2,2), B'(4,2), C'(4,1), D'(2,1). Describe fully the transformation.

Solution The rectangle before and after the transformation is shown in fig 23.22. The rectangle has been rotated clockwise. AB is vertical and A'B' is horizontal, hence the rotation is through 90°. D is fixed, so D is the centre of rotation.

Fig 23.22

Rotation 90° clockwise about (2,1)

(2) The square with vertices A(1,0), B(1,1), C(2,1), D(2,0) is transformed to the square A'(0,-2), B'(0,1), C'(3,1), D'(3,-2). Describe fully the transformation. Find the factor by which the area of the figure has been increased.

Solution The square before and after transformation is shown in fig 23.23. The square has been enlarged by a factor of 3. The midpoint of BC is unchanged, hence this must be the centre of enlargement.

Fig 23.23

An enlargement of scale factor 3 from the point $(1\frac{1}{2},1)$

The increase in area is the square of the increase in lengths.

The area has been multiplied by $3^2 = 9$

23.1.6 Exercises for level 3

(1) A triangle has vertices at A(0,2), B(2,3), C(2,2). After a transformation it has moved to A'(3,1), B'(2,3), C'(3,3). Describe fully the transformation.

(2)   The triangle of question 1 is transformed to A"(0,2), B"(-2,1), C"(-2,2). Describe fully the transformation.

(3)   The rectangle PQRS with vertices P(0,0), Q(0,2), R(1,2), S(1,0) is transformed to P'(1,-1), Q'(3,-1), R'(3,-2), S'(1,-2). Describe the transformation.

(4)   The rectangle of question 3 is transformed to P"(-2,-2), Q"(-2,-4), R"(-3,-4), S"(-3,-2). Describe the transformation.

(5)   The triangle of question 1 is transformed to A'''(0,0), B'''(2,-1), C'''(2,0). Describe the transformation.

(6)   The rectangle of question 3 is transformed to P'''(-4,0), Q'''(-4,2), R'''(-5,2), S'''(-5,0). Describe the transformation.

(7)   The triangle of question 1 is transformed to A''''(-6,-1), B''''(2,3), C''''(2,-1). Describe the transformation. What is the ratio of areas of the two triangles?

(8)   The rectangle of question 3 is transformed to P''''(0,0), Q''''(0,-4), R''''(-2,-4), S''''(-2,0). Describe the transformation. By what factor has the area of the rectangle been multiplied by?

(9)   The triangle of question 1 is reflected in the line x = 3. Give the co-ordinates of the new vertices.

(10)   The triangle of question 1 is reflected in the line y = -1. Give the co-ordinates of the new vertices.

(11)   On graph paper draw the rectangle PQRS of question 3. Rotate the figure through 40° anti-clockwise about the origin (0,0). Write down as accurately as you can the co-ordinates of the new vertices.

(12)   On graph paper draw the triangle ABC of question 1. Rotate the figure through 60° clockwise about A. Write down as accurately as you can the co-ordinates of the new vertices.

23.2 COMBINED TRANSFORMATIONS. Level 3

If a transformation P is followed by a transformation Q, the *combined transformation* is written QP. In particular, if P is done twice then the transformation is written P^2.

If a transformation P takes A to A', the *inverse* transformation which takes A' back to A is written P^{-1}.

23.2.1 Examples for level 3

(1)   The triangle T with vertices at A(1,1), B(1,3), C(2,3) is reflected in the line x = 2, giving the triangle T'. T' is reflected in the line x = -1, to give the triangle T".

Draw the three triangles on graph paper. Describe the transformations which take T to T" and which take T" to T.

Solution   The triangles are shown in fig 23.24. T is taken to T" by a translation, of 6 units to the left. T" is taken to T by a translation of 6 units to the right.

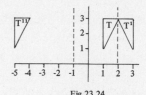

Fig 23.24

(2)   The triangle T of example 1 is rotated through a quarter turn clockwise about (1,1), then through a further quarter turn clockwise about (0,0). To what single transformation is this equivalent?

Solution   The three triangles are shown in fig 23.25. The two rotations of 90° give a rotation of 180°. The centre of rotation is (1,0).

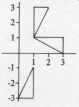

Fig 23.25

23.2.2 Exercises for level 3

(1)   The triangle S with vertices A(1,1), B(2,2), C(4,1) is reflected in the line x = 1 and then in the line x = 2. Find the single transformation equivalent to the two reflections.

(2)   The triangle S of question 1 is reflected in y = 1 and then in y = -1. Find the single transformation equivalent to the two reflections.

(3)   The triangle S of question 1 is reflected in x = 1 and then in y = 1. Describe fully the single transformation.

(4)   The triangle S of question 1 is reflected in y = -1 and then in x = 2. Describe fully the single transformation.

(5)   LMN is the triangle with vertices L(0,2), M(3,2), N(0,1). Describe the single transformation equivalent to the following successive transformations:

        (a) Rotation of 90° clockwise about L.
        (b) Rotation of 90° anti-clockwise about (0,0).

(6)   With ΔLMN as in question 5, describe the single transformation equivalent to the following successive transformations:

        (a) Rotation of 90° clockwise about (0,0).
        (b) Rotation of 90° clockwise about (2,2).

(7)   With ΔLMN as in question 5, describe the single transformation equivalent to the following successive transformations:

        (a) Rotation of 180° about (2,3).
        (b) Rotation of 180° about (1,0).

(8)　Let R be the rectangle with vertices at X(1,1), Y(1,3), Z(2,3), W(2,1). Let P be the operation of rotation through 90° anti-clockwise about (0,0), and let Q be the operation of rotation through 90° clockwise about (1,1). Describe the effect on R of the following combined transformations:

(a) P^2 (b) P^{-1} (c) PQ (d) QP (e) $(PQ)^{-1}$ (f) $(QP)^{-1}$ (g) $P^{-1}Q^{-1}$.

What can you conclude about inverses of combined transformations?

(9)　Let T be the triangle with vertices at J(-1,1), K(0,3), L(2,1). Let F be the operation of reflection in the line x = y, and G the operation of reflection in y = 0. Describe the effect on T of the following transformations:

(a) F^{-1} (b) G^2 (c) FG (d) GF.

COMMON ERRORS

(1)　**Single transformations**

　　　(a) Be careful that you do not confuse the different types of transformation. It is common to mistake reflections for rotations and vice-versa. A quick check is as follows: A reflection reverses the order of lettering of a triangle, so that if it is lettered clockwise before the reflection then it will be anti-clockwise after. A rotation preserves the order.

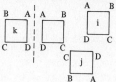

　　　(b) It is very easy to confuse transformations when a square has been transformed. The square of fig 23.26 has been translated to (i), rotated to (j), and reflected to (k).

Fig 23.26

(2)　**Combined transformations**

　　　The order in which transformations are done is important. PQ is not the same as QP. Be sure to remember that PQ means that Q is done first and then P.

24 Matrices

Matrices are rectangular blocks of numbers. They are used to store information, and to represent transformations in the plane.

24.1 MATRIX OPERATIONS. Level 3

A matrix with n rows and m columns is called an *n by m* matrix. Two matrices can be added provided that they have the same numbers of rows and columns.

$$
\begin{pmatrix} 1 & 3 & 4 & 5 \\ 2 & 3 & 4 & 2 \\ 0 & {}^1\!/_2 & 1 & 3 \end{pmatrix} + \begin{pmatrix} 2 & 4 & -3 & 12 \\ 4 & -1 & 3 & 9 \\ 8 & {}^1\!/_2 & -2 & -9 \end{pmatrix} = \begin{pmatrix} 3 & 7 & 1 & 17 \\ 6 & 2 & 7 & 11 \\ 8 & 1 & -1 & -6 \end{pmatrix}
$$

If two matrices do not have same numbers of rows and columns then they cannot be added.

Matrices can be multiplied by an ordinary number. Multiply each term of the matrix by that number.

$$
3 \times \begin{pmatrix} 1 & 2 \\ 2 & -1 \\ {}^1\!/_2 & -3 \end{pmatrix} = \begin{pmatrix} 3 & 6 \\ 6 & -3 \\ 1{}^1\!/_2 & -9 \end{pmatrix}
$$

Matrices can be multiplied together by the following method:

To multiply A by B, each term of AxB corresponds to a row of A multiplied by a column of B.

$$
\begin{pmatrix} 1 & 2 & 3 \\ 0 & 4 & 3 \end{pmatrix} \times \begin{pmatrix} 4 & 5 \\ 1 & 1 \\ 2 & 0 \end{pmatrix} = \begin{pmatrix} 1\mathrm{x}4+2\mathrm{x}1+3\mathrm{x}2 & 1\mathrm{x}5+2\mathrm{x}1+3\mathrm{x}0 \\ 0\mathrm{x}4+4\mathrm{x}1+3\mathrm{x}2 & 0\mathrm{x}5+4\mathrm{x}1+3\mathrm{x}0 \end{pmatrix} = \begin{pmatrix} 12 & 7 \\ 10 & 4 \end{pmatrix}
$$

In order for the product AB to be found, the number of columns of A must equal the number of rows of B.

Two special matrices are as follows.

$$\begin{pmatrix} 0 & 0 \\ 0 & 0 \end{pmatrix}$$

The *zero* matrix contains nothing but zeros.

The *identity* or *unit* matrix I has 1's down the main diagonal and zeros elsewhere.

$$I = \begin{pmatrix} 1 & 0 \\ 0 & 1 \end{pmatrix}$$

I has the property that IxA = AxI = A for all matrices A which can be multiplied by I.

If A is a 2 by 2 matrix, then its *inverse* A^{-1} has the property that:

$$A \times A^{-1} = A^{-1} \times A = I$$

The inverse is given by the following formula:

$$\text{For } A = \begin{pmatrix} a & b \\ c & d \end{pmatrix}, \quad A^{-1} = \frac{1}{ad - bc} \begin{pmatrix} d & -b \\ -c & a \end{pmatrix}$$

The expression ad - bc is called the *determinant* of A.

24.1.1 Examples for level 3

(1)   For the following matrices A, B, C, find where possible (a) A + B, (b) A + C, (c) A + 3B, (d) A x C, (e) C x A.

$$A = \begin{pmatrix} 2 & 4 \\ 1 & -2 \end{pmatrix} \quad B = \begin{pmatrix} 4 & 0 \\ 0 & 2 \end{pmatrix} \quad C = \begin{pmatrix} 2 & 5 & 3 \\ 1 & -1 & 0 \end{pmatrix}$$

Solution   (a) A and B have the same number of rows and columns, hence they can be added.

$$A + B = \begin{pmatrix} 2 & 4 \\ 1 & -2 \end{pmatrix} + \begin{pmatrix} 4 & 0 \\ 0 & 2 \end{pmatrix} = \begin{pmatrix} 6 & 4 \\ 1 & 0 \end{pmatrix}$$

(b) A and C have different numbers of rows and columns. Hence A + C does not make sense.

(c) To find A + 3B first multiply B by 3 and then add.

$$A + 3B = \begin{pmatrix} 2 & 4 \\ 1 & -2 \end{pmatrix} + 3 \times \begin{pmatrix} 4 & 0 \\ 0 & 2 \end{pmatrix} = \begin{pmatrix} 2 & 4 \\ 1 & -2 \end{pmatrix} + \begin{pmatrix} 12 & 0 \\ 0 & 6 \end{pmatrix} = \begin{pmatrix} 14 & 4 \\ 1 & 4 \end{pmatrix}$$

(d) A has 2 columns and C has 2 rows. Hence AC can be found.

$$A \times C = \begin{pmatrix} 2 & 4 \\ 1 & -2 \end{pmatrix} \times \begin{pmatrix} 2 & 5 & 3 \\ 1 & -1 & 0 \end{pmatrix} = \begin{pmatrix} 8 & 6 & 6 \\ 0 & 7 & 3 \end{pmatrix}$$

(e) C has 3 columns and A has 2 rows. Hence CA does not make sense.

(2)   Find the inverse of the matrix A, where:

$$A = \begin{pmatrix} 4 & 2 \\ -1 & 3 \end{pmatrix}$$

Hence solve the equations:

$$4x + 2y = 10$$
$$-x + 3y = 8$$

Solution   Using the formula above:

$$A^{-1} = \frac{1}{4 \times 3 - 2 \times (-1)} \begin{pmatrix} 3 & -2 \\ 1 & 4 \end{pmatrix} = \frac{1}{14} \begin{pmatrix} 3 & -2 \\ 1 & 4 \end{pmatrix} = \begin{pmatrix} 3/14 & -2/14 \\ 1/14 & 4/14 \end{pmatrix}$$

Write the equations in matrix form:

$$\begin{pmatrix} 4 & 2 \\ -1 & 3 \end{pmatrix} \begin{pmatrix} x \\ y \end{pmatrix} = \begin{pmatrix} 10 \\ 8 \end{pmatrix}$$

Multiply both sides by the inverse of A:

$$\begin{pmatrix} x \\ y \end{pmatrix} = \frac{1}{14} \begin{pmatrix} 3 & -2 \\ 1 & 4 \end{pmatrix} \begin{pmatrix} 10 \\ 8 \end{pmatrix} = \frac{1}{14} \begin{pmatrix} 14 \\ 42 \end{pmatrix} = \begin{pmatrix} 1 \\ 3 \end{pmatrix}$$

Hence x = 1 and y = 3

24.1.2 Exercises for level 3

(1)   For the following matrices A and B find (a) A + B (b) A + 2B, (c) 5A - 3B.

$$A = \begin{pmatrix} 1 & 4 \\ 2 & 7 \end{pmatrix} \qquad B = \begin{pmatrix} 5 & -2 \\ -3 & 9 \end{pmatrix}$$

(2)   Evaluate AB and BA for the matrices in question 1.

(3)   For the following matrices C, D, E, state which pairs can be (a) added (b) multiplied.

$$C = \begin{pmatrix} 2 & 3 & 2 \\ 2 & 0 & 9 \\ -2 & 1 & -8 \end{pmatrix} \quad D = \begin{pmatrix} 4 & 6 \\ 3 & 6 \end{pmatrix} \quad E = \begin{pmatrix} -1 & 3 & -2 \\ 7 & 3 & 1 \end{pmatrix}$$

(4)   Evaluate where possible the matrix additions and multiplications of question 3.

(5)   For A and B as in question 1, find a matrix F such that $A + 2F = B$.

(6)   For A and B as in question 1, and D as in in question 3, evaluate A^2, B^2, D^2.

(7)   Why is it not possible to evaluate E^2?

(8)   Find the inverse of the matrix A. Hence solve the equations:

$$x + 4y = -2$$
$$2x + 7y = 7$$

(9)   Find the inverse of the matrix B. Hence solve the equations:

$$5x - 2y = 16$$
$$-3x + 9y = 6$$

(10)   Solve by matrix methods the equations:

$$7x + 4y = -1$$
$$3x + 5y = -7$$

(11)   Solve by matrix methods the equations:

$$2x - 3y = 1$$
$$-3x + 5y = -2$$

(12)   For the matrices A and B of question 1, evaluate (a) $(AB)^{-1}$, (b) $(BA)^{-1}$, (c) $A^{-1}B^{-1}$, (d) $B^{-1}A^{-1}$. What do you notice?

(13)   For the matrices M and N below, verify that $(MN)^{-1} = N^{-1}M^{-1}$.

$$M = \begin{pmatrix} 1 & 4 \\ 2 & -1 \end{pmatrix} \quad N = \begin{pmatrix} 8 & -2 \\ -3 & 7 \end{pmatrix}$$

(14)   For the matrices A and B of question 1, solve the equation $CA = B$.

(15)   For the matrices A and B of question 1, solve the equation $AC = B$.

24.2 INFORMATION MATRICES. Level 3

24.2.1 Examples for level 3

(1) Fig 24.1 shows the roads which connect three villages X, Y and Z. Find the route matrix A for this map. Find A^2. How many two-stage routes are there from Y to X?

Fig 24.1

Solution Each entry gives the number of direct routes from one village to another. There are 3 villages, so the matrix will be 3 by 3. The matrix is as follows:

$$A = \begin{pmatrix} & X & Y & Z & \\ 0 & 1 & 1 & X \\ 1 & 0 & 2 & Y \\ 1 & 2 & 0 & Z \end{pmatrix}$$

Square this matrix to obtain:

$$A^2 = \begin{pmatrix} 0 & 1 & 1 \\ 1 & 0 & 2 \\ 1 & 2 & 0 \end{pmatrix} \times \begin{pmatrix} 0 & 1 & 1 \\ 1 & 0 & 2 \\ 1 & 2 & 0 \end{pmatrix} = \begin{pmatrix} 2 & 2 & 2 \\ 2 & 5 & 1 \\ 2 & 1 & 5 \end{pmatrix}$$

This matrix represents the number of two stage journeys between the villages. Hence there are 2 two stage journeys between Y and X.

(2) Four schools P, Q, R, S play a league of football. The results are given by the following table:

	P	Q	R	S
P	X	L	D	L
Q	W	X	W	L
R	D	L	X	D
S	W	W	D	X

So, for example, School R lost to school Q.

(a) Represent this information as a matrix A, showing the numbers of games won, drawn and lost by each team.

(b) Form the product AV, where V is given by:

$$V = \begin{pmatrix} 3 \\ 1 \\ 0 \end{pmatrix}$$

What does this product represent?

Solution   (a) Allowing one row for each team, the following matrix is obtained:

$$\begin{array}{c} \\ P \\ Q \\ R \\ S \end{array} \begin{pmatrix} W & D & L \\ 0 & 1 & 2 \\ 2 & 0 & 1 \\ 0 & 2 & 1 \\ 2 & 1 & 0 \end{pmatrix} = A$$

(b)

$$A \times V = \begin{pmatrix} 0 & 1 & 2 \\ 2 & 0 & 1 \\ 0 & 2 & 1 \\ 2 & 1 & 0 \end{pmatrix} \times \begin{pmatrix} 3 \\ 1 \\ 0 \end{pmatrix} = \begin{pmatrix} 1 \\ 6 \\ 2 \\ 7 \end{pmatrix}$$

This product represents the total points for each team, on the system of 3 points for a win and 1 for a draw.

24.2.2 Exercises for level 3

(1)   Write down the route matrices for the following maps. In each case find the square of the matrix.

(a)   (b)   (c)

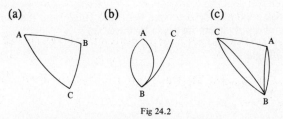

Fig 24.2

(2)   In the following maps the arrows indicate one-way streets. Write down the route matrices for these maps.

(a)   (b)   (c)

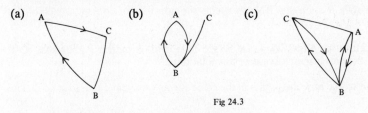

Fig 24.3

(3)   The three towns A, B, C are connected by the map in fig 24.4. The council introduces a system of one-way streets, so that the route matrix is as follows:

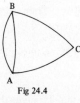

Fig 24.4

$$\begin{array}{c} \\ A \\ B \\ C \end{array} \begin{pmatrix} A & B & C \\ 0 & 2 & 0 \\ 2 & 0 & 1 \\ 1 & 0 & 0 \end{pmatrix}$$

Insert arrows on the map to indicate which streets are now one-way.

263

(4)   4 children Albert, Beatrice, Cynthia and David took exams in Maths, English and French. Their percentage marks were as follows:

	Maths	English	French
Albert	57	63	83
Beatrice	85	54	66
Cynthia	43	75	43
David	73	96	56

Consider this as a 4 by 3 matrix M. Define other matrices by:

$$U = (^1/_4, \ ^1/_4, \ ^1/_4, \ ^1/_4), \qquad V = \begin{pmatrix} 1 \\ 1 \\ 1 \end{pmatrix}$$

(a) Form the products UM and MV. What do they represent?

(b) The marks are scaled so that Maths and English account for 40% each of the total mark, and French accounts for the remaining 20%. Form a 3 by 1 matrix W so that MW gives the total percentage marks.

(5)   Emilia has a job delivering newspapers. She delivers the Post, the Bugle and the News to Acacia Avenue and Beech Close. The numbers of papers are given by the table:

	Post	News	Bugle
Acacia	24	32	18
Beech	42	21	13

Consider this as a 2 by 3 matrix N.

(a) Find a 3 by 1 matrix T such that NT represents the numbers of papers she delivers to the two streets. Evaluate NT.

(b) Find a 1 by 2 matrix R such that RN represents the numbers of copies of each of the three papers. Evaluate RN

(c) The Post costs 20 p, the News 25 p, the Bugle 15 p. Find a matrix P such that RNP represents the total price of all the newspapers. Evaluate this total price.

(6)   A survey was carried out to see how the pupils in the three forms of a year came to school. The results were:

	Car	Bus	Train	Foot	Bike
Form 1	2	6	8	10	2
Form 2	3	3	7	12	1
Form 3	5	4	2	9	4

Call this matrix T.

(a) Find a matrix P so that TP gives the total number of pupils in each form.

(b) Find a matrix Q so that QT gives the total number of pupils who used each type of transport.

(c) Find matrices S and R so that STP and QTR both give the total number of pupils in the year. Verify that these two products give the same answer.

(7)   The prices charged by 3 shops A, B, and C for litres of Milk, Orange and Beer are given by the following matrix P:

$$
\begin{array}{c}
\quad\quad\;\; \text{Milk} \quad \text{Orange} \quad \text{Beer} \\
\begin{array}{c} A \\ B \\ C \end{array}
\left(
\begin{array}{ccc}
50 & 76 & 154 \\
52 & 83 & 164 \\
49 & 74 & 170
\end{array}
\right)
\end{array}
$$

Mrs Smith wants to buy 2 litres of Orange, 5 litres of milk and 6 litres of Beer. Express her shopping list in the form of a matrix L, and find the matrix product PL which will show how much she has to pay in each shop.

(8)   A baker has two sorts of pastry: type A is half butter and half flour, type B is two fifths butter to three fifths flour. Express this as a matrix giving the amounts of butter and flour for 1 kilogram of each sort of pastry.

If he is going to make 5 kg of type A and 3 kg of type B, form a matrix product which will give the total amounts of butter and flour he will use.

(9)   A village cricket team has four bowlers, Smith, Jones, Brown and White. After a season their results are given by the matrix C:

$$
\begin{array}{c}
\quad\quad\quad\;\; \text{Overs} \quad \text{Runs} \quad \text{Wickets} \\
\begin{array}{c} \text{Smith} \\ \text{Jones} \\ \text{Brown} \\ \text{White} \end{array}
\left(
\begin{array}{ccc}
157 & 608 & 30 \\
163 & 793 & 25 \\
97 & 438 & 12 \\
208 & 849 & 35
\end{array}
\right)
\end{array}
$$

Find a matrix product which will give the numbers of overs, runs and wickets for the whole season.

To find its best bowler, the team decides to award 50 points for each wicket and to subtract 1 point for each run. Form a matrix product which will determine who is the best bowler.

24.3 MATRIX TRANSFORMATIONS. Level 3

24.3.1 Examples for level 3

(1)   The points of the triangle T are A(1,1), B(3,1), C(1,2). Draw the triangle T on graph paper.

The points ABC are transformed to A'B'C', forming the triangle T', by the matrix M where:

$$
M = \begin{pmatrix} 0 & -1 \\ 1 & 0 \end{pmatrix}
$$

Draw the triangle T' on the same graph paper.

T' is reflected in the x-axis, to obtain T''. Draw T'' on the graph paper. What single matrix would send T to T''?

Solution   Plot A, B, C on the graph. Apply the matrix M to the points.

$$\begin{pmatrix} 0 & -1 \\ 1 & 0 \end{pmatrix} \overset{\text{A B C}}{\begin{pmatrix} 1 & 3 & 1 \\ 1 & 1 & 2 \end{pmatrix}} = \overset{\text{A' B' C'}}{\begin{pmatrix} -1 & -1 & -2 \\ 1 & 3 & 1 \end{pmatrix}}$$

Plot the points A', B', C'.

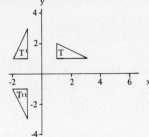

Fig 24.5

Reflect the triangle T' in the x axis. Notice that the y coordinates have all been multiplied by -1. Hence the matrix of this transformation is:

$$N = \begin{pmatrix} 1 & 0 \\ 0 & -1 \end{pmatrix}$$

The matrix which will go from T to T'' is the product of M and N.

$$\begin{pmatrix} 1 & 0 \\ 0 & -1 \end{pmatrix} \begin{pmatrix} 0 & -1 \\ 1 & 0 \end{pmatrix} = \begin{pmatrix} 0 & -1 \\ -1 & 0 \end{pmatrix}$$

(2)   The matrix M takes a point P to (3,4), where:

$$M = \begin{pmatrix} 5 & 3 \\ 3 & 2 \end{pmatrix}$$

Find the coordinates of the point P.

Solution   M takes P to (3,4). Hence M^{-1} takes (3,4) back to P. This gives:

$$P = M^{-1} \begin{pmatrix} 3 \\ 4 \end{pmatrix} = \frac{1}{1} \begin{pmatrix} 2 & -3 \\ -3 & 5 \end{pmatrix} \begin{pmatrix} 3 \\ 4 \end{pmatrix} = \begin{pmatrix} -6 \\ 11 \end{pmatrix}$$

P is at (-6,11)

(3)   Find the matrix which takes (1,0) to (3,5) and (0,1) to (4,-2)

Solution   Notice that for a general matrix:

$$\begin{pmatrix} a & b \\ c & d \end{pmatrix} \begin{pmatrix} 1 \\ 0 \end{pmatrix} = \begin{pmatrix} a \\ c \end{pmatrix}$$

So (1,0) always goes to the left column of the matrix. Similarly (0,1) is sent to the right column of the matrix. The required matrix can now be written down:

$$\begin{pmatrix} 3 & 4 \\ 5 & -2 \end{pmatrix}$$

24.3.2 Exercises for level 3

(1)   Let T be the triangle with vertices L(1,2), M(3,2), N(2,1). Find the image of T after action by the following matrices:

$$A = \begin{pmatrix} 1 & 0 \\ 0 & -1 \end{pmatrix} \quad B = \begin{pmatrix} -1 & 0 \\ 0 & 1 \end{pmatrix} \quad C = \begin{pmatrix} 0 & 1 \\ -1 & 0 \end{pmatrix} \quad D = \begin{pmatrix} 0 & -1 \\ 1 & 0 \end{pmatrix} \quad E = \begin{pmatrix} 0 & -1 \\ -1 & 0 \end{pmatrix}$$

(2)   Repeat question 1, for the rectangle with vertices at P(1,1), Q(1,3), R(2,3), S(2,1).

(3)   Describe the action of the matrices A, B, C, D, E defined in question 1.

(4)   Let S be the unit square with vertices at (0,0), (1,0). (1,1), (0,1). Find the image of S after action by the following matrices:

$$F = \begin{pmatrix} 3 & 0 \\ 0 & 3 \end{pmatrix} \quad G = \begin{pmatrix} 1/2 & 0 \\ 0 & 1/2 \end{pmatrix} \quad H = \begin{pmatrix} 1/2 & 0 \\ 0 & 1 \end{pmatrix} \quad J = \begin{pmatrix} 1 & 2 \\ 0 & 1 \end{pmatrix} \quad K = \begin{pmatrix} 2 & 1 \\ 1 & 2 \end{pmatrix}$$

(5)   Describe the action of the matrices F, G, H, J, K defined in question 4.

(6)   Find the areas of the images of the unit square after the actions of the matrices defined in question 4.

(7)   Find the determinants of the matrices defined in question 4. What do you notice?

(8)   On squared paper plot the triangle T with vertices (1,2), (1,-1), (2,2). T is taken to U by the matrix:

$$A = \begin{pmatrix} 0 & 1 \\ -1 & 0 \end{pmatrix}$$

Plot U on the same squared paper. U is now taken to V by the matrix:

$$B = \begin{pmatrix} -1 & 0 \\ 0 & 1 \end{pmatrix}$$

Plot V on the same paper. What single matrix will take T to V? What matrix will take V to T?

(9) Let S be the square with corners at (1,1), (1,2), (2,2), (2,1). Plot S on squared paper. S is sent to S' by the matrix:

$$C = \begin{pmatrix} 1 & 2 \\ 0 & 1 \end{pmatrix}$$

Plot S' on the paper. S' is taken to S'' by the matrix:

$$D = \begin{pmatrix} 1 & 0 \\ 0 & -1 \end{pmatrix}$$

Plot S'' on the same paper. What single matrix will take S to S''? What matrix will take S' to S?

(10) On squared paper plot the triangle with vertices at (1,4), (5,4), (1,3). Draw the image of the triangle after action by A, A^2, A^3, A^4, where:

$$A = \begin{pmatrix} 0 & 1 \\ -1 & 0 \end{pmatrix}$$

Write down A^{17}.

(11) Let M be the rectangle with corners at (1,1), (1,5), (2,5), (2,1). Plot this rectangle on squared paper, and its image after action by B, C and BC where:

$$B = \begin{pmatrix} 1 & 0 \\ 0 & -1 \end{pmatrix} \qquad\qquad C = \begin{pmatrix} -1 & 0 \\ 0 & 1 \end{pmatrix}$$

(12) Find the matrix which takes (1,0) to (4,-2) and (0,1) to (2,-1).

(13) Find the matrix which takes (2,0) to (4,6) and (0,3) to (9,6).

(14) A certain matrix takes the square with vertices (0,0), (3,0), (3,3), (0,3) to the parallelogram with vertices (0,0), (2,5), (6,7), (4,2). Find the matrix.

(15) Let H be given by:

$$H = \begin{pmatrix} 2 & 4 \\ 1 & 6 \end{pmatrix}$$

Find which point is taken by H to (8,-16).

(16) With H as in question 15, find which point is taken to (-4,2)

(17) With H as in question 15, find the parallelogram which is taken to the unit square. (With vertices (0,0), (1,0), (1,1), (0,1))

(18) Draw the letter b on squared paper, by joining up (1,1) and (1,4), (1,1) and (2,1), (2,1) and (2,2), (2,2) and (1,2). Apply to the letter some of the matrices used in these exercises. In each case find (a) the determinant of the matrix, and (b) whether or not the letter has been reversed to a d. What can you conclude?

COMMON ERRORS

(1)   Operations

(a) Be careful that you do not add together matrices which are not of the same shape.

(b) When multiplying a matrix by a number, do not just multiply the top row. Multiply *every* term of the matrix by the number.

(c) When finding the product of matrices, make sure that they can be multiplied. The number of columns of the left-hand matrix must be the same as the number of rows of the right-hand matrix. Be sure also that they are in the correct order. In general, AB ≠ BA.

(d) There are several mistakes commonly made when finding inverses. The b and c terms are multiplied by -1, even if they are negative already. Do not forget to divide by the determinant ad - bc. Be careful of negative numbers when working out the determinant.

(2)   Information matrices

Make sure that you multiply matrices in the correct order. Not just the result, but the *meaning* of a matrix product depends on the order in which the matrices are multiplied.

(3)   Transformation matrices

(a) Usually, the co-ordinates of a point may be written either horizontally or vertically. But when a matrix is applied to a point, its co-ordinates must be vertical.

(b) Be careful when multiplying matrices to give a combined transformation. The product AB is a transformation which does B first and then A. It is easy to get this the wrong way round.

25 Vectors

Vectors are quantities which have direction as well as magnitude.

25.1 VECTORS AND CO-ORDINATES. Level 3

Vectors can be represented as translations in the plane. If a translation moves x units to the right and y upwards then it is given by:

$$v = \begin{pmatrix} x \\ y \end{pmatrix}$$

Vectors can be added, or multiplied by ordinary numbers.

$$\begin{pmatrix} a \\ b \end{pmatrix} + \begin{pmatrix} c \\ d \end{pmatrix} = \begin{pmatrix} a+c \\ b+d \end{pmatrix} \qquad\qquad m \times \begin{pmatrix} a \\ b \end{pmatrix} = \begin{pmatrix} ma \\ mb \end{pmatrix}$$

The length or *modulus* of a vector is given by:

$$\begin{vmatrix} a \\ b \end{vmatrix} = \sqrt{a^2 + b^2}$$

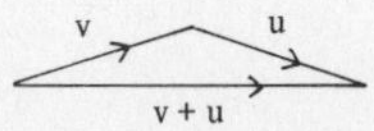

Vectors **v** and **u** can be added by completing the third side of the triangle as in fig 25.1.

Fig 25.1

25.1.1 Examples for level 3

(1)   For the vectors **a, b, c** as defined below find (a) **a + b**, (b) **3a + 2b**, (c) |**c**| (d) numbers x and y such that **xa + yb = c**.

$$\mathbf{a} = \begin{pmatrix} 3 \\ 2 \end{pmatrix} \quad \mathbf{b} = \begin{pmatrix} 5 \\ -3 \end{pmatrix} \quad \mathbf{c} = \begin{pmatrix} 4 \\ 9 \end{pmatrix}$$

Solution   (a) Add the terms of the vectors:

$$\mathbf{a} + \mathbf{b} = \begin{pmatrix} 3 \\ 2 \end{pmatrix} + \begin{pmatrix} 5 \\ -3 \end{pmatrix} = \begin{pmatrix} 8 \\ -1 \end{pmatrix}$$

(b) Multiply the terms of the vectors and add:

$$3\mathbf{a} + 2\mathbf{b} = 3x \begin{pmatrix} 3 \\ 2 \end{pmatrix} + 2x \begin{pmatrix} 5 \\ -3 \end{pmatrix} = \begin{pmatrix} 9 \\ 6 \end{pmatrix} + \begin{pmatrix} 10 \\ -6 \end{pmatrix} = \begin{pmatrix} 19 \\ 0 \end{pmatrix}$$

(c) Use the formula for modulus:

$$|\mathbf{c}| = \left| \begin{pmatrix} 4 \\ 9 \end{pmatrix} \right| = \sqrt{4^2 + 9^2} = \sqrt{16 + 81} = \sqrt{97} = 9.85$$

(d) Re-write the equation $x\mathbf{a} + y\mathbf{b} = \mathbf{c}$ as:

$$x \times \begin{pmatrix} 3 \\ 2 \end{pmatrix} + y \times \begin{pmatrix} 5 \\ -3 \end{pmatrix} = \begin{pmatrix} 4 \\ 9 \end{pmatrix}$$

$$\begin{pmatrix} 3x + 5y \\ 2x - 3y \end{pmatrix} = \begin{pmatrix} 4 \\ 9 \end{pmatrix}$$

This gives the simultaneous equations:

$$3x + 5y = 4$$
$$2x - 3y = 9$$

Solve these to obtain:

x = 3 and y = -1

(2)   Illustrate on a graph the vectors $\mathbf{a}$, $\mathbf{b}$, $\mathbf{a} + \mathbf{b}$, where $\mathbf{a}$ and $\mathbf{b}$ are as in question 1.

Solution   For $\mathbf{a}$ start from the origin and draw a line 3 units to the right and 2 units up. Draw $\mathbf{b}$ similarly.

For $\mathbf{a} + \mathbf{b}$ start at the end of $\mathbf{a}$ and draw a line 5 units to the right and 3 downwards. Join the end of this line to the origin to show $\mathbf{a} + \mathbf{b}$.

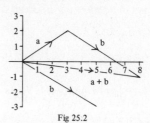

Fig 25.2

25.1.2 Exercises for level 3

(1)   For the vectors $\mathbf{a}$, $\mathbf{b}$, $\mathbf{c}$ defined below find (a) $\mathbf{a} + \mathbf{b}$, (b) $2\mathbf{a} + 5\mathbf{b}$, (c) $2\mathbf{a} - 3\mathbf{b}$, (d) $\mathbf{a} - \mathbf{b}$.

$$\mathbf{a} = \begin{pmatrix} 2 \\ -7 \end{pmatrix} \qquad \mathbf{b} = \begin{pmatrix} 5 \\ 1 \end{pmatrix}$$

(2)   With the vectors **a** and **b** as in question 1, evaluate:

(a) |**a**| (b) |**b**| (c) |**a** + **b**| (d) |2**a** + 3**b**| (e) |**a** - **b**|

(3)   Find x and y from the following vector equation:

$$\begin{pmatrix} x \\ 2 \end{pmatrix} + \begin{pmatrix} 3 \\ y \end{pmatrix} = \begin{pmatrix} 7 \\ 9 \end{pmatrix}$$

(4)   Find x and y from the following vector equation:

$$\begin{pmatrix} 2 \\ 1 \end{pmatrix} + \begin{pmatrix} -1 \\ y \end{pmatrix} = \begin{pmatrix} x \\ 3 \end{pmatrix}$$

(5)   With **a** and **b** as in question 1, solve the following vector equations:

(a) **v** + **a** = **b**

(b) **v** - **b** = 2**a**

(c)
$$x\mathbf{a} + y\mathbf{b} = \begin{pmatrix} 22 \\ -3 \end{pmatrix}$$

(d)
$$\lambda\mathbf{a} + \mu\mathbf{b} = \begin{pmatrix} 9 \\ -13 \end{pmatrix}$$

(6)   On squared paper draw the vectors **a**, **b** of question 1. Shift the tail of **b** to the head of **a**, and hence draw **a** + **b**.

(7)   On squared paper draw **c**, **d**, **c** + **d**, **c** - **d**, where:

$$\mathbf{c} = \begin{pmatrix} 2 \\ 1 \end{pmatrix} \qquad \mathbf{d} = \begin{pmatrix} 3 \\ 5 \end{pmatrix}$$

(8)   Write down in terms of co-ordinates the vectors shown in fig 25.3.

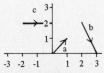

Fig 25.3

(9)   Let **a** and **b** be as in question 1. Write down the co-ordinates of A(1,1) after translation by (a) **a** (b) **b** (c) **a** + **b**.

(10)   Under a translation, A(3,4) is moved to B(4,2). Write down the vector of the translation. What translation takes B to A?

25.2 VECTOR GEOMETRY. Level 3

If A and B are two points in the plane the vector from A to B is written $\underline{AB}$.

The vector which goes from the origin to A is called the *position vector* of A.

Two vectors are parallel when one is a multiple of the other.

25.2.1 Examples for level 3

(1)   A quadrilateral has vertices A(1,1), B(2,3), C(3,4), D(2,2). Show that ABCD is a parallelogram but is not a rhombus.

Solution   The vectors of the sides of the quadrilateral are found by subtracting the co-ordinates.

$$\underline{AB} = \binom{2-1}{3-1} = \binom{1}{2} : \quad \underline{DC} = \binom{3-2}{4-2} = \binom{1}{2}$$

Since AB = DC, opposite sides are equal and parallel. Hence ABCD is a parallelogram.

$$|\underline{AB}| = \sqrt{1^2 + 2^2} = \sqrt{5}$$

$$|\underline{AD}| = \sqrt{1^2 + 1^2} = \sqrt{2}$$

Since adjacent sides are not equal, ABCD is not a rhombus.

(2)   ABC is a triangle, and $\underline{AB} = \mathbf{b}$, $\underline{AC} = \mathbf{c}$. X and Y are the midpoints of AB and AC respectively. Express in terms of **b** and **c**:

(a) $\underline{BC}$ (b) $\underline{AX}$ (c) $\underline{AY}$ (d) $\underline{XY}$.

What can you conclude about BC and XY?

If △ABC has area 10, what is the area of △AXY?

Solution   Go from B to C by way of A. $\underline{BA}$ is minus $\underline{AB}$, which gives:

(a) $\underline{BC} = -\mathbf{b} + \mathbf{c} = \mathbf{c} - \mathbf{b}$.

$\underline{AX}$ is halfway along AB. Hence:

(b) $\underline{AX} = {}^1/2\mathbf{b}$ and (c) $\underline{AY} = {}^1/2\mathbf{c}$.

By similar reasoning to part (a):

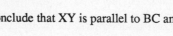
Fig 25.4

(d) $\underline{XY} = {}^1/2\mathbf{c} - {}^1/2\mathbf{b} = {}^1/2(\mathbf{c} - \mathbf{b})$.

Since $\underline{XY} = {}^1/2\underline{BC}$, conclude that XY is parallel to BC and half its length.

The two triangles are similar in the ratio 1:2. Hence the areas are in the ratio $1^2:2^2 = $ 1:4.

The area of △AXY is $^{10}/4 = 2.5$

25.2.2 Exercises for level 3

(1)   Four points in the plane are A(1,2), B(2,4), C(2,-4), D(4,0). Write down the vectors $\underline{AB}$, $\underline{AC}$, $\underline{CD}$, $\underline{BD}$, $\underline{DA}$. Which of these vectors are parallel to each other?

(2)   J(-3,1), K(-2,-2), L(-1,6), M(2,-3) are four points in the plane. Write down the vectors JL, JK, ML, MK, MJ. Which of these vectors are parallel to each other?

(3)   Four points in the plane are A(2,3), B(5,9), C(2,2), D(3,4). Show that AB is parallel to CD. What is the ratio of their lengths? Do the four points form a parallelogram?

(4)   A quadrilateral ABCD has vertices at A(-1,1), B(1,1), C(4,-1), D(3,-1). Show that ABCD is a trapezium. Is it a parallelogram?

(5)   Show that the four points P(1,0), Q(3,3), R(4,2), S(2,-1) form a parallelogram. Find the lengths of the sides. Is PQRS a rhombus?

(6)   A quadrilateral has its vertices at W(1,6), X(1,1), Y(4,5), Z(4,10). Show that WXYZ is a rhombus.

(7)   J(1,2), K(5,4), L(6,2), M(2,0) are the four vertices of a quadrilateral. Show that they form a parallelogram. By considering the lengths of the diagonals show that they form a rectangle.

(8)   W(1,2), X(2,4), Y(5,3), Z(4,1) are the four vertices of a quadrilateral. Prove that they form a parallelogram. Do they form a rectangle? Do they form a rhombus?

(9)   A(-4,4), B(-1,3), C(1,2), D(5,1) are four points in the plane. By considering the vectors between them, find out which three of them lie on a straight line.

(10)   Which three of the points L(-2,-2), M(0,-1), J(2,1) and K(4,1) lie on a straight line?

(11)   Given the three points A(1,1), B(2,3), C(1,2) find the point D such that ABCD is a parallelogram.

(12)   Given L(0,-2), M(-1,3), N(4,4) find the point P so that LMNP is a parallelogram.

(13)   Find x to ensure that A(1,1), B(2,5), C(x,9) lie on a straight line.

(14) ABC is a triangle, and AB = **b**, AC = **c**. D and E lie on AB and AC respectively so that AD = $^1/_4$AB and AE = $^1/_4$AC. Express in terms of **b** and **c**:

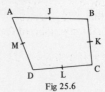

A

D   E

b   c

B   C

Fig 25.5

BC, AD, AE, DE, DB, DC.

Which of these vectors are parallel to each other?

If △ABC has area 32, what is the area of the quadrilateral DBCE?

(15) Let A, B, C, D be four points, and let their position vectors be **a**, **b**, **c**, **d** respectively. Let the midpoints of AB, BC, CD, DA be J, K, L, M respectively.

A   J   B

M   K

D   L   C

Fig 25.6

Find the position vectors of J, K, L, M in terms of **a**, **b**, **c**, **d**. Find the vectors JK, KL, LM, MJ. What can you deduce about the quadrilateral JKLM?

(16) Fig 25.7 shows three equilateral triangles arranged in a row. $\underline{AB}$ = **b**, $\underline{AC}$ = **c**. Express in terms of **b** and **c**:

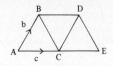

Fig 25.7

$$\underline{BD}, \underline{CD}, \underline{AE}, \underline{BC}$$

(17) OABCDE is a regular hexagon, with the position vectors of A and E relative to O being **a** and **e** respectively. Express in terms of **a** and **e**:

$$\underline{BC}, \underline{CD}, \underline{EB}, \underline{OB}.$$

Fig 25.8

(18) (For draughts players)

Think of a draughts-board as graph paper, in which the x-axis is the base line for White. Each white piece can move one forward diagonally, so its possible movements are described by the vectors:

$$\begin{pmatrix} 1 \\ 1 \end{pmatrix} \quad \text{and} \quad \begin{pmatrix} -1 \\ 1 \end{pmatrix}$$

Write down the possible movements for the black pieces and for the kings.

(19) (For chess-players)

Write down the vectors describing the possible movements for chess pieces.

(20) Which three of the points with position vectors **a**, **b**, **a** + **b**, **a** - **b** lie on the same straight line?

(21) ABCD is a parallelogram: $\underline{AB}$ = **b**, $\underline{AD}$ = **d**. Let X, Y be the midpoints of AB and CD respectively. Express in terms of **b** and **d**:

$$\underline{AC}, \underline{AX}, \underline{AY}, \underline{XY}.$$

What can you conclude about the line XY?

(22) OABC is a parallelogram, in which the position vectors of A, B, C relative to O are **a**, **b**, **c** respectively. Let X and Y be the midpoints of the diagonals OB and AC respectively.

Express **b** in terms of **a** and **c**.

Find expressions for $\underline{AB}, \underline{AY}, \underline{OX}, \underline{OY}$.

What can you say about the points X and Y?

(23) In the triangle ABC, $\underline{AB}$ = **b** and $\underline{AC}$ = **c**. X lies on AB so that AX = $^3/_4$AB. Y is the midpoint of AC, and Z is the midpoint of CX.

Fig 25.9

Find in terms of **b** and **c** the vectors $\underline{AY}, \underline{AX}, \underline{CX}, \underline{CZ}, \underline{AZ}, \underline{ZY}$. What can you say about ZY?

(24) In the triangle OAB, P is the midpoint of AB, and Q lies on OA so that $OQ = (^2/_3)OA$. Extend OB to R so that OB = BR.

Let the position vectors of A and B relative to O be **a** and **b**. Find the position vectors of P, Q, R. Find <u>PQ</u> and <u>PR</u>. What can you say about the points P, Q, R?

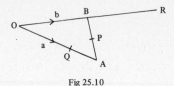

Fig 25.10

COMMON ERRORS

(1) Co-ordinates

Be careful not to confuse vectors with points. (3,4) is a point, and $\begin{pmatrix} 3 \\ 4 \end{pmatrix}$ is a vector which goes from the origin to that point.

The vector will also go from (1,1) to (4,5), or from (-1,-3) to (2,1) and so on. The vector does not have any position: it can start from anywhere in the plane.

(2) Modulus

Be careful when working out the modulus of a vector. If **v** is the vector $\begin{pmatrix} 3 \\ 4 \end{pmatrix}$, then its modulus is given by:

$$|\mathbf{v}| = \sqrt{3^2 + 4^2} = 5.$$

The modulus of **v** is not equal to 3 + 4.

Similarly, when adding vectors do not add the moduli.

$$|\mathbf{a} + \mathbf{b}| \neq |\mathbf{a}| + |\mathbf{b}|.$$

Be careful with negative signs. The modulus of $\begin{pmatrix} -3 \\ 4 \end{pmatrix}$ is 5, not $\sqrt{7}$.

(3) Vector Geometry

(a) A vector is only a position vector if it starts from the origin. Otherwise take account of where it does start.

(b) The vector from (3,2) to (5,7) is $\begin{pmatrix} 2 \\ 5 \end{pmatrix}$, not $\begin{pmatrix} -2 \\ -5 \end{pmatrix}$,

If A has position vector **a**, and B has position vector **b**, then <u>AB</u> is **b** - **a**.

It is not **b**, or **a** + **b**, or **a** - **b**.

26 Sets

26.1 SOLVING PROBLEMS BY SETS. Level 3

Problems involving the numbers of elements in finite sets can often be solved by means of a *Venn diagram*.

26.1.1 Examples for level 3

(1)   26 children were asked about the pets in their families. 11 had a dog, 14 had a cat, and 7 had neither. How many had a cat but not a dog?

Solution   Let E be the set of the families. Let C be the set of those families with a cat, and D the set of the families with a dog. Fig 26.1 shows the sets on a Venn diagram.

Fig 26.1

Put 7 into the region outside the two circles. There are 19 families left. 11 + 14 = 25, and so 6 families must have been counted twice. Put 6 into the common part of the two circles. Now fill up the rest of the diagram.

8 families have a cat but no dog.

(2)   A canteen served lunch consisting of soup, meat and pudding to 100 people. 50 had all three courses, 84 had the main course and 73 had the soup. 62 had both meat and pudding, and 58 had soup and pudding. 62 had soup and meat.

(a) How many had soup only?

(b) How many had pudding only?

Solution   Let S, M, P denote the sets of those people who had soup, meat and pudding respectively. Draw three intersecting circles as shown in fig 26.2. The only region that can be filled in immediately is the central one, for those taking all three courses.

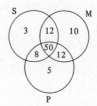

Fig 26.2

62 had meat and pudding. So 12 had meat, pudding but not soup. Put 12 into the appropriate region of fig 26.2. Similarly we find that 8 had soup, pudding but not meat, and 12 had soup, meat but not pudding. Subtract from 73 to find that:

(a) **3 had soup only.**

Fill in all the figures, using the fact that there are 100 customers in total.

(b) **5 had pudding only**

26.1.2 Exercises for level 3

(1)   Out of 20 boys, 15 play darts and 12 play snooker. 10 play both games. How many play darts but not snooker? How many play neither game?

(2)   A cricket eleven contains 7 players who are good at batting, and 6 who are good at bowling. If only one is good at neither, how many are batters but not bowlers?

(3)   Out of a class of 25, 15 are taking French and 11 are taking German. If 4 take neither language, how many are taking both? How many are taking German but not French?

(4)   A newsagent conducts a survey of a street, and finds that out of 50 houses 30 take the Daily News and 37 take the Evening Star. If 3 houses take neither paper, how many take the News but not the Star?

(5)   The police did a check on the safety of bicycles at a school, and found that 34 were without lights, 27 had defective brakes, and 17 had both faults. If 56 bicycles were without fault, how many bicycles were there in total?
        If a bicycle is picked at random what is the probability that it has no fault?

(6)   Out of 40 people, 30 own a car and 25 own their own house. Without any more information, what is the least possible number of people who own both a car and a house? What is the greatest possible of people who own both?
        If you are told that 19 own both, how many own neither a house nor a car?

(7)   A survey is done to find out whether there is a connection between left-handedness and musical ability. Out of 100 children, 23 are left-handed and 35 play a musical instrument. What is the greatest possible number of left-handed children who play an instrument? What is the least possible number?

(8)   A newspaper-boy delivers three newspapers. 43 houses take the News, 42 take the Bugle and 47 take the Globe. 15 take the Globe and the News, 17 take the Bugle and the Globe, 21 take the News and the Bugle. 5 houses take all three papers. How many households does the newspaper-boy deliver to?

(9)   There are 25 books on a shelf, of which 17 are hard-backed thrillers. There are twice as many paper-backed thrillers as hard-backed non-thrillers. Find the possible numbers of thrillers.

(10)   100 people were asked whether they had been to any of the three countries France, Spain or Italy. The answers were as follows: 17 had been to France only, 6 had been to France and Italy only, 7 had been to Italy only. 4 had been to Spain and Italy only, 8 had been to France and Spain only, and 22 had only been to Spain. If 10 had been to none of the three countries, how many had been to all three?

(11)   5 children are fair-haired, blue-eyed and left-handed. These come from a group of 300, of whom 18 are left-handed. If the same number x of the left-handed children are fair-haired as are blue-eyed, and there are 7 left-handers who are neither blue-eyed nor fair-haired, find x.
        80 children are fair-haired, of whom 60 are neither left-handed nor blue-eyed. Find the number of fair-haired blue-eyed children. If there are 90 blue-eyed children, find how many children have none of the three properties.
        If a child is picked at random, what is the probability that he or she has exactly one of the three properties?

26.2 OPERATIONS WITH SETS. Level 3

ξ is the *universal set,* which contains all the elements that are being dealt with.

ø is the *empty set,* which has no elements in it at all.

x ∈ A means *x is a member of A.*

x ∉ A means that *x is not a member of A.*

A∩B is the *intersection* of A and B, which contains the elements in both A and B.

A∪B is the *union* of A and B, which contains the elements in either A or B.

A' is the *complement* of A, which contains the elements not in A.

A⊂B means that *A is a subset of B.*

n(A) is the *number of elements in A.*

26.2.1 Examples for level 3

(1)   Let ξ be the set of whole numbers between 1 and 10 inclusive. Let A be the set of prime numbers, and B the set of even numbers.

(a) Illustrate these sets on a Venn diagram.

(b) List the elements of (i) A∩B, (ii) A∪B, (iii) A', (iv) A'∩B.

Solution   (a) List the elements of ξ, A, B as follows:

ξ = {1,2,3,4,5,6,7,8,9,10}, A = {2,3,5,7}, B = {2,4,6,8,10}

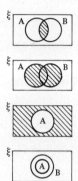

Fig 26.4

Draw circles A, B on the Venn diagram as shown, and put the numbers in the appropriate regions.

(b) The elements of each set can be read off from the Venn diagram as follows:

(i) A∩B = {2} (ii) A∪B = {2,3,4,5,6,7,8,10} (iii) A' = {1,4,6,8,9,10} (iv) A'∩B = {4,6,8,10}.

(2)   Of the households in a village, some but not all have television. All the households with cars have television, but some households with television do not have cars.

Use ξ for the set of households, T and C for the sets of houses which own televisions and cars respectively. Illustrate the above facts on a Venn diagram. Shade the region corresponding to T∩C', and describe this set in words.

Solution   The information given above can be summarized by the following set relations:

Fig 26.5

$T \neq \emptyset$ and $T' \neq \emptyset$.

$C \subset T$ but $T \cap C' \neq \emptyset$.

These facts are shown on the Venn diagram of fig 26.5.

T∩C' is the region in T but not in C. It is shown shaded in fig 26.5. It consists of the households which have television but do not have a car.

(3)   The rectangle ABCD of fig 26.6 represents a field. It is drawn in the scale of 1 cm to 20 m. Let P be the set of points within 20 m. of AB, and Q the set of points within 30 m. of C. Shade the region P∩Q.

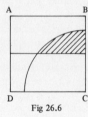

Fig 26.6

Solution   Draw a circle radius $1^{1}/2$ cm. centre C. The inside of the circle represents Q.

Draw a line 1 cm from AB. P is the region between this line and AB.

P∩Q is shown shaded.

26.2.2 Exercises for level 3

(1)   Let $\xi = \{1,2,3,4,5,6,7,8,9,10\}$, let A be the set of the odd numbers in ξ, and B be the set of the members of ξ which are greater than 5.

Draw a clearly labelled Venn diagram, putting each number into its appropriate region. List the elements of the sets:

(a) A' (b) A∩B (c) A∪B (d) A'∩B' (e) A'∪B'.

(2)   Let ξ be the set of the letters of the alphabet. Let A be the set of vowels (including y) and let B be the set of letters which come before m.
Draw a Venn diagram to show these sets, putting each letter into the correct region. List the elements of the sets:

(a) B' (b) A∩B (c) A'∩B' (d) A∩B' (e) (A∪B)'

(3)   Let $\xi = \{1,2,3,4,5,6,7,8,9,10\}$, and let A = $\{1,3,6,10\}$. B is another subset of ξ, and A∩B = $\{1,6\}$, A ∪ B = $\{1,2,3,6,8,10\}$.

Illustrate these sets on a Venn diagram, and find the elements of B.

(4)   Let ξ be the set of cards in a pack (without jokers) and let A be the Aces, B the Queens, C the Hearts. Draw a Venn diagram to show the relationship between these sets.

List the elements of:
(a) A (b) A∩B (c) A∩C (d) (A∪B)∩C.

Find the values of:
(e) n(A) (f) n(A') (g) n(A∪B) (h) n(A'∩B') (i) n(C∩A')

(5)   Let ξ be the set of months. Let A consist of the months with less than 31 days, and let B be the months with an r in them. List the elements of A and of B.

Draw a Venn diagram to illustrate these sets. On your diagram shade the region A'∩B. Describe this set in words.
Let x = June. Place x on your diagram. State in terms of set operations which subset of ξ x belongs to.

(6)   Make 6 copies of the diagram shown, labelling them (a) to (f). Shade the following sets on the appropriate diagrams:

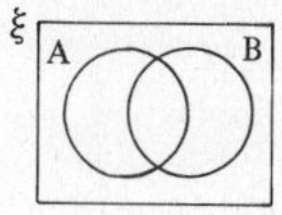

Fig 26.7

(a) A' (b) B' (c) A∩B (d) A'∪B' (e) A'∩B' (f) (A∩B)'. What conclusions can you reach?

(7)   Let ξ be the set of all integers, and let A = {x : x ≤ 3}, B = {x : x > -2}.
Illustrate these sets on a number line. Describe in the form {x : ... } the following sets:

(a) A' (b) B' (c) A∩B (d) A∪B (e) A'∩B' (f) A'∩B (g) A'∪B (h) A∩B' (i) A∪B'.

Simplify these sets where possible.

(8)   Let ξ be the set of all triangles, let I be the set of Isosceles triangles, E the set of Equilateral triangles, R the set of Right-angled triangles.

Draw a Venn diagram to show the relationship between these sets. Shade the region R∪(I∩E').
If the triangle ABC has sides of length 7, 7, 3, mark the position of ΔABC on the Venn diagram.

(9)   Let ξ be the set of all quadrilaterals. Let T be set of all trapezia, P the set of all parallelograms, Re the set of all rectangles, Rh the set of all Rhombi, and S the set of all squares.

Write in terms of set notation the relationship between P and T. Describe S in terms of Re and Rh. Draw a Venn diagram to illustrate the relations between these sets, and shade the region (Re∪Rh)∩S'.

(10)   In a sixth form several subjects are available, but with various restrictions. Anyone taking Physics must also take Mathematics, and a pupil cannot take Spanish without also taking French. The combination of Spanish with Mathematics is impossible.

Illustrate these restrictions on a Venn diagram, taking ξ for the set of all the pupils, P, M, F, S for the sets of those taking Physics, Mathematics, French, Spanish respectively..

(11) In the Venn diagram of fig 26.8 describe, using set notation, the relationships between (a) A and B (b) B and C.

(12) In the Venn diagram of fig 26.9, the universal set represents all the pupils in a class. A consists of those who play tennis, B consists of those who play squash, and C consists of those who play soccer. Describe in words the relationships between the sets A, B and C.

Fig 26.8

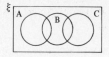

Fig 26.9

(13) A, B and C are subsets of the universal set ξ. You are told that (a) A⊂B (b) B∩C = ø (c) B∪C ≠ ξ.
    Illustrate this information by drawing the three sets A, B, C on a Venn diagram.

(14) A and B are subsets of the universal set ξ. n(A) = 45, n(B) = 63, n(A∩B) = 22. Find n(A∪B).

(15) A and B are subsets of the universal set ξ. n(A) = 12, n(B) = 17, n(A∪B) = 21. Find n(A∩B).

(16) The square PQRS of fig 26.10 is of side 3 cm. Let A be the points less than 2 cm from P, and B the points less than $2^1/2$ cm from Q. Shade the region A∩B.

(17) Draw a square ABCD, 3 cm by 3 cm. Sets X, Y, Z are given by:

X = {points nearer AB than CD}
Y = {points nearer C than A}
Z = {points more than 3 cm from D}

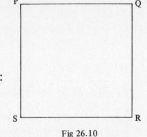

Fig 26.10

Mark the sets X, Y, Z on your diagram. Shade the region X∩Y∩Z'.

(18) Draw points P and Q 7 cm apart. X consists of those points x cm from P, and Y consists of those points y cm from Q.

   (a) If x = 2 and y = 3, draw X and Y on your diagram.

   (b) If x = 5 and y = 6, what is n(X∩Y)?

   (c) Let x = 4. If n(X∩Y) = 1, what is y?

COMMON ERRORS

(1) **Solving problems with sets**
       Make sure that you read the information carefully. If you have a problem involving 2 sets A and B, then do not confuse the statements:

There are 10 elements in A. (And maybe in B as well).

There are 10 elements in A *only*. (Not in B).

(2) **Set symbols**

(a) Do not mistake ⊂ and ∈. ⊂ is a relationship between sets. A⊂B means that A is a subset of B. But ∈ is a relationship between an element and a set. x ∈ A means that x is an element of A. For example:

$$\{1,2\} \subset \{1,2,3,4\} \quad \text{but} \quad 2 \in \{1,2\}.$$

(b) Do not mistake ∪ for ∩ . The union of two sets A and B is a large set which includes them both. The intersection of A and B is the small set inside them both. Be sure that you know the difference between them.

(c) Do not confuse ⊂ and ∩ . A⊂B is a *statement*, it tells us that A is contained in B. But A∩B is a set, it is the set of things common to both A and B. Be sure that you do not mix up these two symbols.

TEST PAPERS

If you have worked through the examples and exercises at the appropriate level then you are ready to find out how well you are likely to do in your examination.

In the following pages are 19 test papers modelled in the style, syllabus and mark scheme of the specimen papers of the four English examining groups. They are grouped by board and, for each paper, the following is indicated:

1. The equivalent **level** (1,2,3)
2. The information and formulae provided for you at your examination.
3. The time allowed for the test paper.
4. The marks allocated (shown in brackets) for each question.

You are advised to test yourself under as near as examination conditions as you can, i.e. set aside the appropriate time, be in a quiet room etc., and then check your answers against the solutions shown at the back of the book.

Contents:-

London and East Anglian Group. Paper 1. (Level 1.)

INFORMATION AND FORMULAE
(Provided on the examination paper)

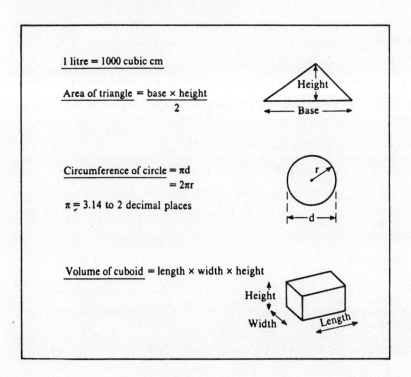

1 litre = 1000 cubic cm

Area of triangle $= \dfrac{\text{base} \times \text{height}}{2}$

Circumference of circle $= \pi d$
$\qquad\qquad\qquad\quad = 2\pi r$

$\pi = 3.14$ to 2 decimal places

Volume of cuboid $=$ length $\times$ width $\times$ height

Answer all the questions. Time 1^1/$_2$ hours

1/   Find the cost of 12 loaves of bread at 56 p each.   [1]

2/   What fraction of the shape is shaded?   [1]

3/   A film lasts 93 minutes. If it began at 4.35, when will it end?   [1]

4/   Write in figures: fifty six thousand two hundred and seventeen.   [1]

5/   A television programme began at 7.45 and ended at 8.20. How long did it last?   [1]

6/   A raffle has 100 tickets. Algernon, Berenice and Chloe each buy 6 tickets. What is the probability that one of them wins?   [2]

7/   Arrange in increasing order:

      1 ^{1}/$_3$, 1.3, 1 ^{2}/$_7$, ^{7}/$_5$.   [2]

8/   Donna runs three times round a circular running track of diameter 50 metres. Taking π to be 3.14, find how far she has run.   [2]

9/   Eggs are packed in boxes of 6. How many boxes can be filled from 100 eggs, and how many will be left over?   [2]

10/   (a) Find the next two terms in the sequence 2, 5, 8, 11, 14.   [1]

      (b) Find two prime numbers from this sequence.   [1]

11/   How much is left from 23 ounces of cheese after 5.32 ounces have been eaten?   [2]

12/   (a) Express 0.4 as a fraction.   [1]

      (b) Write ^{23}/$_{50}$ as a decimal.   [1]

13/   (a) Mark the lines of symmetry of the equilateral triangle.   [2]

      (b) How many lines of symmetry does the isosceles triangle have?   [1]

14/ The rectangle is reflected in the dotted line shown. Shade in the new rectangle. [2]

15/ The dots of the diagram are 1 cm apart. The net is folded to make a solid.

    (a) What is its name? [1]

    (b) What is its volume? [2]

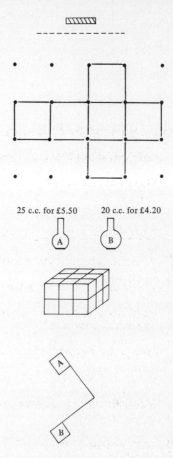

16/ How much is perfume, in pence per c.c., for the two bottles shown? [2]

    Which is cheaper? [1]

25 c.c. for £5.50   20 c.c. for £4.20

A   B

17/ A cuboid is made out of cube blocks as shown. How many blocks are there? [1]

    What is the least number of blocks we could add to make the cuboid into a cube? [3]

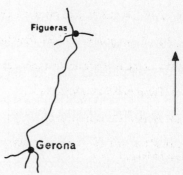

18/ (a) What is the rotation which takes (A) to (B)? [2]

    (b) Complete the diagram so that it has rotational symmetry. [2]

19/ The map is in the scale of 1 cm per 10 km.

    (a) What is the distance from Gerona to Figueras? [2]

    (b) What is the bearing of Figueras from Gerona? [2]

    If the bearing of A from B is 45°, what is the bearing of B from A? [2]

Figueras

Gerona

287

20/ The lines of the grid shown are 1 cm apart.

(a) What are the co-ordinates of A? [1]

(b) Plot D at (3,1). [1]

(c) What figure is ABCD? [2]

(d) What is the area of ABCD? [2]

21/ A pizza parlour sells four types of pizza, A, B, C, D. During an evening the sales were:

B B A C C A D D D A A B A C D B A D A A B B D C D B A A D C A A B C A D

Complete the following table: [2]

Type of pizza	A	B	C	D
Number sold				

Complete the bar-chart shown. [2]

What is the most popular pizza? [1]

If a prize is given to one customer in the evening, what is the probability that it is won by someone who orders an A pizza? [2]

22/ A population of bacteria doubles every hour. Fill in the table for the number of bacteria.

Time	12	1	2	3	4
Number	800	1600			

[3]

How many bacteria were there at 11 o'clock? [1]

How many bacteria were there at 10 o'clock? [1]

23/ Measure the sides of the triangle shown. [3]

What is the perimeter of the triangle? [1]

24/ Two maths exams were taken, and the total mark is given by $2X/5 + 3Y/5$, where X is the mark in the first exam and Y the mark in the second. Find the total mark for:

(a) Liz, with 40 in the first exam and 70 in the second. [2]

(b) Kath, with 60 in the first exam and 48 in the second. [2]

288

London and East Anglian Group. Paper 2. (Levels 1 & 2.)

INFORMATION AND FORMULAE
(Provided on the examination paper)

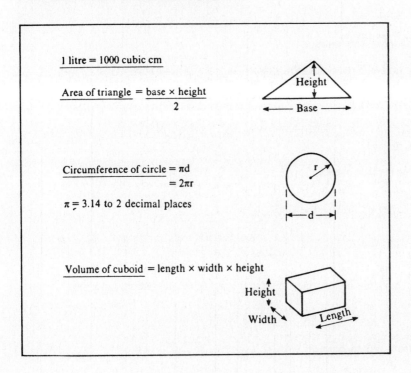

1 litre = 1000 cubic cm

$$\text{Area of triangle} = \frac{\text{base} \times \text{height}}{2}$$

Circumference of circle = πd
$\qquad\qquad\qquad\qquad = 2\pi$r

$\pi = 3.14$ to 2 decimal places

Volume of cuboid = length × width × height

Answer all the questions. Time 1^1/$_2$ hours

1/   A builder is paid £4.75 per hour. How much is he paid for a 42 hour week?   [1]

2/   The temperature at night was -4°. By noon the following day it had risen 7°. What was the temperature at noon?   [1]

3/   The income tax rate goes up from 30% to 35%. How much more will a man with a taxable income of £12,000 have to pay?   [2]

4/   (a) Read the meter dials shown. [1]

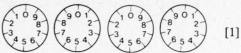

           (b) Fill in a reading of 3487 on the dials.

                                                    [1]

5/   A driver sees that there are 44 miles to go to the end of the motorway. How many minutes will this take at 66 m.p.h.?   [2]

6/   A map grid is lettered A, B, C from left to right, and 1, 2, 3 from top to bottom. How many squares share a side with B2?   [2]

7/   (a) If I arrive at North Sheen at 7.40, when do I get to Waterloo?   [1]

           (b) When must I arrive at Barnes to ensure that I get to Waterloo before 8.55?   [1]

North Sheen	d	07 35	——	07 53	08 00	
Mortlake	d	07 37	——	07 55	08 02	
Barnes	d	07 40	07 52	07 58	08 05	
Waterloo	a	07 57	08 09	08 15	08 23	

——	——	——	08 23	08 31	08 39	——
——	08 16	——	08 25	08 33	08 41	——
08 12	——	08 22	08 28	08 36	08 44	08 53
08 28	08 31	08 39	08 45	08 54	09 00	09 10

8/   Jocelyn drives for ^{3}/$_4$ hour at 56 m.p.h. How far has he gone? Give your answer to the nearest 10 miles.   [2]

9/   (a) How much does it cost to process 3 rolls each of 36 prints?   [1]

           (b) How much will it cost if a 10% discount is given?   [1]

CHARGES PER ROLL	
12 EXP.	£1.30
24 EXP.	£2.50
36 EXP.	£3.10

10/ The recipe for Hollandaise sauce requires:

3 egg yolks, 3 tablespoons of lemon juice, 6 ounces of butter.

Write out the ingredients if 1 egg yolk is used. [2]

11/ A job is advertised at £6,340, with a London allowance of £642. What is the total salary? [1]

The salary increases at £402 per year. How much will it be after 4 years? [2]

12/ The pass grades of an exam are A, B, C, D. The results of the students who passed are shown in the pie chart.

(a) How many students were there? [3]

(b) 120 students got grade A. What is the angle in the A sector? [2]

(c) Equal number of students got C and D. How many got C? [2]

(d) If 960 students in all took the exam, what was the proportion who failed? [2]

13/ (a) Divide 112 by 2 until you get an odd number. How many times have you divided by 2? [2]

(b) What is the largest odd number which divides 120? [2]

14/ The prices per person for a 14 day skiing holiday are shown. How much will the holiday cost Mr and Mrs Jones and their two children:

(a) If they all share a room? [2]

(b) If the adults share a room and the children share a separate room. [2]

14 DAY HOLIDAY CHARGES	
4 sharing a room £295 10% reduction for children	2 sharing a room £350 15% reduction for children

15/ Bert walks to the shops, buys his food and walks home. The graph shows his distance against time.

(a) How far away are the shops? [1]

(b) How long did he spend at the shops? [1]

(c) What was his speed while he returned home? [2]

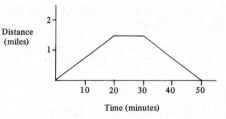

291

16/ Peter at A sees a tree T on a bearing of 30°. He walks 100 metres East to B, and it is now on a bearing of 350°.

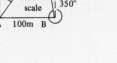

Make a diagram of the triangle ABT, using a scale of 1 cm for 10 m. [3]

How far is B from the tree? [2]

17/ In a certain year August 4th is a Saturday.

(a) What day of the week is August 20th? [2]

(b) What day of the week is July 31st? [2]

(c) What day of the week is September 3rd? [3]

(d) What is the date of the Saturday before August 4th? [3]

18/ 3 newspapers are delivered to a street. The bar-chart shows the numbers of each newspaper. Fill in the frequency table. [2]

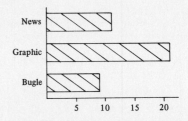

Paper	News	Graphic	Bugle
Number			

What is the total number sold? [2]

Which is the most popular paper? [1]

19/ The quarterly rental for a telephone is £15, and each unit call costs 2p. Find the total bill for (a) 100 units (b) 200 units (c) 300 units. [3]

Draw a graph to show these figures, using 1 cm to represent 20 units along the x-axis, and 1 cm to represent £5 up the y-axis. [3]

If the bill is £27, how many unit calls have been made? [2]

20/ The lines of the grid shown are 1 cm apart.

(a) Find the area of ∆ABC. [2]

(b) What is <ABC? [2]

(c) Mark D on the grid to make ABCD a parallelogram.

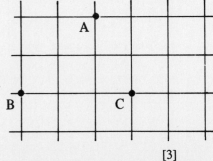

[3]

21/ A "Manana" car has a turning circle of radius 20 m. How long is the circle? (Take π to be 3.14)

[2]

How long does the car take to drive round a full circle at 2 metres per second? [2]

At 3 m. per second the car can safely go round a circle of radius 40 m. How long will it take to go round this circle at the higher speed? [3]

London and East Anglian Group. Paper 3a. (Levels 2 & 3.)

INFORMATION AND FORMULAE
(Provided on the examination paper)

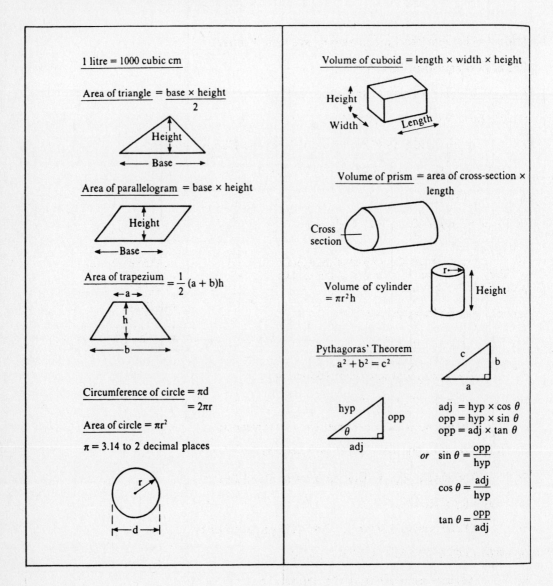

1 litre = 1000 cubic cm

Area of triangle $= \dfrac{\text{base} \times \text{height}}{2}$

Area of parallelogram $=$ base $\times$ height

Area of trapezium $= \dfrac{1}{2}(a + b)h$

Circumference of circle $= \pi d$
$\qquad\qquad\qquad\qquad\ = 2\pi r$

Area of circle $= \pi r^2$

$\pi = 3.14$ to 2 decimal places

Volume of cuboid $=$ length $\times$ width $\times$ height

Volume of prism $=$ area of cross-section $\times$ length

Volume of cylinder $= \pi r^2 h$

Pythagoras' Theorem
$a^2 + b^2 = c^2$

adj $=$ hyp $\times \cos \theta$
opp $=$ hyp $\times \sin \theta$
opp $=$ adj $\times \tan \theta$

or $\quad \sin \theta = \dfrac{\text{opp}}{\text{hyp}}$

$\cos \theta = \dfrac{\text{adj}}{\text{hyp}}$

$\tan \theta = \dfrac{\text{opp}}{\text{adj}}$

Attempt all the questions. Circle the answer you think is correct.

One mark per question. Time $^3/_4$ hour.

1/ The product of -0.2 and -0.5 is:

(a) -0.7 (b) 1 (c) -1 (d) 0.1 (e) -0.1

2/ A man earning £9,000 receives a 12% pay rise. His new salary is:

(a) £9,012 (b) £10,227 (c) £9,000 (d) £10,200 (e) £10,080

3/ If 60k = 24, then k is:

(a) $^1/_3$ (b) $^2/_5$ (c) 1440 (d) $2^1/_2$ (e) -36

4/ James makes a bar-chart of the species of birds which come to his bird-table. The proportion of sparrows is:

(a) $^1/_2$ (b) $^1/_3$ (c) $^1/_4$ (d) $^1/_5$ (e) $^1/_6$

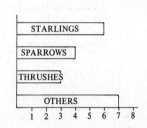

5/ The highest common factor of 30 and 24 is:

(a) 3 (b) 120 (c) 6 (d) 12 (e) 54

6/ The number of cubic centimetres in a cubic metre is:

(a) 10 (b) 100 (c) 1,000 (d) 100,000 (e) 1,000,000

7/ The number of prime numbers between 10 and 20 is:

(a) 1 (b) 2 (c) 3 (d) 4 (e) 5

8/ £1,200 is divided in the ratio 3:4:5. The smallest share is:

(a) £900 (b) £300 (c) £400 (d) £100 (e) £250

9/ A square has area 100 cm^2. Its perimeter is:

(a) 100 (b) 20 (c) 40 (d) 10 (e) 50

10/   A figure is both a rhombus and a rectangle. It is a:

(a) kite (b) square (c) parallelogram (d) trapezium (e) quadrilateral

11/   When simplified, (2 - 4x) - (3 - 7x) is:

(a) 3x - 1 (b) 5 + 11x (c) 1 + 3x (d) -1 - 3x (e) -1 - 11x

12/   Aaron is twice as old as Barry. Barry is 3 years younger than Charles. If Charles is x years, then Aaron is:

(a) 2x - 6 (b) $^{1}/2$x - 3 (c) 2x - 3 (d) 2x + 3 (e) 3x - 2

13/   A speed of 24 metres per second is equivalent in kilometres per hour to:

(a) 6 $^{2}/3$ (b) 86.4 (c) 24 (d) 36 (e) 12

14/   A cube of side 4 cm is cut up into cubes of side 1 cm. The number of small cubes is:

(a) 4 (b) 64 (c) 12 (d) 16 (e) 256

15/ $3x10^4 + 4x10^3$ in standard form is equal to:

(a) $34x10^3$ (b) $7x10^3$ (c) $1.2x10^7$ (d) $7x10^4$ (e) $3.4x10^4$

16/   The sides of a rectangle are 3.8 and 4.3, where both numbers are given to 1 decimal place. The greatest possible area of the rectangle is:

(a) 16 (b) 16.4 (c) 16.8 (d) 20.3 (e) 16.75

17/   I stand 60 m from the base of a 50 m tower. The angle of elevation of the top is:

(a) 33.5° (b) 50.2° (c) 75° (d) 56.4° (e) 39.8°

18/   In the triangle ABC, AE = 4, EB = 2, AF = 3, FC = 5. The ratio EF:BC is:

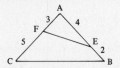

(a) 1:2 (b) 2:3 (c) 3:2 (d) 1:1 (e) 3:4

19/   A cylinder with base radius 3 cm contains 300 c.c. of water. The depth of water in centimetres is:

(a) 8,482.3 (b) 100 (c) 10.6 (d) 42.4 (e) 5.6

20/   A bag contains 7 red and 6 green marbles. 2 are drawn out. The probability that the second is red is:

(a) $^{7}/12$ (b) $^{6}/13$ (c) $^{6}/7$ (d) $^{1}/2$ (e) $^{7}/13$

21/ If $a^2 = b^2 + c^2$, then c is equal to:

(a) $(a^2 - b^2)^2$ (b) $\sqrt{a^2 - b^2}$ (c) a - b (d) $b^2 - a^2$ (e) $\sqrt{a} - \sqrt{b}$

22/ Below are several nets. Which make a cube?

(i) (ii) (iii)

(a) all (b) (i) only (c) (ii) only (d) none (e) (ii) and (iii) only

23/ Which graph represents $y = \cos x$ for x between 0° and 90°?

(a) (b) (c) (d) (e)

24/ I walk 10 miles South and 10 miles West. My bearing from my starting point is:

(a) 090° (b) 045° (c) 225° (d) 135° (e) 210°

25/ $27 \times 9^2 \times 3^3$ is the following power of 3:

(a) 12 (b) 10 (c) 9 (d) 7 (e) 6

26/ In 10 matches, a team scores goals with the following frequencies:

Number of goals	0	1	2	3
Frequency	2	4	3	1

The mean number of goals is:

(a) 1 (b) $1^5/_8$ (c) $3^1/_4$ (d) $1^1/_2$ (e) 1.3

27/ I buy 10 biros at 8 p each, then 15 more at 12 p each. The average cost of the biros is:

(a) 10.4 (b) 10 (c) 9.84 (d) 11 (e) 10.5

28/ A dealer sells a car for £1,000, making a 25% profit. The original price was:

(a) £1,250 (b) £750 (c) £666.67 (d) £800 (e) £975

29/ The transformation of ABCD to A'B'C'D' is:

(a) rotation of 180° about B (b) reflection in y = x
(c) rotation of 90° about (0,0) (d) translation of
(1,-1) (e) reflection in x = 1

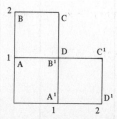

30/   AB is a diameter of the circle, and TB is a tangent. Which of the following must be true?

(a) y = z (b) x + z = 90° (c) x = z (d) x + y + z = 180° (e) x = y

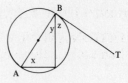

London and East Anglian Group. Paper 3b. (Levels 2 & 3.)

INFORMATION AND FORMULAE
(Provided on the examination paper)

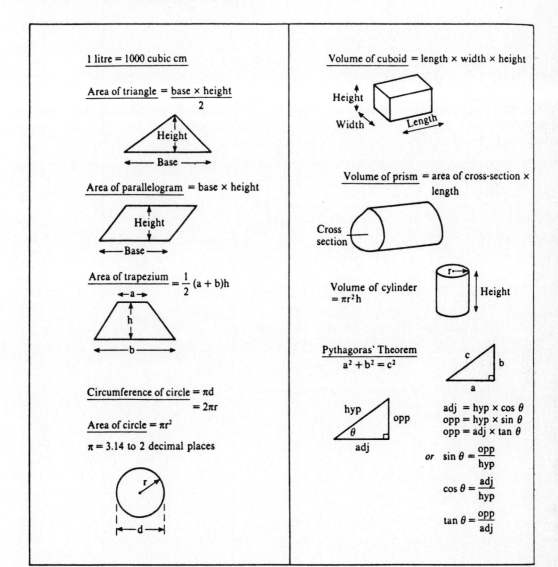

1 litre = 1000 cubic cm

Area of triangle = $\dfrac{\text{base} \times \text{height}}{2}$

Area of parallelogram = base × height

Area of trapezium = $\dfrac{1}{2}(a + b)h$

Circumference of circle = πd
$\qquad\qquad\qquad\qquad = 2\pi r$

Area of circle = πr^2

$\pi = 3.14$ to 2 decimal places

Volume of cuboid = length × width × height

Volume of prism = area of cross-section × length

Volume of cylinder = $\pi r^2 h$

Pythagoras' Theorem
$a^2 + b^2 = c^2$

adj = hyp × cos θ
opp = hyp × sin θ
opp = adj × tan θ

or $\quad \sin\theta = \dfrac{\text{opp}}{\text{hyp}}$

$\cos\theta = \dfrac{\text{adj}}{\text{hyp}}$

$\tan\theta = \dfrac{\text{opp}}{\text{adj}}$

Time 1¹/₂ hours. Answer all the questions.

1/ Mr Record took his car to be serviced. Fill in the blanks in the bill.

(a) .... [3]

(b) .... [3]

(c) .... [3]

(d) .... [3]

Oil filter	6.50
Air filter	5.45
Spark Plugs	5.80
Contacts	a
Material Total	25.62
Labour	b
Total	43.12
VAT @ 15%	c
TOTAL	d

2/ Fred reckons that he spends his day as follows:

Activity	Sleep	Work	Play	Meals
Time in hours	8	8	6	2

(a) Construct a bar-chart to illustrate these figures. [4]

(b) Construct a pie-chart to illustrate these figures. [4]

(c) If he measured the times in minutes, what would be the angle in the pie-chart corresponding to sleep? [1]

3/ The diagram shows two adjacent fields.
<A = <BCD = 90°. AC = 50 m, AB = 40 m and
CD = 30 m.

(a) Find BC to 3 significant figures. [3]

(b) Find <ACB to the nearest degree. [3]

(c) Find <D to the nearest degree. [3]

(d) Find the total area of the fields, giving
your answer to 2 decimal places. [4]

300

4/ A tessellation consists of equilateral triangles as shown. What transformation would take:

(a) ΔABC to ΔDCE?   [2]

(b) ΔABC to ΔCDA?   [2]

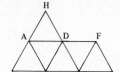

(c) ΔABC to ΔHBE?   [2]

(d) ΔABC to ΔFGE?   [2]

5/ A garden is 50 ft by 30 ft. A paved courtyard occupies the first 5 ft. A 2 ft wide path surrounds the lawn. A circular flower bed of radius 3 ft is in the centre of the lawn.

(a) Find the area of the courtyard.   [2]

(b) Find the area of the path.   [4]

(c) Find the area of the flower bed, giving your answer to 3 significant figures.   [3]

(d) Find the area of lawn.   [2]

6/ The diagram represents a ship S sailing due West. X represents a lighthouse, 30 miles South and 100 miles West of S. The lighthouse is visible for 50 miles.

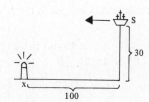

(a) Make an accurate scale drawing of the course of the ship, using a scale of 1 cm per 10 miles.   [4]

(b) Shade the region North of the lighthouse within which it is visible. What is its area?   [5]

(c) If the ship is sailing at 8 m.p.h., when will it be able to see the lighthouse? For how long will it be able to see the lighthouse?   [8]

London and East Anglian Group. Paper 4. (Level 3.)
INFORMATION AND FORMULAE
(Provided on the examination paper)

1 litre = 1000 cubic cm

Area of triangle $= \dfrac{\text{base} \times \text{height}}{2}$

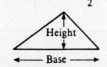

Area of parallelogram = base × height

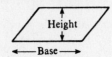

Area of trapezium $= \dfrac{1}{2}(a + b)h$.

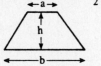

Circumference of circle = πd
$= 2\pi$r

Area of circle = πr^2

Volume of cuboid = length × width × height

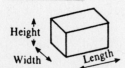

Volume of prism = area of cross-section × length

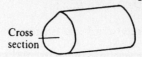

Cross section

Volume of cylinder $= \pi r^2 h$

Height

Volume of sphere $= \frac{4}{3}\pi r^3$

Surface area of sphere $= 4\pi r^2$

Volume of pyramid (including cone)

$= \frac{1}{3} \times$ area of base × height

Volume of cone = $\frac{1}{3}\pi r^2 h$

Pythagoras' Theorem
$a^2 + b^2 = c^2$

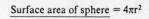

hyp

opp

θ

adj

adj = hyp × cos θ
opp = hyp × sin θ
opp = adj × tan θ

or $\sin \theta = \dfrac{\text{opp}}{\text{hyp}}$

$\cos \theta = \dfrac{\text{adj}}{\text{hyp}}$

$\tan \theta = \dfrac{\text{opp}}{\text{adj}}$

Quadratic Equations

The solutions of $ax^2 + bx + c = 0$
where $a \neq 0$, are given by

$$x = \dfrac{-b \pm \sqrt{(b^2 - 4ac)}}{2a}$$

302

Answer all the questions. Time $2^1/_2$ hours

1/   F varies inversely as the square of d. F = 4 when d = 5.

         (a) Find an equation giving F in terms of d.   [2]

         (b) Find the value of d when F = 400.   [2]

2/   A bag contains 4 red and 5 blue counters. Two are drawn out. What are the probabilities
that: that:

         (a) Both are red.   [2]

         (b) Both are the same colour.   [2]

         (c) Exactly one is blue.   [2]

3/   The speed of a car was measured as follows:

Time in secs.| 0 | 1 | 2 | 3 | 4 | 5 | 6 |
Speed in m/s.| 10 | 17 | 21 | 24 | 27 | 29 | 30 |

       Plot these points on a graph, using 1 cm per second along the x-axis, and 1 cm per 5 m/sec
along the y-axis. Join the points as smoothly as you can.   [4]

       By approximating the region under the curve by two trapezia, estimate the distance covered by
the car during this time.   [4]

4/   Let a, b, c be positive numbers related by the formula $a^2 = b^2 + (2c)^2$.

         (a) Find a when b = 7 and c = 12.   [1]

         (b) Find b when a = 8 and c = 1, giving your answer to 3 decimal places.   [2]

         (c) If a = 11, find all the possible whole number values of c.   [5]

5/   In the figure shown AD = DB = BC. <EBC = 117°. Find <BCD.   [3]

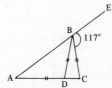

6/ (a) Expand $(5x + 3y)(2x - 7y)$, simplifying your answer as much as possible. [2]

(b) Make x the subject of $a = \sqrt{2 + x}$. [3]

(c) Solve the equation $x^2 + 3x - 3 = 0$, giving your answers to 3 significant figures. [4]

7/ (a) Let E = {1,2,3,4,5,6,7,8,9,10}, A = {2,4,6,8}, B = {3,6,9}. Illustrate these sets on a Venn diagram. [4]

(b) Make 3 copies (i), (ii), (iii) of the diagram shown. Use them to shade the regions:

(i) P∩Q [1]
(ii) P∩Q' [2]
(iii) (P∩Q')∪(P'∩Q) [3]

8/ A cylindrical jar has base radius 3 cm. It contains water to a depth of 8 cm.

(a) A long cylindrical rod with diameter 2 cm is placed vertically in the jar. By how much does the water rise? [4]

(b) A sphere of radius $1^1/2$ cm is dropped in. By how much does the water rise? [4]

9/ Let the matrix A be given by:

$$A = \begin{pmatrix} 4 & 2 \\ 3 & 2 \end{pmatrix}$$

(a) Find the inverse A^{-1}. [2]

(b) Find the values of x and y for which $A \begin{pmatrix} x \\ y \end{pmatrix} = \begin{pmatrix} 8 \\ 7 \end{pmatrix}$.

Hence solve the equations:

$4x + 2y = 16$
$3x + 2y = 14$ [3]

(c) Plot the points P(0,0), Q(1,0), R(1,1), S(0,1) on graph paper. Plot its image P'Q'R'S' after applying the matrix A. What sort of quadrilateral is the image? [4]

10/ A tree AB is 30 m high. P stands 100 m south of the tree, and Q stands 200 m west of the tree.

(a) Find the distance PQ. Give your answer to 3 significant figures. [2]

(b) Find the angles of elevation of the top of the tree from P and from Q. [4]

(c) Find the volume of the region enclosed by the lines AB, AQ, AP, BP, BQ, PQ. [4]

11/   In $\triangle$ABC $\underline{AB}$ =**b** and $\underline{AC}$ = **c**. X,Y,Z are the midpoints of BC, AC,AB respectively.

    (a) Express in terms of **b** and **c**, $\underline{AZ}$, $\underline{BC}$, $\underline{BX}$, $\underline{AX}$.   [4]

    (b) What is the relationship of $\triangle$ABC to $\triangle$XYZ?   [2]

    (c) List the three triangles which are congruent to $\triangle$XYZ   .[3]

    (d) $\triangle$ABC has area 24. What is the area of $\triangle$XYZ?[2]

12/   Complete the table for the function y = $x^3/8$ - $x^2/4$ + 2.

x	-3	-2	-1	0	1	2	3
y	-3⅝				1⅞		

Draw the graph of this curve, taking 1 cm per unit along both the x and the y axes.   [6]

By drawing a suitable line on your graph, solve the equation:

$x^3/8$ - $x^2/4$ + 2 = 1 - $\tfrac{1}{2}$x.   [4]

Draw a tangent to the curve at x = 2, and hence find the gradient of the curve at that point.[4]

Midland Group. Paper 1. (Level 1.)
INFORMATION AND FORMULAE
(Provided on the examination paper)

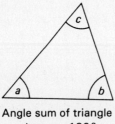

Angle sum of triangle
$a + b + c = 180°$

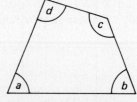

Angle sum of quadrilateral
$a + b + c + d = 360°$

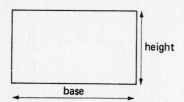

Area of rectangle = base × height

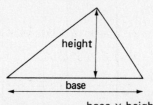

Area of triangle $= \dfrac{\text{base} \times \text{height}}{2}$

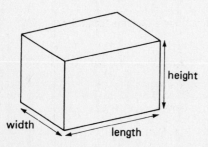

Volume of cuboid = length × width × height

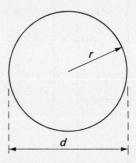

Circumference of circle = $2\pi r$ or πd

Answer all the questions. Time 1^1/$_2$ hours.

1/   What fraction of the shape is shaded?   [1]

2/   Mr Eliot's bank sends him a statement as shown. (DR indicates that he owes money to the bank.)

	Balance
	£123
Cheque Drawn	£——
	£27 DR

       How much was the cheque for?   [1]

3/   Find 17% of £850.   [2]

4/   Anne is 5 feet and six inches high. How tall is she in centimetres?
(1 inch = 2.54 cm.)   [3]

5/   Sid buys a jacket at £23 and two shirts at £9 each. He is given a discount of 10 p in the £.
How much does he pay?   [2]

6/   (a) What is the cost of 5 pounds of flour at 35 p per pound?   [1]

               (b) The flour is used to make 50 cakes. How much flour is there in each cake?   [1]

7/   "Dial a Meal" charges 10% for delivery of its meals. What would be the total cost for the items ringed on the list?   [2]

Moussaka	£5.25
Klefliko	£6.85
Dolmades	£4.50
Afelia	£5.55

8/   Draw dotted lines to show the lines of symmetry of the rectangle shown.   [2]

9/ z is given in terms of x by z = 3x - 7.

    (a) Find z when x = 4.   [1]

    (b) Find z when x = 3 $^2/_3$.   [1]

10/ A film on television began at 7.35, and ended at 9.03.

    (a) How long was the film?   [1]

    (b) Rachel saw only the last 37 minutes. When did she start watching?   [1]

11/ A fair die is rolled. What is the probability that the score is greater than 4?   [2]

12/ Continue the tessellation shown with 6 more triangles.

    [3]

13/   (a) Find the number which when added to 3 gives 17.   [1]

    (b) How many biros at 11 p each can be bought for 90 p?
How much will be left over?   [2]

14/ The dots shown are 1 cm apart.

    (a) Find the area of the square.   [1]

    (b) Find the area of the triangle.   [2]

15/   (a) What is the cost of 12 cakes at 15 p each?   [1]

    (b) Joan, Jackie and Liz share a bill of £15.60. How much does each pay?   [2]

16/ The figure shows a pyramid. Write down the number of
(a) vertices (b) edges (c) faces.   [3]

17/ Out of 60 children, 45 are boys. Express the proportion of boys as (a) a fraction (b) a decimal (c) a percentage.   [3]

18/ A car does 40 miles per gallon.

(a) How far will it go on 5 gallons? [1]

(b) How many gallons will be needed for 120 miles? How much will this cost at £1.70 per gallon? [2]

19/ A swimming bath is open for 7 hours per day from Monday to Friday, for 8 hours on Saturday and for 2 hours on Sunday. Alf and Bert are the attendants: if Alf is present for 27 hours for how many hours must Bert be there? [4]

20/ A cricketer scores 5, 0, 40, 0, 61, 14 in six successive innings. What is his average? [2]

In his next match he scores 13. What is his new average? [2]

21/ A map is in the scale of 1 cm per 2 km. If two towns are 10 km apart, how far apart are they on the map? [1]

The map is 50 cm by 50 cm. What area of land does it represent? [3]

22/ (a) Lorna donates $^1/_{10}$ of her pocket money to charity. If she receives £5 per week, how much does she give? [1]

(b) Fred gives 60 p per week. If his pocket money is £4 per week, what fraction does he give to charity? [3]

23/ ABC is an isosceles triangle, with AB = AC. <B = 66°.

A

66°

B   C   D

Find (a) <ACB (b) <A (c) <ACD   [4]

24/ The pie-chart shows how 30 books are classified.

(a) How many travel books are there? [2]

(b) What is the angle in the Sci-Fi slice? [2]

Sci-Fi
3 Travel
60°
Fiction Thrillers
Romance

25/ The diagram of the M5 motorway is not to scale.

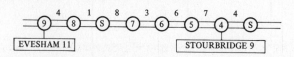

(a) How far is it between service areas? [1]

(b) How far is it from Evesham to Stourbridge via the motorway? [2]

(c) How long will it take between junctions 6 and 8 at 60 m.p.h.? [2]

26/ £1 is worth 10.5 French Francs or $1.45.
(a) How many French Francs do I get for £120? [1]

(b) An American tourist changes $232 into pounds. How many does he get? [2]

(c) How many dollars is 840 French Francs worth? [3]

27/ The graph shows the height of a lift in terms of time.

(a) How high is each floor? [2]

(b) How long does the lift take to rise to the top? [2]

(c) How long did the lift wait at the second floor? [2]

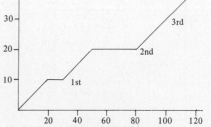

28/ The lines of the grid shown are 1 cm apart.

(a) Find the perimeter of ABCD. [2]

(b) Find the area of PQR. [2]

(c) Give the co-ordinates of the point S such that PQRS is a rectangle. [3]

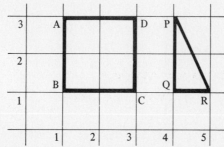

29/ A year is a *leap year* if it is divisible by 4, unless it is also divisible by 100.

(a) Find the leap years between 1895 and 1905. [4]

(b) How many leap years are there between 1930 and 1987? [2]

(c) How many leap years are there between 1850 and 1933? [4]

Midland Group. Paper 2. (Level 2.)
INFORMATION AND FORMULAE
(Provided on the examination paper)

MENSURATION

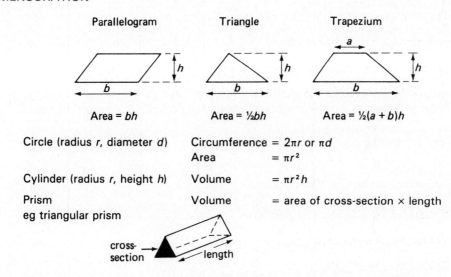

Parallelogram	Triangle	Trapezium
Area = bh	Area = $\frac{1}{2}bh$	Area = $\frac{1}{2}(a + b)h$

Circle (radius r, diameter d)   Circumference = $2\pi r$ or πd

Area   = πr^2

Cylinder (radius r, height h)   Volume   = $\pi r^2 h$

Prism   Volume   = area of cross-section × length

eg triangular prism

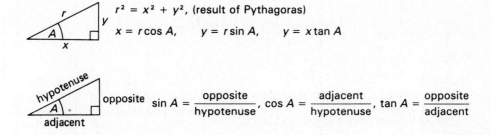

cross-section   length

TRIGONOMETRY

Right-angled triangle

$r^2 = x^2 + y^2$, (result of Pythagoras)

$x = r\cos A$, $y = r\sin A$, $y = x\tan A$

$\sin A = \dfrac{\text{opposite}}{\text{hypotenuse}}$, $\cos A = \dfrac{\text{adjacent}}{\text{hypotenuse}}$, $\tan A = \dfrac{\text{opposite}}{\text{adjacent}}$

NUMBER
Standard form is $a \times 10^n$ where $1 \leqslant a < 10$ and n is an integer.

Answer all the questions. Time 2 hours.

1/ $^3/_4$ of the members of a golf club are men. If there are 73 women, how many members are there in all? [3]

2/ A model aircraft is 15 cm long. If the real aircraft is 12 m long, what is the scale of the model? [3]

3/ A jug with capacity 1.35 litres is filled from a 30 litre barrel. How many times can the jug be filled? [2]

How much is left over? [2]

4/ Belinda wants 16 driving lessons. How much will she save if she books two sets of 8, instead of paying for them singly? [2]

| LESSONS £5.50 each |
| OR |
| £41 for a course of 8 |

5/ (a) What is the whole number before 100,000? [1]

(b) What is the least whole number greater than -1,000,000? [1]

(c) Multiply your answers together and give the answer in standard form to 3 significant figures. [3]

6/ A house is divided into 3 flats, and expenses are shared in the ratio 10:9:7. If it costs £520 to repaint the house, how much are the three shares? [3]

7/ Jane got 72% in her first exam. In the second exam she lost twice as many marks. What was her percentage overall? [3]

8/ Solve the equations:

(a) $4x - 3 = 17$ [1]   (b) $3(x + 1) - 2(x - 3) = 8$ [2]

9/ The standard European week is 7 days long. The ancient Chinese week was 10 days long. How frequently does the first day of the European week fall on the same day as the first day of the Chinese week? [3]

10/ Fill in the blanks in the following sums:

```
    134              158
+  -1-           +  -6-
=  3-7  [2]       =  5-3   [2]
```

11/ A number is divisible by 9 if the sum of its digits is divisible by 9. What is the first number after 2135 which is divisible by 9? [2]

12/ In 10 innings, Geoffrey scores 0, 45, 55, 23, 0, 43, 6, 34, 49, 10.

What is his average score? [2]

13/ The figure shows a river of width 20 metres. Angie stands directly opposite Bert. If Bert now walks 30 metres along the river, how far is he from Angie? [2]

What angle does the line between them make with the bank?[2]

14/ Alfred and Beatrice play darts: each must throw a double before they can start scoring. Alfred has probability $1/10$ of throwing a double, and Beatrice has a chance of $1/8$. Find the probabilities that:

(a) Alfred does not start with his first dart. [1]

(b) Neither starts with their first dart. [2]

(c) Beatrice starts with her first dart but Alfred does not. [2]

15/ The figure shows two triangles.

(a) What transformation has taken S to T? [2]

(b) Draw the triangle which is obtained by reflecting S in the x-axis.[2]

16/ The figure shows a garden shed.

(a) Find the area of a side wall. [2]

(b) Find the volume of the shed. [1]

17/ A cylindrical tank of height 1 metre is filled up at a steady rate from a tap, so that it is full in $1/2$ hour. It is left for $1/4$ hour, then a plug at the bottom is removed so that the water drains away in $1/4$ hour. Draw a graph of the level of the water in the tank against time. [4]

18/ In a borough there were 12 magistrates, of whom 3 were doctors, 2 were farmers, 3 were teachers and 4 had other professions. Draw up a pie-chart to illustrate these figures. [4]

19/ F is given in terms of C by $F = {}^{9C}/5 + 32$. Find F when C takes the values 0,10,-25. [2]

Find a formula giving C in terms of F. [3]

20/ A man is 5 times as old as his son. 2 years ago he was 7 times as old. Let x be the son's age now. Form an equation in x and solve it. [5]

313

21/ 30 cars were in a garage to be repaired so that they would pass the MOT test. 15 had one fault, 12 had two faults and 3 had three faults.

(a) Draw a histogram to show these figures. [2]

(b) What is the mean number of faults per car? [2]

22/ The shape S is transformed to T by reflection in the line x = 1. Draw the shape T. [2]

The shape T is transformed to U by reflection in the line x = 0. Draw U. [2]

Describe the single transformation which will take S to U. [2]

23/ Write down 2 points which the line y = 3x - 2 passes through. [2]

Find the gradient of this line. [2]

24/ A room is 10 ft by 12 ft. The only power point is at X, midway along one of the shorter walls. The flex of a fire is 6 ft long.

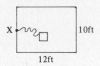

Make an accurate scale drawing of the room, to a scale of 1 cm per 2 ft, and shade the region within which the fire may be put. [4]

25/ Find the angles x and y in the figures shown. O marks the centres of the circles.

[2]

[2]

26/ In $\triangle ABC$, BC = 8 cm, <B = 53°, <C = 64°. Make an accurate scale diagram of the triangle.[3]

What is the length of AB? [2]

If the diagram is a map in the scale of 1:50,000, which represents three towns A, B, C, what is the distance of B from C? [2]

Midland Group. Paper 3. (Level 3.)

INFORMATION AND FORMULAE
(Provided on the examination paper)

MENSURATION

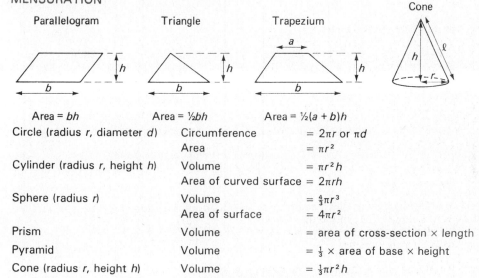

Parallelogram Triangle Trapezium Cone

Area = bh Area = $\frac{1}{2}bh$ Area = $\frac{1}{2}(a + b)h$

Circle (radius r, diameter d)	Circumference	$= 2\pi r$ or πd
	Area	$= \pi r^2$
Cylinder (radius r, height h)	Volume	$= \pi r^2 h$
	Area of curved surface	$= 2\pi rh$
Sphere (radius r)	Volume	$= \frac{4}{3}\pi r^3$
	Area of surface	$= 4\pi r^2$
Prism	Volume	$=$ area of cross-section $\times$ length
Pyramid	Volume	$= \frac{1}{3} \times$ area of base $\times$ height
Cone (radius r, height h)	Volume	$= \frac{1}{3}\pi r^2 h$
	Area of curved surface	$= \pi r \ell$

where ℓ = slant height = $\sqrt{h^2 + r^2}$

TRIGONOMETRY

Right-angled triangle

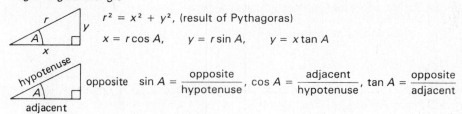

$r^2 = x^2 + y^2$, (result of Pythagoras)

$x = r\cos A, \qquad y = r\sin A, \qquad y = x\tan A$

opposite $\sin A = \dfrac{\text{opposite}}{\text{hypotenuse}}$, $\cos A = \dfrac{\text{adjacent}}{\text{hypotenuse}}$, $\tan A = \dfrac{\text{opposite}}{\text{adjacent}}$

Any triangle

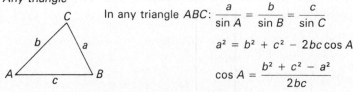

In any triangle ABC: $\dfrac{a}{\sin A} = \dfrac{b}{\sin B} = \dfrac{c}{\sin C}$

$a^2 = b^2 + c^2 - 2bc\cos A$

$\cos A = \dfrac{b^2 + c^2 - a^2}{2bc}$

Area of triangle $ABC = \frac{1}{2}ab\sin C$

NUMBER   Standard form is $a \times 10^n$ where $1 \leqslant a < 10$ and n is an integer.

Compound Interest formula $A = P\left(1 + \dfrac{r}{100}\right)^n$

ALGEBRA   The quadratic equation $ax^2 + bx + c = 0$ has solutions

$$x = \frac{-b \pm \sqrt{b^2 - 4ac}}{2a}$$

Answer all the questions. Time 2 hours.

1/ In 10 games of golf Enid has an average of 95. How much must she score in her next game to bring her average down to 93? [2]

2/ After a pay rise of 12%, Robert is earning £17,248. How much was he earning before? [2]

3/ A grocer mixes Chinese and Indian tea in the ratio 1:5 to make his special blend. How much Indian tea is there in 18 kg of the mixture? [2]

4/ Mr King buys a car for £8,000, and reckons that every year its value decreases by 10%. What is its value after 5 years? [3]

5/ The sides of a right-angled triangle are x, 3x + 3, 3x + 4. Write down Pythagoras' Theorem for the triangle, and show that it reduces to:

$$x^2 - 6x - 7 = 0$$ [3]

Solve this equation to find the sides of the triangle. [2]

6/ 5 cakes and 7 buns cost £3.04. 4 cakes and 9 buns cost £3.35. Let x p be the price of a cake, and y p be the price of a bun. Form two equations in x and y and solve them. [5]

7/ (a) Find the area of the front wall of the house shown. [2]

(b) Find the volume of the house. [1]

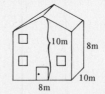

8/ Solve the equation $(x + 2)/(x - 1) = 2$. [3]

9/ (a) If $a = \sqrt{b^2 + (2c)^2}$ find a when b = 3 and c = 2. [2]

(b) Make x the subject of the formula $x + 2y = 3x - 4z$. [2]

10/ A rectangle is 5.3 cm long by 4.2 cm wide, where both figures are rounded to 1 decimal place. What is the greatest possible value of the length? [2]

What is the greatest possible value of the area? [2]

11/ In the triangle shown, OA = 8, OB = 12, OX = 2 and OY = 3. Find the ratio XY:AB. [2]

If △OAB has area 36, find the area of the quadrilateral ABXY. [2]

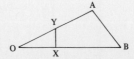

12/ (a) Find the equation of the straight line which goes through (1,1) and (5,2). [3]

(b) Find where the line $2y + 3x = 6$ crosses the axes. Find the gradient of the line. [3]

13/ Draw an equilateral triangle ABC of side 4 cm. Shade the region of those points which are within 2 cm of A. [3]

14/ Express the following as fractions in their simplest form:

(a) $^2/p + 3/p^2$ [2]

(b) $^5q/r \times q^2/10r^2$ [2]

15/ The square ABCD has side 2 cm. The shaded region is enclosed by two circles of radius 2 cm and centres A and C.

(a) Find the area of the triangle ABD. [1]

(b) Find the area of the sector ABD. [2]

(c) Find the area of the shaded region. [2]

16/ A trapdoor is of length 2 ft. It is propped open by a stick as shown.

(a) If the stick is of length 1 ft, what is the angle does the door make with the horizontal? [3]

(b) What length should the stick be to keep the door open at 45° to the horizontal? [3]

17/ ABCD is a parallelogram. X and Y are the midpoints of BC and AD. Let $\underline{AB} = \mathbf{b}$ and $\underline{AD} = \mathbf{d}$.

Express, in terms of $\mathbf{b}$ and $\mathbf{d}$, $\underline{BC}$, $\underline{AC}$, $\underline{BD}$, $\underline{AX}$, $\underline{XY}$. [4]

What can you say about ABXY? [2]

18/ One week in January it is noticed that 20% of the pupils in a school have colds, and 5% have toothaches. 2% have both a cold and toothache. What percentage of the school has neither complaint? [3]

19/ The prism shown is of length 20 cm, and its cross-section is an isosceles triangle of sides 8 cm, 8cm, 14 cm. Find:

(a) The length of the perpendicular from A to BC. [2]

(b) The area of cross-section. [2]

(c) The volume. [1]

317

20/ The matrix A is given by:

$$A = \begin{pmatrix} 3 & 1 \\ 2 & 1 \end{pmatrix}$$

Find the inverse matrix A^{-1}.   [3]

21/ The prices of 100 cars at an auction were given by the following table.

Price in £1,000's |0-1 |1-2|2-3|3-4|4-5|
Frequency | 33 | 27| 16| 14| 10|

Plot a histogram to illustrate these figures.   [3]

Estimate the proportion of cars which cost more than £2,750.   [2]

22/ Plot the graph of y = sin x, taking x from 0° to 180°.   [3]

Use your graph to find both solutions of the equation sin x = 0.6.   [2]

23/ The triangle T is transformed to the triangle
T' by an enlargement of scale factor 3 from the
point (0,0).

(a) Draw the transformed triangle T' on the
same graph.   [2]
(b) Write down the matrix which will
perform this transformation.   [2]

24/ A train accelerates at 2 m/sec² for 15 seconds, then it travels at a constant speed for 60 seconds,
then it brakes to a halt at 3 m/sec².

(a) Draw a graph of the speed against time.   [4]

(b) Find the greatest velocity.   [1]

(c) Find the time taken and the total distance covered.   [3]

Midland Group. Paper 4. (Level 1.)

INFORMATION AND FORMULAE
(Provided on the examination paper)

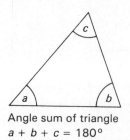

Angle sum of triangle
$a + b + c = 180°$

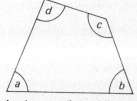

Angle sum of quadrilateral
$a + b + c + d = 360°$

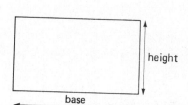

Area of rectangle = base × height

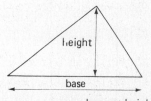

Area of triangle = $\dfrac{\text{base} \times \text{height}}{2}$

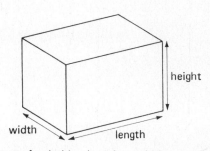

Volume of cuboid = length × width × height

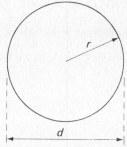

Circumference of circle = $2\pi r$ or πd

Answer all the questions. Time 1^{1}/$_{2}$ hours

1/ A cube is built up out of smaller cubes each of side 1 cm, as shown.

(a) How many smaller cubes are there? [2]

(b) How many cubes are completely hidden? [2]

(c) How many cubes have 3 faces on the outside? [2]

(d) How many cubes have 1 edge on the outside? [2]

(e) How many cubes share a face with 5 other cubes? [2]

2/ The figure shows a pattern of triangles.

(a) Draw the next triangle .[4]

(b) Complete the table below:

Length of side |1|2|3|4|5|
Number of triangles |1|4| | | | [2]

(c) Describe the sequence of the second row. [2]

Find the next two terms of the second row. [2]

3/ The map shown is in the scale of 1 inch to 5 miles.

(a) How far is it from Bradford to Brighouse? [3]

(b) What is the bearing of Dewsbury from Huddersfield? [3]

(c) A car breaks down 5 miles from Huddersfield on the A62. Where is it? [2]

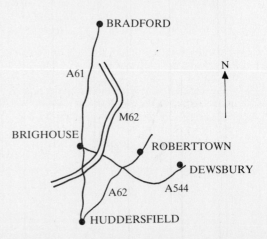

4/ A telephone bill consists of £13.50 standing charge and 2 p per unit call. Mrs Knight makes 93 unit calls: how much is her bill? [2]

Mr Young paid a bill of £25. How many unit calls did he make? [3]

The standing charge is increased by 10%, and the unit call charge increased by 20%. If a bill is £33.50 before, how much would it be after the increase? [3]

5/ (a) 11 pupils score an average of 23 in a test. What was the total of all the marks? [2]

(b) When the marks of 3 more pupils are included the average decreases to 21. What was the total for the new pupils? [4]

6/ 5 schools play a league. 3 points are awarded for a win, 1 for a draw and none for a loss. If two teams have equal points at the top, the winner is the one with the greater difference between goals for and goals against. When there is only one match left to play the table is as shown.

	Played	Won	Drew	lost	goals for	goals against	Points
A	3	0	2	1	1	3	
B	4	1	1	2	3	4	
C	4	2	1	1	5	3	
D	4	2	1	1	3	3	
E	3	1	1	1	5	4	

(a) Fill in the points gained so far. [2]

(b) If A is not to lose the league, what must it get when it plays E? [2]

(c) If E is to win the league, what must it get when it plays A? [3]

7/ The diagram is not to scale. Make an accurate scale drawing of ΔABC. [8]

What is <A? [4]

8/ Damian is a fan of Camford United. He keeps a record of the goals they score and the goals scored against them.

The bar chart shows the goals they score. Fill in the frequency table below. [4]

Number of goals	0	1	2	3	4	5
Frequency						

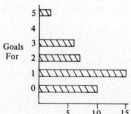

The frequency table below shows the goals against.
Fill in the bar chart. [4]

Number of goals | 0 | 1 | 2 | 3 | 4 | 5 |
Frequency | 14| 12| 8 | 3 | 2 | 1 |

Goals Against (chart y-axis 0–5, x-axis 5, 10, 15)

What was the average number of goals scored against? [2]

What was the average number of goals in a match? [2]

What was the most common number of goals that the team scored? [1]

9/   A taxi firm charges 40 p per mile for the first 10 miles, then 35 p per mile for the next 30 miles, then 30 p per mile thereafter. Fill in the following table. [4]

Miles | 5 | 10 | 20 | 30 | 40 | 50 | 60 | 100 |
Charge |£2 | |£7.5| | |£17.5| | |

Draw a graph of cost against distance, using 1 cm per 10 miles along the x-axis, and 1 cm per £5 along the y-axis. [4]

What is the cost per mile, over (a) 10 miles (b) 100 miles? [3]

How far can I get for £11? [2]

How far, to the nearest mile, can I get for £5? [2]

10/ A local borough charges rates at 185 p in the £.

(a) Mrs Ashley's house has a rateable value of £832. How much does she pay in rates?[2]

(b) Ms Broomfield pays £962 in rates. What is the rateable value of her house? [3]

(c) In the next borough, Mr Cholmondely pays £784 on a house with rateable value £448. What are the rates in his borough? [3]

(d) The total rateable value of the property in a borough is £34 million. The borough needs to raise £52.7 million. What rate should it set? [3]

Midland Group. Paper 5. (Level 2.)

INFORMATION AND FORMULAE
(Provided on the examination paper)

MENSURATION

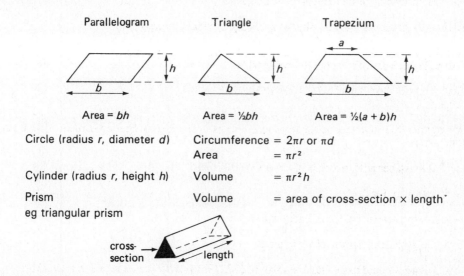

Parallelogram — Area = bh

Triangle — Area = $\frac{1}{2}bh$

Trapezium — Area = $\frac{1}{2}(a + b)h$

Circle (radius r, diameter d) Circumference $= 2\pi r$ or πd
Area $= \pi r^2$

Cylinder (radius r, height h) Volume $= \pi r^2 h$

Prism Volume $=$ area of cross-section $\times$ length
eg triangular prism

cross-section → ← length

TRIGONOMETRY

Right-angled triangle

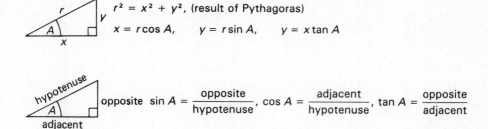

$r^2 = x^2 + y^2$, (result of Pythagoras)

$x = r \cos A$, $y = r \sin A$, $y = x \tan A$

opposite $\sin A = \dfrac{\text{opposite}}{\text{hypotenuse}}$, $\cos A = \dfrac{\text{adjacent}}{\text{hypotenuse}}$, $\tan A = \dfrac{\text{opposite}}{\text{adjacent}}$

NUMBER
Standard form is $a \times 10^n$ where $1 \leqslant a < 10$ and n is an integer.

Answer all the questions in Section A, and 4 questions from Section B

Time 2 hours

Section A

1/ (a) Mr Wilson is paid £3.60 per hour for a 40 hour week. How much is he paid for a week in which he worked an extra 4 hours of overtime at time and a half? [3]

(b) Mr Duncan is paid at the same rate. If he receives £176.40 in a week, how much overtime did he do? [4]

2/ The angle of elevation of a tower is 14°, when I am 300 m from its base. How high is the tower? [3]

If I now walk 100 m towards the tower, what is the new angle of elevation? [2]

How far am I now from the top of the tower? [2]

3/ X and Y are the centres of circles with radii 6 cm and 5 cm respectively. <BAD = 40°.

(a) What is <BDC? [2]

(b) Find AD. [3]

(c) Find <CAD. [3]

4/ Mr Leppard moved house, and received the bill shown. Complete it.

Removal charge	£198		Storage	£12
VAT at 15%	£ 29.70		Vat at 15%	£ (c)
Insurance at 7%	£ (a)		Total	£ (d)
Total	£ (b)			

Grand total £ (e) [9]

5/ An electricity company offers its customers two tariffs.

Tariff A: £25 per quarter and 4 p per unit.
Tariff B: £10 per quarter and 6 p per unit.

Let £PA and £PB be the prices charged when u units of electricity have been used. Find formulae for PA and PB in terms of u. [4]

On the same sheet of paper draw graphs of PA and PB against u, taking u from 0 to 1,000.[4]

At what rate of consumption is it cheaper to pay by Tariff A? [2]

6/ In a racing sport the winner, 2nd and 3rd of each race gain 5 points, 3 points and 1 point respectively. At the end of the season all the points are added up and an overall champion is declared. When there is only one race to be run A and B have 23 and 20 points respectively, and their nearest rival has 10 points.

If B wins the last race and A comes second, what are the final totals? [1]

If A comes third in the last race, what must B do to win the championship? [2]

Write down all the possible results of the last race which will ensure that A wins the championship. [4]

Section B

7/ Fill up the following table for the function $y = x^2 - 2x + 2$.

x	−1	0	$1/2$	1	$1^1/2$	2	3
y							

[3]

Draw the graph of the function, indicating the axis of symmetry. [6]

Solve the equations

$x^2 - 2x + 2 = 2$ [2]

$x^2 - 2x + 2 = 1^1/2$ [2]

8/ The figure shows a regular hexagon ABCDEF. O is the centre. Describe the transformations which take:

(a) ∆ABO to ∆BCO. [2]

(b) ∆AOF to ∆COD. [2]

(c) ∆ABO to ∆DEO. [2]

What is the name of the figure AFEB? [2]

Find on the figure a rhombus and a rectangle. [5]

9/ *Pascal's triangle* is the following triangle of numbers, in which term is obtained by adding the two immediately above.

(a) Find the next two rows. [3]

(b) Find the sum of the terms in each row. [2]

(c) Find the sum of the terms in the tenth row. [2]

(d) How many numbers are there in the n'th row? [2]

(e) What are the first two terms in the n'th row? [2]

(f) What is the sum of the terms in the n'th row? [2]

10/ Jeff thinks a die might be unfair. He throws it 60 times, and records the frequency of each score.

Score| 1 | 2 | 3 | 4 | 5 | 6 |

//// //// //// //// //// ////

//// / //// //// / ////

//// ///

(a) Make up a frequency table for the scores. [2]

(b) Draw a bar-chart for the scores. [5]

(c) Find the average score. [3]

(d) What conclusion would you reach? give *brief* reasons. [3]

11/ (a) A cube has volume 20 cm^3. What is the length of the side? Give your answer to 3 significant figures. [3]

(b) How many cubes of side 1 cm can be cut from the original cube? [3]

(c) A cylinder with base radius 4 cm has the same volume as the original cube. What is its height? [3]

(d) A cylinder with height 2 cm has half the volume of the original cube. What is its base radius? [4]

Midland Group. Paper 6. (Level 3.)
INFORMATION AND FORMULAE
(Provided on the examination paper)

MENSURATION

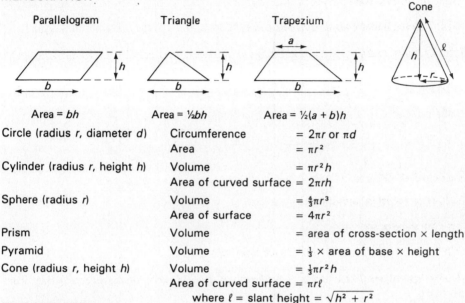

Parallelogram | Triangle | Trapezium | Cone

Area = bh Area = $\frac{1}{2}bh$ Area = $\frac{1}{2}(a + b)h$

Circle (radius r, diameter d)	Circumference	$= 2\pi r$ or πd
	Area	$= \pi r^2$
Cylinder (radius r, height h)	Volume	$= \pi r^2 h$
	Area of curved surface	$= 2\pi rh$
Sphere (radius r)	Volume	$= \frac{4}{3}\pi r^3$
	Area of surface	$= 4\pi r^2$
Prism	Volume	$=$ area of cross-section $\times$ length
Pyramid	Volume	$= \frac{1}{3} \times$ area of base $\times$ height
Cone (radius r, height h)	Volume	$= \frac{1}{3}\pi r^2 h$
	Area of curved surface	$= \pi r\ell$

where ℓ = slant height = $\sqrt{h^2 + r^2}$

TRIGONOMETRY

Right-angled triangle

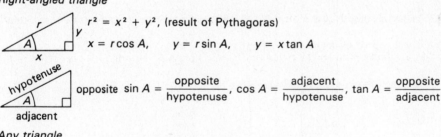

$r^2 = x^2 + y^2$, (result of Pythagoras)

$x = r\cos A,\qquad y = r\sin A,\qquad y = x\tan A$

$\text{opposite } \sin A = \dfrac{\text{opposite}}{\text{hypotenuse}},\ \cos A = \dfrac{\text{adjacent}}{\text{hypotenuse}},\ \tan A = \dfrac{\text{opposite}}{\text{adjacent}}$

Any triangle

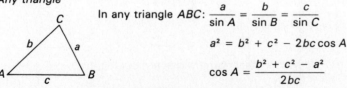

In any triangle ABC: $\dfrac{a}{\sin A} = \dfrac{b}{\sin B} = \dfrac{c}{\sin C}$

$$a^2 = b^2 + c^2 - 2bc\cos A$$

$$\cos A = \frac{b^2 + c^2 - a^2}{2bc}$$

Area of triangle $ABC = \frac{1}{2}ab\sin C$

NUMBER   Standard form is $a \times 10^n$ where $1 \leqslant a < 10$ and n is an integer.

Compound Interest formula $A = P\left(1 + \dfrac{r}{100}\right)^n$

ALGEBRA   The quadratic equation $ax^2 + bx + c = 0$ has solutions

$$x = \frac{-b \pm \sqrt{b^2 - 4ac}}{2a}$$

327

Time 2^{1}/2 hours. Answer all the questions in Section A, and 4 of the questions in Section B.

Section A

1/ My car does 30 miles to the gallon, and I must drive 90 miles. A gallon of petrol costs £1.70, and I have £6.

Let x be the number of gallons I buy. Find 2 inequalities in x. [6]

Find the integer values of x which satisfy both inequalities. [3]

2/ (a) The internal angle of a regular polygon is 120° greater than the external angle. Find the number of sides of the polygon. [4]

(b) A point is 10 cm away from the centre of a circle with radius 3 cm. Find the length of the tangent from the point to the circle, giving your answer to 3 significant figures. [4]

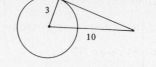

3/ When Mr Dixon leaves for work in the morning, he takes his umbrella with probability $^{1}/_{4}$ if it is fine, and with probability $^{7}/_{8}$ if it is wet. The probability that a day is fine is $^{2}/_{3}$. Find the probabilities that:

(a) It is fine and he takes his umbrella. [2]

(b) He takes his umbrella. [2]

(c) It is wet or he does not take his umbrella. [2]

4/ The graph shows the level of water in a rain-barrel.

(a) When did it start raining? [1]

(b) Draw a tangent to the curve at the point where it was raining most heavily. What was the flow in cm/hour? [4]

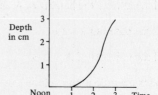

(c) At 3 o'clock a plug in the bottom of the barrel was pulled out, and the barrel drained in 10 minutes. Extend the graph to show this. [3]

5/   (a) Lorna tells Martha to think of a nur.ber, then to add 5, then to square it. Martha's answer is 1. What are the two possible values for Martha's original number?   [4]

(b) Solve the equation $x^2 + 3x - 2 = 0$, giving your answers to 3 decimal places.   [4]

6/   The triangle ABC is at (-1,1), (1,1), (1,3). Draw this triangle on graph paper, and plot its image after transformation by the matrix A, where:

$$A = \begin{pmatrix} 0 & -1 \\ 1 & 0 \end{pmatrix}$$

[5]

Describe the action of this matrix.   [2]

What point is taken to (-1,1) by A?   [2]

Section B

7/   5 schools play each other football in a league. 2 points are awarded for a win, 1 for a draw and 0 for a defeat. The results of the first 8 matches are:

A beat B: C drew with D: E drew with A: A beat D: B beat C: E drew with D: E beat C: A beat C.

After 10 games B had 2 points. What were the results of the last 2 games?   [3]

Fill in the points table below:

```
Team  |A|B|C|D|E|
Points|  |  |  |  |  |
```
[3]

What is the total number of points?   [1]

If 6 schools play each other, how many games will there be?   [3]

If 7 schools play each other, what will be the total number of points?   [3]

8/   A cone C is of height 12 cm, and its base radius is 6 cm. A cut is made 4 cm from the top. The top cone is A, and the lower frustrum is B.

(a) Find the ratio of the heights of A and C.   [2]

(b) Find the base radius of A.   [3]

(c) Find the volumes of A and B.   [4]

(d) What is the ratio of the volumes of A and B?   [2]

(e) What is the ratio of the curved surface areas of A and B?   [2]

9/   Complete the following table for the function $y = x^2 - 2/x$

```
x |1/4 |1/2 |3/4 |1 |11/2|2 |3 |
y |    |    |    |  |    |  |  |
```
[3]

329

Draw the curve, using 2 cm per unit along the x-axis and 1 cm per 2 units along the y-axis. [5]

Solve the equation $x^2 - {}^2/x = 1$. [2]

By drawing an appropriate line, solve the equation:

$$x^2 - {}^2/x = 1 - x.$$ [3]

10/ Richard walks 10 miles North. Then he turns through 140°, and walks a further 11 miles.

(a) How far is he from his starting point? [5]

(b) What is his bearing from his starting point? [4]

(c) He now returns directly to his starting point. What is the area of the triangle enclosed by the three parts of his journey? [4]

11/ In $\triangle$ ABC, $\underline{AB} = \mathbf{b}$ and $\underline{AC} = \mathbf{c}$. X is the midpoint of BC.

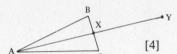

(a) Find $\underline{BC}$, $\underline{BX}$, $\underline{AX}$ in terms of $\mathbf{b}$ and $\mathbf{c}$. [4]

(b) AX is extended to Y, where AY = 2AX. Find $\underline{AY}$, $\underline{BY}$, $\underline{CY}$ in terms of $\mathbf{b}$ and $\mathbf{c}$. [4]

(c) What sort of quadrilateral is ABYC? [2]

(d) $\triangle$ ABC has area 16. What is the area of ABYC? [3]

12/ The heights of 100 girls were given by the following table:

Height in cm|120-130|130-140|140-150|150-160|
Frequency | 12 | 33 | 38 | 17 |

(a) Find the cumulative frequencies.[2]

(b) Draw a cumulative frequency graph, using a scale of 2 cm for 20 cm along the x-axis, and 8 cm for the 100 girls along the y-axis. [5]

(c) Find the median and the interquartile range. [4]

(d) What proportion of girls were taller than 145 cm? [2]

Northern Group. Paper 1. (Level 1.)

INFORMATION AND FORMULAE
(Provided on the examination paper)

Mensuration

Area of a trapezium $= \frac{1}{2} \times$ (sum of parallel sides) $\times$ (distance between)

Volume of a prism $=$ (area of cross section) $\times$ (height)

Volume of a pyramid $= \frac{1}{3} \times$ (area of base) $\times$ (height)

Circumference of a circle $= \pi \times$ (diameter)

Area of a circle $= \pi \times$ (radius)^2

Length of a circular arc $= \dfrac{\theta}{360} \times$ (circumference)

Area of a sector of a circle $= \dfrac{\theta}{360} \times$ (area of circle)

Volume of a cylinder $= \pi \times$ (radius)$^2 \times$ (height)

Curved surface area of a cylinder $=$ (circumference) $\times$ (height)

Volume of a cone $= \frac{1}{3} \times$ (area of base) $\times$ (height)

Curved surface area of a cone $= \pi \times$ (radius of base) $\times$ (slant height)

Volume of a sphere $= \frac{4}{3}\pi \times$ (radius)^3

Surface area of a sphere $= 4\pi \times$ (radius)^2

Algebra

$$a^{-m} = \frac{1}{a^m} \qquad a^{1/m} = \sqrt[m]{a}.$$

Standard form $a \times 10^n$, $1 \leqslant a < 10$, n an integer.

Trigonometry

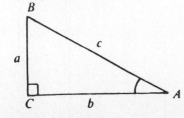

$$\cos A = \frac{b}{c}, \quad \sin A = \frac{a}{c}, \quad \tan A = \frac{a}{b}$$

Matrices and transformations

Inverse under multiplication of $\begin{pmatrix} a & b \\ c & d \end{pmatrix}$ is $\dfrac{1}{(ad-bc)} \begin{pmatrix} d & -b \\ -c & a \end{pmatrix}$.

The image (x', y') of the point (x, y) under the transformation given by

$\begin{pmatrix} a & b \\ c & d \end{pmatrix}$ is given by $\begin{pmatrix} x' \\ y' \end{pmatrix} = \begin{pmatrix} a & b \\ c & d \end{pmatrix} \begin{pmatrix} x \\ y \end{pmatrix}$.

Functions

For two functions $f(x)$, $g(x)$

$$fg(x) = f(g(x)).$$

Statistics

Probability of an event for equally likely outcomes is

$$\frac{\text{number of favourable outcomes.}}{\text{total possible outcomes}}$$

Mean of a set of N numbers $= \dfrac{\Sigma x}{N}$.

Mean of a frequency distribution $= \dfrac{\Sigma fx}{\Sigma f}$.

Mean of a grouped interval frequency distribution $= \dfrac{\Sigma fx}{\Sigma f}$, where x is the

mid-interval value.

Answer all the questions. Time 1^{1}/2 hours

1/ George has a cassette which lasts for 30 minutes. He records a 26 minute piece of music. How much is there left? [1]

2/ A gannet is flying 50 feet above the water. It spots a fish and dives 10 feet below the surface. How far has it fallen? [1]

3/ The figure shows three lines meeting at a point. What is the angle labelled x? [1]

4/ What fraction of the shape is shaded? [1]

5/ From Ayville to Bridgetown is 60 miles. How long does it take at 40 m.p.h.? [1]

6/ (a) Write two thousand four hundred and one in figures. [1]

(b) What is the number before 3,500? [1]

7/ Fill in the boxes for the equation:

$$3^2 + 2^2 = 9 + \square = \square$$ [2]

8/ A motor-bike uses petrol and oil in the ratio 16:1. How much oil should be added to 12 litres of petrol? [2]

9/ A tea-pot holds 1,200 c.c., and a mug holds 350 c.c. How many mugs can be filled from the pot, and how much is left over? [2]

10/ Find the distance between junctions 32 and 35 on the motorway map below .[1]

```
—(32)———(S)—(33)——(34)——(35)—
     11      2      6      4
```

If the journey takes ^{1}/2 hour, what is the speed? [1]

11/ The rules of a tennis club are that at least ^{2}/5 of the members must be women. If there are 150 men, what is the least possible number of women? [2]

12/ How much will it cost Jane and her two children to join the squash club for one year? [2]

13/ The pattern below uses rotations, reflections and translations. Complete the pattern. [2]

14/ X is given in terms of Y by $X = 7 - 3Y$.

(a) Find X when $Y = 2$. [1]

(b) Find X when $Y = -3$. [1]

15/ The radius of a circular running track is 70 m. Diane runs round 4 times. How far has she run? (Take π to be $^{22}/_7$.) [2]

16/ The television programmes for Wednesday 4th October are shown.

10.30	International Hockey
11.25	Film: "Revenge of the Sea Troll"
12.47	Closedown

(a) How long is the film? [1]

(b) Nancy watches only the first 43 minutes of the film. When does she switch off? [1]

(c) The film is repeated on the following Tuesday. What will the date be then? [1]

17/ Find x in the following sequences:

(a) 1, 2, 4, x, 16. [1]

(b) 7, 3, -1, x, -9, -13. [2]

18/ For a journey 20 litres of petrol at 43.2 p per litre are used. Find the cost of the petrol. [1]

The driver pays half the cost, and the two passengers share the rest of the cost equally. How much does each passenger pay? [1]

19/ ABC is a triangle, in which AB = 8 cm and BC = 7 cm.
What is the upper limit for AC? [1]

334

20/ An office block is five stories high, and each floor has nine rooms in a 3 by 3 square.

   (a) How many rooms have no windows? [2]

   (b) How many rooms have exactly 2 outside walls? [2]

21/ Measure the angles of the triangle ABC. [3]

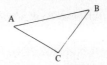

What is the sum of the angles? [1]

22/ On the dotted paper, complete the net of the pyramid. [3]

   If the dots are 1 cm apart, what is the base area of the pyramid? [1]

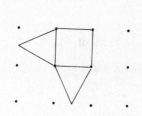

23/ Find the bearing of Nakuru from Nairobi. [2]

   If the scale of the map is 1 cm for 50 km, what is the distance between the towns? [2]

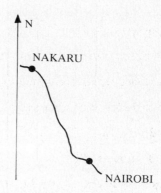

24/   (a) What is the cost of 5 records at £3.98 each?   [2]

      (b) What would the price be if a discount of 10% is given?   [2]

25/   The dwellings in Eckstown are classified by
the pie-chart shown.

      (a) How many dwellings are there?   [2]

      (b) What is the angle for the detached
houses sector?   [2]

      (c) How many flats are there?   [2]

26/   Two dice are rolled. Complete the table for the total score.

```
            first die | 1 | 2 | 3 | 4 | 5 | 6 |
               1      | 2 | 3 | 4 |   |   | 7 |
               2      | 3 | 4 |   |   |   |   |
Second die     3      | 4 |   |   |   |   |   |
               4      |   |   | 7 |   |   |   |
               5      |   | 7 |   |   | 10|   |
               6      |   | 8 |   |   | 11|   |
```
                                                     [4]

What is the probability that the total score is 12?   [2]

What is the most likely total, and what is its probability?   [3]

27/   Bertram is a bowler for his school team. In 20 matches he takes the following number of wickets:

    2, 1, 0, 3, 2, 2, 1, 3, 1, 0, 2, 0, 1, 4, 0, 1, 2, 1, 2, 2.

Complete the following frequency table   .[3]

```
Number of wickets| 0 | 1 | 2 | 3 | 4 |
Frequency        |   |   |   |   |   |
```

Complete the bar chart.   [3]

What was the average number of wickets that he took?   [2]

What was his most common total of wickets in a match?[1]

Northern Group. Paper 2. (Levels 1 & 2.)

INFORMATION AND FORMULAE
(Provided on the examination paper)

Mensuration

Area of a trapezium $= \frac{1}{2} \times$ (sum of parallel sides) $\times$ (distance between)

Volume of a prism $=$ (area of cross section) $\times$ (height)

Volume of a pyramid $= \frac{1}{3} \times$ (area of base) $\times$ (height)

Circumference of a circle $= \pi \times$ (diameter)

Area of a circle $= \pi \times$ (radius)^2

Length of a circular arc $= \frac{\theta}{360} \times$ (circumference)

Area of a sector of a circle $= \frac{\theta}{360} \times$ (area of circle)

Volume of a cylinder $= \pi \times$ (radius)$^2 \times$ (height)

Curved surface area of a cylinder $=$ (circumference) $\times$ (height)

Volume of a cone $= \frac{1}{3} \times$ (area of base) $\times$ (height)

Curved surface area of a cone $= \pi \times$ (radius of base) $\times$ (slant height)

Volume of a sphere $= \frac{4}{3}\pi \times$ (radius)^3

Surface area of a sphere $- 4\pi \times$ (radius)^2

Algebra

$$a^{-m} = \frac{1}{a^m} \qquad a^{1/m} = \sqrt[m]{a}.$$

Standard form $a \times 10^n, \quad 1 \leqslant a < 10, \quad n$ an integer.

Trigonometry

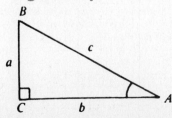

$$\cos A - \frac{b}{c}, \quad \sin A - \frac{a}{c}, \quad \tan A - \frac{a}{b}$$

Matrices and transformations

Inverse under multiplication of $\begin{pmatrix} a & b \\ c & d \end{pmatrix}$ is $\dfrac{1}{(ad-bc)} \begin{pmatrix} d & -b \\ -c & a \end{pmatrix}$.

The image (x', y') of the point (x, y) under the transformation given by

$\begin{pmatrix} a & b \\ c & d \end{pmatrix}$ is given by $\begin{pmatrix} x' \\ y' \end{pmatrix} = \begin{pmatrix} a & b \\ c & d \end{pmatrix} \begin{pmatrix} x \\ y \end{pmatrix}$.

Functions

For two functions $f(x)$, $g(x)$

$$fg(x) = f(g(x)).$$

Statistics

Probability of an event for equally likely outcomes is

$$\frac{\text{number of favourable outcomes.}}{\text{total possible outcomes}}$$

Mean of a set of N numbers $= \dfrac{\Sigma x}{N}$.

Mean of a frequency distribution $= \dfrac{\Sigma fx}{\Sigma f}$.

Mean of a grouped interval frequency distribution $= \dfrac{\Sigma fx}{\Sigma f}$, where x is the

mid-interval value.

Answer all the questions. Time 1^1/$_2$ hours

1/   A stack of coins is 7 cm high, and each coin is ^{1}/$_5$ cm thick. How many coins are there?[2]

2/   A typist has a speed of 30 words per minute. How long will it take her to type 11 pages with an average of 360 words per page?   [2]

3/   The emperor Augustus died in 14 AD at the age of 77. When was he born?   [2]

4/   A holiday was advertised as costing £320. The pound was devalued and a surcharge of 10% was added. What is the new cost?   [2]

5/   How can 82 p be made up from 12 p and 17 p stamps?   [2]

6/   Henry has two jackets and one pair of trousers and one suit to be cleaned. How much will it cost him?   [2]

Jackets	£2.00
Trousers	£1.50
Suits	£3.20

7/   Helen saves £120 in a building society at 8%. How much does she have after one year?[2]

8/   A sprinter completes 100 metres in 15 seconds. What is his speed in km per hour?[3]

9/   A new computer is priced at £200. One shop sells it at a 20% discount, another sells it at a reduction of £35. What is the difference between the two prices?   [3]

10/   After her holiday Debbie has 2 rolls of 24 exposures and 1 roll of 36 exposures. How much will it cost her to have them processed?   [2]

How much change will she get from £10?   [1]

Processing cost per roll	
24 Exposures	£2.70
36 Exposures	£3.50

11/   The King in chess can move one square in any direction. How many squares can the King move to in position (a) in the centre, or in position (b) in a corner.   [4]

Mark on the board a position from which the King can move to exactly 5 squares.   [2]

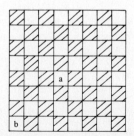

12/ Nick writes down some equations. He forgets to put brackets in. Insert brackets to make the equations correct.

(a) 15 - 3 + 2 = 10 [2] (b) 4 x 2 + 1 = 12 [2]

13/ The bill is for a meal in a Chinese restaurant.

2 Soups	95p each
Sweet & Sour Pork	£2.20
Chicken in Black Bean Sauce	£1.90
1 Boiled Rice	50p
2 Tea	30p each

(a) What is the total? [2]

(b) If 10% is added for service, what is the total cost? [2]

14/ What solid is formed from the net shown? [1]

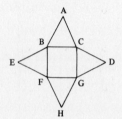

How many edges does it have? [2]

Which points will be joined with A? [2]

15/ The chart shown converts gallons to litres.

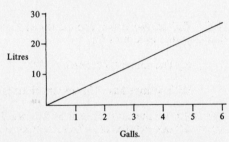

(a) How many litres are there in 5.2 gallons? [2]

(b) How many gallons are there in 10.3 litres? [2]

16/ (a) A room is 2 m high, 3¹/2 m long and 3 m wide. There is a window which is 1 m by 1¹/2 m. What is the area of walls and ceiling? [3]

(b) One litre of paint will cover a surface of 20 m². How much paint is needed for two coats of paint? [2]

(c) How much will the paint cost at £3.50 per litre? [1]

(d) What is the cost of the paint per square metre? [2]

17/ The dots shown are 1 cm apart. Draw a net for a cuboid which is 2 cm by 2 cm by 1 cm.[4]

What is the volume of the cuboid?[1]

What is its surface area?   [2]

18/ Zoe has twice as much money as Yvonne. Yvonne has £20 more than Xanthippe. If Xanthippe has £x, write in terms of x:

(a) The amount Yvonne has [2]   (b) The amount Zoe has. [3]

If Xanthippe has £15, how much does Zoe have?   [2]

19/ An income tax scheme is as follows. The first £3,000 of income is tax free. Income over £3,000 up to £20,000 is taxed at 30%. Income over £20,000 is taxed at 40%.

Find the tax on incomes of £2,900, £7,500, £15,000, £27,000.   [4]

Using a scale of 1 cm per £5,000, draw a graph of tax against income.   [4]

20/ Mr Jones will cash cheques for you, taking a commission of £5. Mr McKay will also cash cheques, but he gives 85 p per £. Complete the following table: [3]

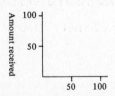

Amount of cheque	£10	£30	£50	£70
Mr Jones pays	£5			£65
Mr McKay pays	£8.50		£42.5	

Draw lines on the graph to show how much money the two dealers give.   [3]

For what size of cheque do they give the same amount?   [2]

21/ Building work of 144 hours needs to be done.
     (a) How long will it take 3 men, working an 8 hour day? [1]

     (b) If they are paid £3.20 per hour each, how much will it cost? [2]

     (c) The work must be done in 5 days. How many hours of overtime must be worked? [2]

     (d) If overtime is paid at time and a half, how much will the work now cost? [3]

22/ Jane walks for 3 km in a straight line. She
then turns through 150° and walks a further 4 km.
Make a diagram of her journey, using a scale of 1
cm per km.   [4]

     How far is she from her starting point? [3]

     If the first part of her journey was due
North, what is her bearing from her starting point?
                                              [3]

Northern Group. Paper 3. (Levels 2 & 3.)

INFORMATION AND FORMULAE
(Provided on the examination paper)

Mensuration

Area of a trapezium $= \frac{1}{2} \times$ (sum of parallel sides) $\times$ (distance between)

Volume of a prism $=$ (area of cross section) $\times$ (height)

Volume of a pyramid $= \frac{1}{3} \times$ (area of base) $\times$ (height)

Circumference of a circle $= \pi \times$ (diameter)

Area of a circle $= \pi \times$ (radius)^2

Length of a circular arc $= \dfrac{\theta}{360} \times$ (circumference)

Area of a sector of a circle $= \dfrac{\theta}{360} \times$ (area of circle)

Volume of a cylinder $= \pi \times$ (radius)$^2 \times$ (height)

Curved surface area of a cylinder $=$ (circumference) $\times$ (height)

Volume of a cone $= \frac{1}{3} \times$ (area of base) $\times$ (height)

Curved surface area of a cone $= \pi \times$ (radius of base) $\times$ (slant height)

Volume of a sphere $= \frac{4}{3}\pi \times$ (radius)^3

Surface area of a sphere $= 4\pi \times$ (radius)^2

Algebra

$$a^{-m} = \frac{1}{a^m} \qquad a^{1/m} = {}^m\!\sqrt{a}.$$

Standard form $a \times 10^n$, $1 \leqslant a < 10$, n an integer.

Trigonometry

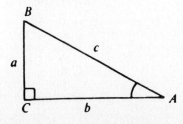

$$\cos A = \frac{b}{c}, \quad \sin A = \frac{a}{c}, \quad \tan A = \frac{a}{b}$$

Matrices and transformations

Inverse under multiplication of $\begin{pmatrix} a & b \\ c & d \end{pmatrix}$ is $\dfrac{1}{(ad-bc)} \begin{pmatrix} d & -b \\ -c & a \end{pmatrix}$.

The image (x', y') of the point (x, y) under the transformation given by

$\begin{pmatrix} a & b \\ c & d \end{pmatrix}$ is given by $\begin{pmatrix} x' \\ y' \end{pmatrix} = \begin{pmatrix} a & b \\ c & d \end{pmatrix} \begin{pmatrix} x \\ y \end{pmatrix}$.

Functions

For two functions $f(x)$, $g(x)$

$$fg(x) = f(g(x)).$$

Statistics

Probability of an event for equally likely outcomes is

$$\frac{\text{number of favourable outcomes.}}{\text{total possible outcomes}}$$

Mean of a set of N numbers $= \dfrac{\Sigma x}{N}$.

Mean of a frequency distribution $= \dfrac{\Sigma fx}{\Sigma f}$.

Mean of a grouped interval frequency distribution $= \dfrac{\Sigma fx}{\Sigma f}$, where x is the mid-interval value.

Answer all the questions. Time 2 hours

1/ What is the cost of 15 metres of string at 23 p per metre? [1]

2/ O is the centre of the circle shown and XT and XR are both tangents. If <X = 50° find <O. [2]

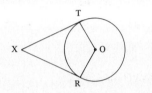

3/ Solve the equation $3x - 7 = 20$. [2]

4/ An aunt has 2 nephews and 1 niece. She leaves them money in the ratio 2:3:4. If she leaves £2,700, how much is each share? [3]

5/ (a) The area of a square table is 1.69 m². What is the side of the table? [2]

(b) A round table has the same area. What, to 3 decimal places, is the radius of the table?[3]

6/ The weight of a hydrogen atom is 1.7×10^{-23} grams. How many atoms are there in 85 grams of hydrogen? Leave your answer in standard form. [3]

7/ X is the centre of the cube shown. If X is joined to A, B, C, D, what is the name of the solid formed? [2]

How many such solids will make up the whole cube? [2]

8/ The figure shows a mountain which is 2,000 feet high and 10,000 feet across.

(a) What is the angle of the slope? [2]

(b) How far is it to the top? [2]

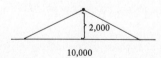

9/ Write the following as integers:

(a) 3^2 (b) 2^{10} (c) $(1/4)^{-1}$ [3]

345

10/ A shop is open for the hours shown.

(a) For how many hours is it open? [3]

(b) A shop assistant is paid £2.30 per hour, and time and a half for Saturdays. How much does she receive per week? [2]

MON	8.30–1	2–5.30
TUE	8.30–1	2–5.30
WED	8.30–1	2–5.30
THUR	8.30–1	Closed
FRI	8.30–1	2–5.30
SAT	9–12	Closed

11/ In a survey of 50 cars it was found that 30 had radios, 25 had clocks and 17 had neither. How many had both? [3]

12/ A population is 15 million and is increasing at $1\frac{1}{2}\%$. How many people will there be:

(a) In one years time? [1] (b) In two years time [2]

13/ Describe the transformation which takes ABCD to APQR. [2]

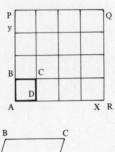

Give the co-ordinates of the point which is taken to (2,0). [2]

14/ (a) ABCD is a parallelogram. Write down two pairs of equal vectors. [2]

(b) PQRS is a quadrilateral in which $\underline{PQ} = 2\underline{SR}$. What Is the name for PQRS? [2]

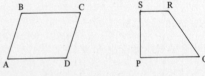

15/ John walks North from his home. The graph shows his distance from home against time. What is his speed? [1]

A quarter of an hour later Jane sets off, walking at 4 m.p.h. Draw a line on the graph to show her distance from home. [3]

When does she catch up with him? [3]

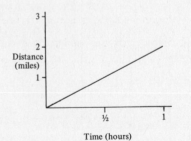

16/ Dawn goes through her books classifying them. The results are shown in the tally table below.

Romance Crime Science Fiction Historical

⊬⊦	///	⊬⊦	//
//		⊬⊦	

(a) Draw up a frequency table. [2]

(b) Draw a bar chart. [3]

17/ The triangle S is reflected in y = 2 to S'. Draw S' on the graph. [2]

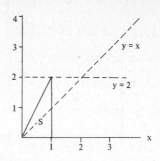

S is reflected in x = y to S". Draw S" on the graph. [2]

What transformation would take S' to S? [2]

18/ What is the distance between the points (1,1) and (5,6)? [2]

What angle does the line joining these points make with the x-axis? [2]

19/ Make an accurate drawing of a triangle ABC with AB = 7 cm, BC = 8 cm, CA = 9 cm. [3]

Construct the perpendicular bisectors of AB and BC. Let them cross at G. Measure the length BG. [3]

20/ A coin is tossed 3 times. Complete the following list of the possible results.

   HHH   HHT   ---   ---
   ---   ---   ---   TTT.   [3]

What are the probabilities of:

(a) All tails. [1]

(b) Exactly one head. [2]

(c) At least one head. [2]

21/ Malcolm walks from home, arriving at his friend's house 2 miles away ¹/2 hour later. He stays for ¹/4 hour, then walks home in ³/4 hour. Complete the travel graph to show his journey. [3]

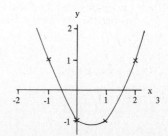

22/ The figure shows the graph of $y = x^2 - x - 1$.

(a) Draw the line of symmetry of the graph. [2]

(b) Draw a tangent to the curve at x = 2 and find its gradient. [4]

(c) Find the solutions of the equation $x^2 - x - 1 = 0$ [4]

347

23/ Certain gold coins are $^1/_8$ inch thick and $^3/_4$ inch in diameter. What is the volume of each coin? [2]

How many of these coins could be made from 2 cubic inches of gold? How much would be left over? [3]

100 of these coins are melted down and recast as a cube. What is the side of the cube? [2]

One cubic inch of gold weighs 11.6 ounces. What is the weight of 50 coins? [2]

Gold costs £300 per ounce. A forger buys gold to make 50 coins, which he then sells for £240 each. How much profit does he make? [4]

Northern Group. Paper 4. (Level 3.)

INFORMATION AND FORMULAE
(Provided on the examination paper)

Mensuration

Area of a trapezium   = $\frac{1}{2}$ × (sum of parallel sides) × (distance between)

Volume of a prism   = (area of cross section) × (height)

Volume of a pyramid = $\frac{1}{3}$ × (area of base) × (height)

Circumference of a circle   = π × (diameter)

Area of a circle   = π × (radius)²

Length of a circular arc   = $\dfrac{\theta}{360}$ × (circumference)

Area of a sector of a circle = $\dfrac{\theta}{360}$ × (area of circle)

Volume of a cylinder   = π × (radius)² × (height)

Curved surface area of a cylinder = (circumference) × (height)

Volume of a cone   = $\frac{1}{3}$ × (area of base) × (height)

Curved surface area of a cone   = π × (radius of base) × (slant height)

Volume of a sphere   = $\frac{4}{3}\pi$ × (radius)³

Surface area of a sphere – 4π × (radius)²

Algebra

$$a^{-m} = \frac{1}{a^m} \qquad a^{1/m} = \sqrt[m]{a}.$$

Standard form $a \times 10^n$, $1 \leqslant a < 10$, n an integer.

Trigonometry

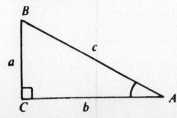

$$\cos A - \frac{b}{c}, \quad \sin A - \frac{a}{c}, \quad \tan A - \frac{a}{b}$$

Matrices and transformations

Inverse under multiplication of $\begin{pmatrix} a & b \\ c & d \end{pmatrix}$ is $\dfrac{1}{(ad-bc)}\begin{pmatrix} d & -b \\ -c & a \end{pmatrix}$.

The image (x', y') of the point (x, y) under the transformation given by

$$\begin{pmatrix} a & b \\ c & d \end{pmatrix} \text{ is given by } \begin{pmatrix} x' \\ y' \end{pmatrix} = \begin{pmatrix} a & b \\ c & d \end{pmatrix}\begin{pmatrix} x \\ y \end{pmatrix}.$$

Functions

For two functions $f(x)$, $g(x)$

$$fg(x) = f(g(x)).$$

Statistics

Probability of an event for equally likely outcomes is

$$\frac{\text{number of favourable outcomes}}{\text{total possible outcomes}}$$

Mean of a set of N numbers $= \dfrac{\Sigma x}{N}$.

Mean of a frequency distribution $= \dfrac{\Sigma fx}{\Sigma f}$.

Mean of a grouped interval frequency distribution $= \dfrac{\Sigma fx}{\Sigma f}$, where x is the

mid-interval value.

Time 2 ¹/₂ hours. Answer all the questions.

1/   A quarterly telephone bill consists of a standing charge of £12 and x p per unit call. Mr Yates's bill was £18.05, after making 275 unit calls. Find x.   [2]

2/   The figure shows an equilateral triangle on top of a square. Find <BDA.   [2]

3/   (a) Factorize $x^2 - 3x - 70$.   [2]

(b) Solve the equation $(x - 1)(x - 2) = 72$.   [3]

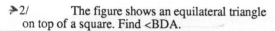

4/   A journey is made up of two stages A to B and B to C as shown. A driver takes ³/₄ hour for the first stage. How fast must he drive in the second stage to ensure that his average speed for the whole journey is 60 m.p.h.?   [2]

```
        30 miles              50 miles
    |-------------+-----------------------|
    A             B                       C
```

5/   A pond is a cuboid 50 cm wide by 70 cm long by 12 cm deep. During a cold spell the top 1 cm freezes.

(a) What volume is unfrozen?   [1]

(b) If the relative density of ice is 0.9, what is the volume of the ice?   [2]

6/   (a) Find the inverse of the matrix A defined below.   [2]

(b) Hence solve the equation AC = B, where B is defined below.

$$A = \begin{pmatrix} 2 & 2 \\ 1 & 2 \end{pmatrix} \qquad B = \begin{pmatrix} 2 & 1 \\ 1 & 7 \end{pmatrix}$$ [2]

7/   Below are 4 graphs and 4 equations. Which graph fits which equation?   [3]

(a) $y = 3 - x^2$ (b) $y = x + 1$ (c) $y = x^2 + 1$ (d) $y = 1 - x$

(i)   (ii)   (iii)   (iv)

  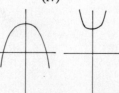

8/ A sheet of cardboard is 10 cm by 20 cm. Squares of side x cm are cut out from each corner and the sheet is folded along the dotted line to make a tray. Find in terms of x:

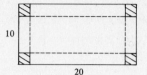

(a) The area of cardboard. [2]

(b) The base area of the tray. [2]

(c) The volume of the tray. [2]

9/ The shaded region is bounded by the lines x = 1, y = 2, 2x + 3y = 12. Describe the shaded region in terms of 3 inequalities. [3]

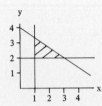

(10) In the triangle OAB, $\underline{OA}$ = **a**, $\underline{OB}$ = **b** and OX = $^{OA}/3$.

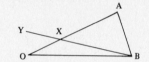

(a) Find $\underline{OX}$, $\underline{BA}$ and $\underline{BX}$ in terms of **a** and **b**. [3]

(b) BX is extended to Y, so that XY = $^1/2$BX. Find $\underline{OY}$ and $\underline{YA}$ in terms of **a** and **b**. [3]

(c) What can you say about OY and BA? [2]

(11) One third of the tickets in a tombola are blank, and two thirds win prizes. Ann, Bert and Charles each buy a ticket. Find the probabilities that:

(a) Ann wins a prize. [1]

(b) Ann and Bert both lose. [2]

(c) Charles is the only one to lose. [3]

(12) Let f:x → 3x + 1 and g: x → 1 - 2x.

(a) Find f(2), g(3), fg(2), gf(2). [4]

(b) Express fg in the form fg: x → .... [2]

(c) What number is sent to 5 by f? [2]

(d) If g(x) = y, express x in terms of y. [3]

(13) A cone is made out of paper. It is 12 cm high and the base radius is 5 cm. Find the slant height AX.   [2]

The cone is now cut along AX and the paper is laid flat as the sector of a circle.

Find the length of the curved arc of the sector.   [2]

Find the angle of the sector.   [2]

Find the area of the paper.   [2]

(14) (a) Express $9^{3/2}$ as an integer.   [2]

(b) Solve the equation $2^{3x+1} = 4^{x+2}$.   [2]

(c) Express as a single power of x: $x^2 \times x^3 \div x^6$.   [2]

(15) Describe, in set notation, the relationships between the sets S and T in the diagrams shown.   [4]

(16) ABCD is a rectangle 6 cm by 4 cm. P = {points nearer BC than AD}, and Q = {points within 3 cm of D}.

Make an accurate scale drawing of ABCD, and of the sets P and Q. Shade the region P∪Q.[4]

(17) The triangle ABC is at (1,1), (3,1), (3,2). The matrix M is given by:

$$M = \begin{pmatrix} 0.6 & 0.8 \\ -0.8 & 0.6 \end{pmatrix}$$

(a) Plot ΔABC on graph paper.   [2]

(b) Apply M to the triangle, and plot the transformed triangle.   [3]

(c) Describe the transformation, as fully as you can.   [2]

(d) Describe the inverse transformation. What point will M take to A?   [4]

(18) A ship is slowing down. The values of its speed against time are given by the following table:

Time in minutes	0	1	2	3	4	5
Speed in m/sec	8	7.1	6.2	5.0	4.9	4.1

(a) Plot these points on a graph, and draw a straight line through them.   [2]

(b) Find the equation of your straight line.   [2]

353

(c) When will the ship come to a halt? [2]

(d) How far has the ship travelled over these 5 minutes? [2]

(19) The matrix P below gives the prices of honey and jam in 3 shops A, B, C.

$$
\begin{array}{c}
 \begin{array}{ccc} A & B & C \end{array} \\
\begin{array}{c} \text{Honey} \\ \text{Jam} \end{array}
\begin{pmatrix} 89 & 95 & 85 \\ 43 & 52 & 41 \end{pmatrix}
\end{array}
$$

(a) Mr Smith wishes to buy 2 jars of honey and 3 of jam. Write down a matrix product which will show how much he has to pay in each of the shops. Evaluate the product. [4]

(b) Find a matrix product which will give the average price of each of the two commodities. [3]

(20) An exam is taken by 200 students. The results are shown in the following table.

Mark	0-29	30-39	40-49	50-54	55-59	60-69	70-79	80-100
Frequency	30	24	35	25	29	21	14	22

(a) Plot a histogram for these frequencies. [5]

(b) Extimate the mean mark. [2]

(c) Find the cumulative frequencies. [2]

(d) Plot a cumulative frequency curve. [4]

(e) Find the median and the quartiles. [3]

(f) 70% of the candidates passed. What was the pass-mark? [2]

Southern Group. Paper 1. (Level 1.)

INFORMATION AND FORMULAE
(Provided on the examination paper)

Volume of cuboid = length × width × height

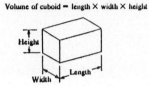

Circumference of circle = πd
= $2\pi r$

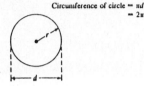

Area of triangle = $\dfrac{\text{base} \times \text{height}}{2}$

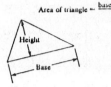

Answer all the questions. Time $1^1/_2$ hours

1/ 43 boys at a camp are to play 5 a side soccer. How many teams will there be? How many will be left over? [2]

2/ Jules prepares a tomato for a salad by cutting it in half, then by cutting each half into 3 equal parts. What fraction of a tomato is each half? [2]

3/ A spinner is twirled. What is the probability that it lands on an odd numbered side? [2]

Ron and Raj play a game with the spinner: Ron pays 4 p to Raj, and Ron receives the score on the spinner in pence. What is the probability that Ron makes a profit? [3]

4/ A housewife bought 6 pounds of meat for £7.80. What was the cost per pound? [2]

5/ Construct an enlargement of scale factor 2 of the shape shown. [3]

6/ Find the next two terms of each of the number sequences:

(a) 1, 4, 7, 10 [2]   (b) 1, 2, 4, 8 [2]

7/ The chart gives the conversion from miles to kilometres.

(a) Convert 5 miles to kilometres. [2]

(b) Convert 12 kilometres to miles. [2]

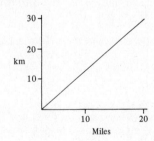

8/ Measure the diameter of the circle shown. [3]

What is the circumference of the circle, taking π to be 3.14? [2]

The area of a circle is given by $\pi d^2/4$, where d is the diameter of the circle. What is the area of this circle? [2]

9/   A bill at an Indian restaurant is shown. Find the total cost.   [2]

1 meat Dhansak	£2.45
1 Bombay Chicken Curry	£2.65
2 Rice	50p each
2 Popadoms	20p each

If the bill is paid with a whole number of pound coins, how much is the change?   [2]

If the change is left as a tip, what is the tip as a fraction of the bill?   [2]

10/   A cuboid is 3 cm high, and has a square base of side 4 cm.

(a) How many 3 by 4 faces are there?   [2]

(b) How many 4 by 4 faces are there?   [2]

(c) What is the volume of the cuboid?   [1]

(d) What is the surface area of the cuboid?   [2]

11/   Use the flow chart below to evaluate the function $(x + 1)^2 + 7$, for $x = 1, 3, -2$.   [3]

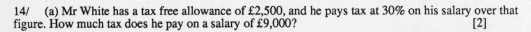

read x → add 1 → square → add 7 → print y

What would you alter to evaluate the function $(x - 3) + 7$?   [2]

12/   (a) What fraction of the circle is shaded?   [2]

(b) Shade parts of the triangle so that $1/3$ of it is shaded.   [2]

13/   Measure the sides of the triangle as accurately as you can.   [4]

What is the perimeter of the triangle?   [2]

What is angle <A?   [2]

14/   (a) Mr White has a tax free allowance of £2,500, and he pays tax at 30% on his salary over that figure. How much tax does he pay on a salary of £9,000?   [2]

(b) Mrs Black has a tax free allowance of £3,000, but pays tax at the same rate. If she pays £1,500 in tax, what is her salary?   [3]

(c) Ms Brown pays £2,100 tax on a salary of £10,500. What is her tax free allowance?[3]

15/ The chart gives the fares between 4 towns.

(a) How much does it cost to go from A to B? [1]

(b) How far can one go from D with only 50 p? [1]

(c) How much does it cost to go from A to B via C? [2]

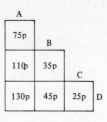

(d) Granny is entitled to a 25 % reduction in fares. How much does it cost her to go from A to C and then back via B? [4]

16/ (a) What is the bearing of Ashdon from Ashqelon? [3]

(b) Measure the distance between these towns on the map. [3]

(c) If the actual distance between these towns is 17.5 km, what is the scale of the map in km per cm?

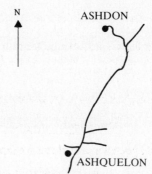

17/ An "L" shape is made from a rectangle of cardboard as shown. The lengths are in cm.

Find the areas of the rectangles AEGH and HFCD. Find their sum. [5]

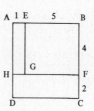

Find the areas of the rectangles ABCD and EBFG. Find their difference. [5]

What do you notice? [1]

If the cardboard weighs 0.04 grams per cm^2, what is the weight of the shape? [2]

18/   Samantha keeps a record over 10 weeks of how much she spends per week. She prepares a
frequency table and a bar chart. Complete them both.   [6]

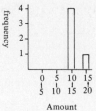

Amount |£0-5|£5-10|£10-15|£15-20|
Frequency | 3 | 2 | | |

What is the greatest possible amount she can have spent?   [4]

Over the next 5 weeks she spends £4, £16, £12, £6, £13. What has been her average
expenditure?   [3]

19/ A wall is made from bricks which are 3x3x6
inches. A brick which is laid along the wall is a
stretcher, and a brick which is laid across the wall
is a *header*.

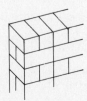

(a) If the wall is 6 inches thick and 12 feet long, how many bricks are there in each course?[2]

(b) If the wall is 4 feet high, how many bricks does it contain?   [2]

(c) What is the volume of the wall?   [2]

(d) If the courses are alternately of headers and stretchers, how many bricks do we see on the
front of the wall?   [3]

(e) How many bricks do we see at each end of the wall?   [3]

A floor is to be covered with wooden tiles which are 2 inches by 4 inches. Extend the following
patterns by drawing at least 8 more tiles in each.
(i)   (ii)

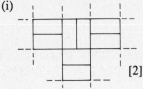

[2]   [3]

If the floor is 8 ft by 12 ft, what is the number of tiles needed in pattern (i)?   [3]

If each tile is $^1/2$ inch thick, what is the volume of wood?   [2]

If the wood weighs 0.07 pounds per cubic inch, what is the weight of the tiles?   [2]

If each tile costs 15 p, what is the cost of the tiles?   [2]

Southern Group. Paper 2. (Levels 1 & 2.)

INFORMATION AND FORMULAE
(Provided on the examination paper)

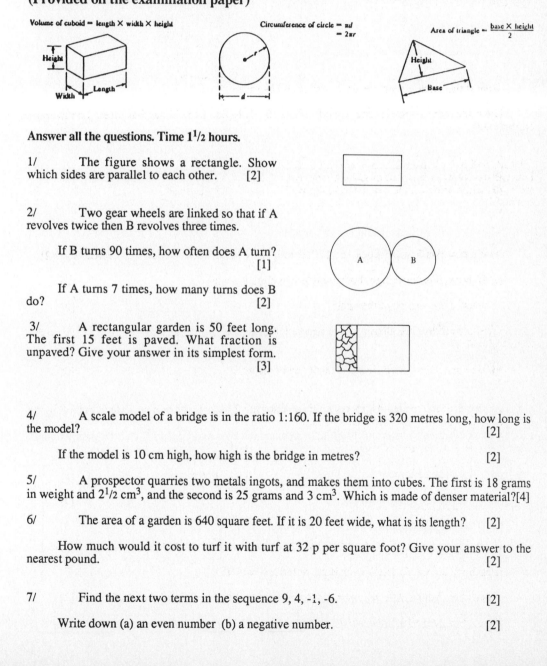

Volume of cuboid = length × width × height

Circumference of circle = πd
= 2πr

Area of triangle = $\frac{base \times height}{2}$

Answer all the questions. Time 1¹/₂ hours.

1/ The figure shows a rectangle. Show which sides are parallel to each other. [2]

2/ Two gear wheels are linked so that if A revolves twice then B revolves three times.

If B turns 90 times, how often does A turn? [1]

If A turns 7 times, how many turns does B do? [2]

3/ A rectangular garden is 50 feet long. The first 15 feet is paved. What fraction is unpaved? Give your answer in its simplest form. [3]

4/ A scale model of a bridge is in the ratio 1:160. If the bridge is 320 metres long, how long is the model? [2]

If the model is 10 cm high, how high is the bridge in metres? [2]

5/ A prospector quarries two metals ingots, and makes them into cubes. The first is 18 grams in weight and 2¹/₂ cm³, and the second is 25 grams and 3 cm³. Which is made of denser material?[4]

6/ The area of a garden is 640 square feet. If it is 20 feet wide, what is its length? [2]

How much would it cost to turf it with turf at 32 p per square foot? Give your answer to the nearest pound. [2]

7/ Find the next two terms in the sequence 9, 4, -1, -6. [2]

Write down (a) an even number (b) a negative number. [2]

8/ When Mr Sampson's car did not start, he had to call in the garage. The bill was as shown. Find:

Labour		Parts	
Breakdown Fee	£8	Plugs	£4.80
Battery Charge	£2	Air-filter	£4.93
Tune Engine	£13.50	Contact Set	£2.97

(a) The labour total. [2]

(b) The parts total. [2]

(c) The total bill after VAT at 15% has been added. [2]

9/ Use your calculator to complete the following:

(a) 3.27 x 6.14 = [2]

(b) (3.1 + 4) x 6 = [2]

(c) (5 - 0.2) + 0.01 = [2]

(d) (5x3x7.2)/(2.1x4.6x0.13) [2]

10/ Building work will take 160 hours to complete.

(a) How long will it take 4 men? [1]

(b) If they are paid £3.50 per hour, how much will each man get? [2]

(c) What will be the total bill for labour, after VAT at 15% has been added? [3]

11/ ABCDEF is a regular figure. X is its centre. What are the names for:

(a) ABCDEF [1]

(b) ABDE [2]

(c) ABXF [2] (d) ABCF [2] (e) ABE [2]

12/ A class contains 10 boys and 12 girls. In an exam the boys averaged 61% and the girls averaged 72%.

(a) What is the total score for the boys? [2]

(b) What is the total score for the girls? [2]

(c) What is the average score for the whole class? [3]

13/ The chart converts pounds to kilograms.

(a) How many pounds for 5 kilograms? [2]

(b) How many kilograms for 7 pounds? [2]

(c) How many pounds for 15 kilograms? [3]

14/   Mark the lines of symmetry of (a)   [2]   (b)   [2]

Which shape has rotational symmetry?   [1]

What is the order of rotational symmetry?   [2]

Draw a shape which has rotational symmetry of order 3.   [3]

Draw the lines of symmetry of your shape.   [2]

15/   Athens is 2 hours ahead of London. When a plane leaves London the time is 1600, and when it arrives in Athens the (local) time is 2200. How long has the flight been?   [2]

The flight back leaves Athens at 0900. If it flies at the same speed, what is the time in London when it arrives?   [2]

A flight leaves London at 1200, and arrives 8 hours later in New York, where the local time is 1500. What is the time difference between London and New York?   [3]

A flight leaves London at 0800, and arrives in Moscow where the time is 1500. The return flight leaves Moscow at 2000 and arrives in London at 2100. What is the time difference between Moscow and London?   [3]

16/   A solid is built up of prisms, whose cross-sections are right-angled triangles. How many prisms are there in the solid?   [2]

What is the area of a triangular face of a prism?   [2]

What is the volume of each prism?   [2]

What is the volume of the solid?   [1]

How many prisms must be added to make the solid into a cuboid?   [3]

What is the volume of this cuboid?   [2]

17/   Ben starts off to school walking, then realizes he might be late and starts to run. The graph of his distance against time is shown.

(a) How far did he walk?   [2] (b) For how long was he walking?   [2]

(c) How far is the school?   [2]

(d) How long did the whole journey take?   [2]

(e) How fast did he walk?   [2] (f) How fast did he run?   [2]

(g) What was his average speed for the whole journey?   [2]

361

18/ Mark is standing 20 metres due West of Tom. Mark walks North at 1 m/sec, and Tom walks North at 1.2 m/sec. After 1 minute they both stop. Make a diagram of their journeys, using a scale of 1 cm for 10 metres.[8]

How far apart are they after 1 minute?   [5]

What is the bearing of Tom from Mark?   [3]

What is the bearing of Mark from Tom?   [2]

If Tom walks to Mark, how long will it take him at 0.8 m/sec?   [2]

Southern Group. Paper 3. (Levels 2 & 3.)

INFORMATION AND FORMULAE
(Provided on the examination paper)

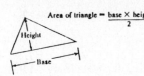

Area of triangle = base × height / 2

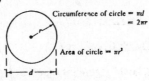

Circumference of circle = πd
= 2πr

Area of circle = πr²

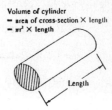

Volume of cylinder
= area of cross-section × length
= πr² × length

Answer all the questions. Time 2 ¹/₄ hours.

1/ Arrange the following in increasing order, using the < symbol.
¹/₅, 0.22, ³/₁₇, ²/₉. [2]

2/ There are 800 children in a school, of whom 450 are boys. What percentage are boys? [2]

3/ Find $3 \times 10^7 \div 4 \times 10^{19}$, leaving your answer in standard form. [3]

4/ R is given in terms of S by R = 5S - 3.

(a) Find R when S = 2. [1]

(b) Find S when R = 22. [2]

5/ A "Variety" stamp packet mixes British and European stamps in the ratio 3:5. How many British stamps are there in a packet containing 32 stamps? [2]

6/ A rectangular lawn is 20 ft by 45 ft. It contains a circular flower bed of radius 3 ft.

(a) Find the area of the flower bed, giving your answer to 3 significant figures. [2]

(b) Find the area of lawn outside the flower bed, giving your answer to 2 decimal places. [3]

7/ In the figure <ABD = <DBC and <ACD = <DCB.
<D = 108°. Find <A. [4]

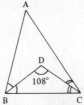

8/ A kite flies 30 ft high, on a string which is 60 ft long. What angle does the string of the kite make with the ground? [3]

9/ Kate is 4 times as old as her son Bill. In 16 years time she will be twice as old. Letting x be Bill's age, find expressions for Bill's age in 16 years time, and for Kate's age in 16 years time. [3]

Find an equation in x. [3]

Solve this equation. [3]

363

10/   In the 6th form of a school, $^3/5$ of the pupils are boys. $^2/3$ of the boys study Mathematics, but only $^1/6$ of the girls study Mathematics.

   (a) If a pupil is picked at random, what is the probability that a female Mathematics student is picked?   [3]

   (b) If there are 300 pupils in the 6th form, how many boys do not study Mathematics?[3]

11/   A cylinder has volume 1,000 cm^3.

   (a) If the base radius is 2 cm, find the height. Give your answer to 2 decimal places.   [3]

   (b) If the height is 4 cm, find the base radius. Give your answer to 2 decimal places.   [4]

12/   Describe the transformation which takes S to S' in the picture shown.   [2]

   S is rotated through 90° clockwise about (0,0). Draw the result on the graph.   [3]

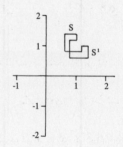

13/   Bristol is 120 miles West of London. Mr Jones starts from Bristol travelling East at 60 m.p.h. At the same time Mrs Smith starts from London, travelling West at 40 m.p.h.

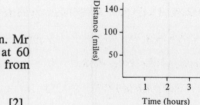

   (a) How long do they take?   [2]

   (b) Draw two lines on the same graph to show their distances from Bristol.   [5]

   (c) After how long do they pass each other? How far are they from London when they pass each other?   [4]

14/ The dots of the diagram are 1 cm apart. The net shown is of a pyramid.

When the pyramid is assembled, how many faces, edges and vertices will it have? [3]

At how many of the vertices will 3 edges meet? [3]

What is the surface area of the pyramid? [4]

Another pyramid is to be made, with a square base of side 1 cm and total surface area 9 cm^2.

What is the surface area of each triangular face? [2]

Draw the net for this pyramid. [6]

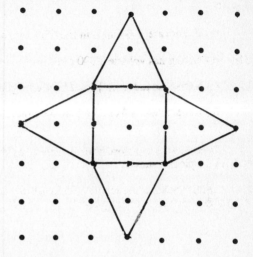

15/ The population of a new town is given by the following figures.

Year	1980	1981	1982	1983	1984	1985	1986
Population in 1,000's	16	20	24	28	32	36	40

Plot these figures on a graph. Join them up by a straight line. [5]

Find the gradient of the line. [3]

What will the population be in 1988? [2]

What was the population in 1978? [2]

When was the town founded? [3]

16/ The figure shows a right-angled triangle.
Write down the relationship between the sides a, b, c [2]

If a = 7 and b = 24 find c. [2]

If a = 4 and c = 7 find b. Give your answer to 3 significant figures. [3]

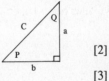

365

Fill in the following table:

a	b	c
3	4	5
5	12	
7	24	
9		41

[4]

If a = 39 find whole numbers b and c such that $a^2 + b^2 = c^2$. [3]

Fill in the following table to give sin, cos and tan of P and Q in terms of a, b, c.

sin P	cos P	tan P || sin Q	cos Q	tan Q
a/c		||		b/a

[3]

Write sin P, cos P and tan P in terms of functions of Q.

sin P =   cos P =   tan P = [3]

17/ Lawrence asks the members of his form how many magazines they read regularly. His results are in the following tally table:

0 mags.	1 mag.	2 mags.	3 mags.
~~HHH~~ ~~HHH~~	~~HHH~~ /	~~HHH~~	///

(a) How many people did he ask? [2]

(b) Draw up a frequency table for these figures. [3]

(c) Illustrate these figures on a bar chart. [5]

(d) Illustrate these figures on a pie chart. [6]

(e) What is the mode number of magazines read? [1]

(f) What is the mean number of magazines read? [3]

Southern Group. Paper 4. (Level 3.)

INFORMATION AND FORMULAE
(Provided on the examination paper)

Volume of prism = area of cross-section × length

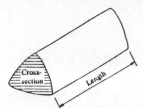

Volume of pyramid = $\dfrac{\text{base area} \times \text{height}}{3}$

If $ax^2 + bx + c = 0$ $(a \neq 0)$

$x = \dfrac{-b \pm \sqrt{(b^2 - 4ac)}}{2a}$

Curved surface of area of cone = $\pi r l$

Sphere: Surface area = $4\pi r^2$

Volume = $\frac{4}{3}\pi r^3$

Answer all the questions. Time 2 hours

1/ A man drives at 50 m.p.h., (to the nearest 5 m.p.h.) over a distance of 80 miles. (To the nearest 10 miles.) What is the greatest time he could have taken? [4]

2/ An insurance company charges £10 plus 25 p for each £100 of cover. If a woman pays £20 premium, how much cover does she have? [3]

3/ (a) A square has the same area as a circle of radius 2 cm. What is the side of the square? [3]

(b) A sphere has the same volume as a cube of side 3 cm. What is the radius of the sphere? [3]

4/ A sphere of radius 2 cm is immersed in a cylinder of base radius 3 cm. By how much does the level of the water rise? [3]

5/ (a) n! is defined as $n \times (n-1) \times (n-2) \times \ldots \times 3 \times 2 \times 1$. Find 3! and 5!. [2]

(b) Follow through the flow-chart for the group of numbers 5,4,6,4,6,3,-2.

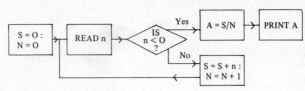

What value does the flow-chart print? [4]

What is the purpose of the flow-chart? [3]

6/ (a) Solve the equation $2^{3x} = 4^{x+1}$. [3]

(b) Express $5^3 \times 25^2 \div 125^2$ as a power of 5. [3]

7/ A box contains 10 screws and 7 nails. Two are drawn out at random. Find the probabilities that:

(a) Both are nails. [2]

(b) There is one screw and one nail. [3]

8/   It is thought that T is proportional to a power of S. The following values are obtained:

$$S\ |\ 2\ |\ 3\ |\ 4\ |$$
$$T\ |\ 6\ |\ 13.5|\ 24\ |$$

(a) Find the power.   [2]

(b) Find T when S = 5.   [2]

(c) Find S when T = $^1/_6$.   [2]

9/   X is the centre of the circle.
<A = 50° and <BXP = 44°. Find:

(a) <PXC   [2]   (b) <BAP   [2]   (c) <BPC   [2]

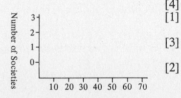

10/   There are 3 societies in the 6th form of a school: Jazz, Film and Debating. There are 140 pupils in the 6th form. 43 belong to all three societies.

There are 64, 78, 84 members of the Jazz, Film and Debating societies respectively. The Jazz and Film societies have 50 members in common, as do the Jazz and Debating societies. There are 58 pupils in both the Film and Debating societies.

Fill in a Venn diagram to show this information.   [4]
(a) How many pupils are in the Jazz society only?   [1]

(b) How many pupils belong to exactly one society?   [3]

(c) How many pupils belong to no societies?   [2]

Complete the bar chart to show the number
of pupils who belong to 0, 1, 2, 3 societies.   [4]

What is the mean number of societies that the pupils belong to?   [4]

11/   Complete the following table for the function y = x^2 - $^3/x$.

$$x\ |\ ^1/_2\ |\ ^3/_4\ |\ 1\ |\ 1^1/_4\ |\ 1^1/_2\ |\ 1^3/_4\ |\ 2\ |$$
$$y\ |\ \ \ |\ \ \ |\ \ |\ \ \ \ |\ \ \ \ |\ \ \ \ |\ \ |$$ [4]

Draw the graph of the function.   [4]

Solve the equations:

(a) x^2 - $^3/x$ = 0   [1]

(b) x^3 - 3 = x   [3]

(c) x^2 - $^3/x$ = x - 4   [3]

12/ Let M be the matrix $\begin{pmatrix} -1 & 0 \\ 0 & 1 \end{pmatrix}$

Let T be the triangle with vertices A(1,1), B(3,1), C(2,5).

Plot T on graph paper. [2]

T' is obtained by applying M to T. Find the vertices of T', and plot them on the same graph paper. [3]

Describe the transformation which has taken T to T'. [2]

T is reflected in the x-axis to T''. Draw T''. [2]

Find the matrix which will take T to T''. [3]

Find the matrix which transforms T' to T''. [3]

13/ A regular octagon is drawn inside a circle of radius 4 cm.

(a) Find the angle each side subtends at the centre. [2]

(b) Find the length of the perpendicular from the centre to a side. [4]

(c) Find the length of the sides of the octagon. [4]

(d) Find the area of the octagon. [4]

14/ An enemy aircraft is flying directly over a missile site. Its speed is 200 m/sec due North, and it is 1,000 m. high. The instant that the aircraft is above the site a missile is fired at 500 m/sec.

(a) If the missile is fired vertically upward, find how long it takes to reach the height of the aircraft and the distance by which it misses. [4]

(b) If the missile is fired at θ to the vertical, its vertical velocity is 500xcos θ Find in terms of θ how long it takes to reach the height of the aircraft. [3]

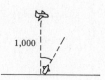

1,000

(c) Find how far North the aircraft will have flown in this time. [3]

(d) The horizontal velocity of the missile is 500xsin θ. Find how far North the missile will have flown. [3]

(e) By equating your answers to (c) and (d) find the value of θ which will enable the missile to intercept the aircraft. [4]

15/ (a) If interest is chargeable at 10%, how much will a man have to pay on a loan of £26,000? [2]

(b) If the man pays income tax at 30%, how much does he have to earn in order to pay the interest? [3]

(c) The interest rate rises to 12%, and the tax rate falls to 25%. What is the percentage change in the amount he has to earn to pay his interest payments? [3]

Solutions to Exercises

Chapter 1

Exercise 1.1.2 Page 2
(1) (a) 12+17 (b) 3x12 (c) 2x12+17 (d) 3x12+2x17 (2) 3x10+5+2x2 or 2x10+3x5+2x2 (3) (a) 3B (b) A+2B (c) A+4B (4) £7 (5) 1965 (6) £125 (7) 58 (8) 119, 120 (9) (a) 13, 1 (b) 6, 6 (c) 14, 2 (10) 7, 2 (11) 6, 11p (12) 36,5 (13) 14, 6 (14) (a) 4,363 (b) 227,412 (c) 500,062 (15) (a) 2,902 (b) 3,981 (c) 4,850 (d) 2,000 (16) (a) 212 (b) 4,209 (c) 1,999 (d) 8,999,999 (17) (a) 9 (b) 7 (c) 4 (d) 12 (18) (a) 241+162=403 (b) 5235+1734=6969 (c) 2431+9389=11820 (d) 746-324=422 (e) 5293-3537=1756 (f) 13783-9457=4326 (19) a,c,e,g. 6=2x3:25=5x5:35=5x7.

Exercise 1.1.4 Page 4
(1) (a) 2 (b) 2 (c) 3 (d) 6 (e) 1 (2) (a) 30 (b) 28 (c) 45 (d) 60 (e) 210 (3) 10p (4) 15 pints (6) 24 secs (7) 30 grams (8) 30x30 cm.

Exercise 1.2.2 Page 6
(1) (a) 11, 13 (b) 26,31 (c) 22,26 (d) 32, 64 (e) 81,243 (f) 48,96 (g) 8, 4 (h) 6,2 (i) 36, 49 (j) 125, 216 (k) 50,72 (l) 37, 50 (2) 9, 3 (3) 10, 5 (4) 1+6+21 (5) 5+10+26 (6) (a) 5x6+1=31, 6x7+1=43 (b) 4x6-1=23, 5x7-1=34 (c) 2x2x2x2x2=32, 2x2x2x2x2x2=64 (d) 1x2x3x4x5=120, 1x2x3x4x5x6=720 (7) (a) 1,4,9,16,25,36 (b)2,6,12,20,30 (c) 3,5,7,9,11 (8) 3,6,9,12,15 (9) (a) 4,8,12,16,20 (b) 3,7,12,18,25 (10) 13,21,34.

Exercise 1.3.2 Page 8
(2) (a) -5 (b) Friday (3) 150 miles (4) £205 (5) 10 hours (6) 53 m. (7) (a) Dead Sea (b) 1,770 m (c) 398 m (d) 1,802 m (8) 2512 miles (9) (a) 1 in, -15 in (b) 12 in (c) 16 in (10) 520, 184 BC (11) 35 AD (12) 74 BC (13) 32 BC (14) (a) 1797, 1774 (b) 11, -43, -2002.

Exercise 1.4.2 Page 11
(1) $1/2$, $1/3$, $1/4$, $3/4$, $5/8$ (2) fifth, ninth, five sixths, seven eights (3) 0.5, 0.25, 0.375,0.7 (4) $1/2$, $1/4$, $7/10$,$7/20$ (5) $1/2$, $1/3$, $1/3$, $1/2$, $1/4$ (6) (a) $2/4$ (b) $4/6$ (c) $3/4$ (d) $1/7$ (7) $1/5$ (8) $3/8$, 0.3 (9) (a) $1/2$ (b) $1/4$ (c) $3/8$ (d) $4/9$ (10) $1/10$p, $1/8$p (11) 5 gallons. At the $3/4$ mark. (12) (a) 30, 45 (b) $1/6$, $1/12$ (13) $9/20$ (14) 1,200

Exercise 1.4.4 Page 13
(1) (a) $1\,1/4$ (b) $2\,1/3$ (c) $6\,2/3$ (d) $1\,1/2$ (e) $4\,1/2$ (2) (a) $3/2$ (b) $11/4$ (c) $27/8$ (d) $19/5$ (e) $16/7$ (f) $53/10$ (3) (a) 0.02 (b) 0.08 (c) 0.7 (d) 0.75 (4) (a) $3/2$, $1\,1/2$ (b) $23/10$, $2\,3/10$ (c) $13/4$, $3\,1/4$ (d) $9/8$, $1\,1/8$ (5) 24 (6) 55 (7) $3/8$, £2,000.

Chapter 2

Exercise 2.1.2 Page 16
(1) (a) 9.4 (b) 22.7 (c) 79.92 (d) 3.9 (e) 3.78 (f) 1.5 (g) 12.3225 (h) 1932.832 (i) 110.7 (j) 4.7 (k) 3.5 (l) 11.5 (m) 20 (2) £8.81 (3) 479 g (4) 20.1 km (5) £2.11 (6) 6.35 m (7) 96 g (8) £7.62 (9) £11.64 (10) £10.02 (11) £1.71 (12) £9.80 (13) 37 p (14) £1.43 (15) £667 (16) 7 (17) (a) £7 (b) £9.62 (c) £10.38 (18) (a) £18.60 (b) £45.60 (19) (a) £27.55 (b) £37.28 (20) 169,344, 294 (21) (a)

13 (b) 25 (c) 16 (d) 6 (e) 7 (f) 3 $^{1}/4$ (g) 12 (h) 2 (22) (a) 3x(4+6)=30 (b) (12-5)x2=14 (c) (7-1)÷2=3 (d) 6÷(17-14)=2

Exercise 2.1.4 Page 18
(1) (a) 49.1 (b) 99.3 (c) 68.1549 (d) 238.76 (2) (a) 24.42 (b) 2.9502 (c) 512.7432 (d) 0.224 (3) £87.95 (4) £24.19 (5) 16

Exercise 2.2.2 Page 19
(1) (a) $^{11}/12$ (b) $^{23}/30$ (c) $^{17}/21$ (d) $^{7}/20$ (e) $^{3}/4$ (f) $^{7}/10$ (g) $^{8}/21$ (h) $^{27}/40$ (2) $^{5}/6$ (3) 1 $^{5}/12$ yards (4) 12 $^{3}/4$ pints (5) $^{1}/20$

Exercise 2.2.4 Page 20
(1) (a) $^{8}/21$ (b) $^{1}/9$ (c) $^{9}/10$ (d) $^{10}/21$ (e) 3 (f) 3 $^{7}/12$ (g) $^{8}/15$ (h) $^{-3}/8$ (i) - 4 $^{3}/20$ (j) 1 $^{1}/6$ (k) $^{5}/6$ (l) $^{5}/27$ (m) 22 (2) (a) 12.8 (b) 2 (c) 9 (d) -0.45 (e) 24 (f) -12 (g) 9 (h) -32 (3) 40 (4) 24 (5) 6 hours (6) $^{1}/2$ mile (7) £210 (8) 125 m (9) 6 (10) $^{7}/30$ (11) $^{11}/240$ lb

Chapter 3
Exercise 3.1.2 Page 23
(1) (a) 4 (b) 4 (c) 130 (d) 987 (e) 6 (f) -8 (g) -77 (h) 1 (i) 0 (j) -1 (k) 0 (l) 9 (m) 76 (n) 1 (2) (a) 1,097,000 (b) 1,097,400 (c) 1,097,380 (3) (a) 460 (b) 500 (c) 0 (4) (a) -67 (b) -70 (c) -100 (5) (a) 90 (b) 1 (c) 13 (d) 5 (6) (a) 17 (b) 86 (c) 13 (d) 7 (e) 5 (f) 1 (7) £8.13 (8) £107 (9) £13.74 (10) 2 p (11) 190 miles (12) 3 hours (13) £590

Exercise 3.2.2 Page 24
(1) (a) 2,300 (b) 660,000 (c) 0.0047 (d) 0.025 (e) -3.5 (f) 1.0 (g) 2.1 (h) 500 (i) 600 (j) 5.1 (k) 4.0 (2) (a) 4,740 (b) 4,700 (c) 5,000 (3) (a) 3.576 (b) 5.235 (c) 23.038 (d) 65.476 (e) 0.346 (f) 0.123 (g) 0.005 (h) 0.000 (i) 0.120 (j) 0.270 (4) (a) 0.1358 (b) 0.136 (c) 0.14 (5) (a) 6.94 (b) 1.51 (c) 7.97 (d) 0.71 (e) 2.72 (f) 1.82 (g) 0.00 (6) 1.47 g/c.c. (7) 660 g (8) £13 (9) (a) 3.571 (b) 3.57 (10) 34.5°, 33.5° (11) 915 kg (12) 85,000, 75,000 (13) 0.545, 0.535 (14) 325 miles, 315 miles (15) (a) 10,200 m^2 (b) 10,000 m^2 (c) 10,000m^2

Exercise 3.3.2 Page 26
(1) 1.3 in (2) 360 g (3) 70 g (4) 600,000,000 miles (5) (a) 21 (b) 7 (c) 30 (d) 2,860 (e) 30 (f) 2.8

Chapter 4

Exercise 4.1.2 Page 29
(1) (a) 9 (b) 64 (c) 4 (d) 81 (e) 100 (f) 144 (g) 10,000 (2) (a) 2 (b) 3 (c) 4 (d) 8 (e) 10 (f) 11 (3) (a) 8 (b) 64 (c) 1,000 (d) 2 (e) $^{1}/4$ (f) $^{1}/9$ (g) 0.125 (4) (a) 100 (b) 9 (c) 2 (d) 1 (e) $^{1}/2$ (f) $^{3}/2$ (5) (a) 36, 85 (b) 16,9 (c) 9 (d) 2 (6) 144 sq. ft. (7) 1,600 cm^2 (8) 2 m (9) 5 cm (10) 27 c.c. (11) 4 cm

Exercise 4.2.2 Page 30
(1) (a) 81 (b) 256 (c) 1 (d) 1,024 (e) $^{1}/8$ (f) 0.027 (g) 3.375 (h) -8 (2) (a) $^{1}/2$ (b) $^{1}/27$ (c) $^{1}/125$ (d) $^{1}/64$ (e) 1 (f) 2 (g) 9 (3) (a) 3^6 (b) 2^7 (c) 4 (d) 3 (e) 7 (4) (a) 4^5 (b) 3^6 (c) 1 (d) (3/2)^9 (5) 2^1 (6) 3^{-3} (7) (a) 1.414 (b) 1.732 (c) 8.062 (d) 0.447 (e) 0.007 (8) (a) 30 (b) 32 (c) 31.623 (9) (a) 3.73 (b) 32.35 (c) 1.22 (d) 6.97

Exercise 4.3.2 Page 32
(1) (a) 3 (b) 10 (c) 2 (d) $^{1}/7$ (e) $^{1}/4$ (f) $^{1}/10$ (2) (a) 4 (b) 27 (c) 100 (d) 16 (e) 2 (f) 10 (3) (a) 2^3 (b) 3^3 (c) 10 (4) (a) x (b) y (c) 3^0 (5) (a) 2 (b) 4 (c) 1 $^2/3$ (6) (a) 2^3 (b) 2^{2}x3 (c) 3x5^2 (d) 2^{2}x3^2 (e) 2^{10} (f) 2^{2}x3^{2}x5 (7) 3 (8) 45

Exercise 4.4.2 Page 34
(1) (a) 2.3×10^4 (b) 8.76×10^8 (c) 1.23×10^{-4} (d) 1×10^{-3} (e) 3.2×10^6 (2) (a) 1×10^5 (b) 2.1×10^9 (c) 9×10^{-2} (d) 8×10^7 (e) 1.04×10^{-12} (f) 3×10^2 (g) 2×10^3 (h) 2.5×10^7 (i) 1.6×10^{17} (j) 3.1×10^{-10} (k) 6.4×10^7 (3) (a) 8.4×10^6 (b) 3.57×10^6 (c) 1.03×10^6 (d) 6.4×10^7 (e) -2.6×10^9 (f) 9.8×10^{-3} (4) 6.696×10^8 miles, 5.9×10^{12} miles (5) 5×10^{27}

Chapter 5

Exercise 5.1.2 Page 36
(1) 4:3 (2) 5:6 (3) 4:1 (4) 2:1 (5) 2:1 (6) 3:1 (7) 10 litres (8) 40 kg (9) 30 kg (10) £32 (11) 3 km (12) 5 cm (13) $^1/_2$ m (14) 25 m (15) 1:100 (16) 9 cm

Exercise 5.1.4 Page 38
(1) £9,000 (2) 4 kg (3) 26 cm (4) 150 (5) 2:1, £4,000 (6) £10,000 (7) 40 (8) £6,000, £8,000, £10,000 (9) 240 c.c. (10) 8 hours (11) 10 min, 4 min, 6 min

Exercise 5.1.6 Page 39
(1) $^1/_4$ km^2 (2) 20 cm^2 (3) $^3/_{25}$ km^2 (4) 5 cm^2 (5) 0.6 m^3 (6) (a) 1:100,000 (b) 1:5,000,000 (c) 1:250,000 (7) (a) 1 km (b) $^1/_2$ km (c) $^1/_4$ km (8) 1 cm to $^1/_4$ km. 1:25,000 (9) 1 cm to 2 km. 1:200,000 (10) 4,000 m^3 (11) (a) 480 cm (b) 27 $^7/_9$ cm^2 (c) 15.12 m^3 (d) 2 (12) (a) 16 $^2/_3$ cm (b) 16.2 m^2 (c) 110 g (13) (a) 1:20 (b) 20 cm^2 (c) 10 (14) 22.5 cm^2 (15) 1 cm to $^1/_2$ km, 1:50,000

Exercise 5.2.2 Page 41
(1) 300 ml (2) £12 (3) 272 miles (4) 90 g (5) £48 (8) $1^1/_2$ hours (9) £6 (10) 2 days (11) 2 hours

Exercise 5.2.4 Page 42
(1) $y=^1/_2 x$, $2^1/_2$ (2) (a) 72 (b) 30 (3) (a) $3^1/_2$ amps (b) 120 volts (4) 100 km (5) $p=^{70}/_q$, $17^1/_2$ (6) (a) 4 (b) 16 (7) $V=^{200}/_P$, 5 m^3 (8) (a) 50,000 cps (b) 250 m

Exercise 5.2.6 Page 43
(1) (a) $y=4x^2$ (b) 36 (c) 1 (2) (a) $T=7s^2/9$ (b) 28 (c) 9 (3) $62^1/_2$ (4) $P=Q^3/432$, 2 (5) $k=^4/_3\sqrt{j}$, $^{28}/_3$ (6) $m=^5/_2 \sqrt[3]{n}$, 32.768 (7) $M=4l^3$, 108 g, 1.5 cm (8) $E=2s^2$, 10 m/sec (9) $P=8I^2/3$, $^{200}/_3$ Watts (10) $E \propto e^2$, $E=7e^2/9$, $^7/_9$ (11) $P=3.5\sqrt{l}$, 1.75 secs, 1.44 m (12) $12^1/_2$ km (13) (a) $T=245/R^2$ (b) 6.81 (c) 1.4 (14) $B=2/c^2$, 8, $^1/_2$ (15) $1^7/_9$ (16) $R=^3/4/r^2$, 12 ohms, 1 cm (17) $2^2/_3$ cm (18) 689 days (19) 2, 108 (20) 3, 27

Chapter 6

Exercise 6.2 Page 47
(1) (a) $^1/_{10}$ (b) $^1/_5$ (c) $^3/_5$ (d) $^1/_{20}$ (e) $^1/_8$ (f) $^3/_2$ (g) 1 (2) (a) 50% (b) 25% (c) 75% (d) 60% (e) 5% (f) 85% (g) 175% (3) (a) 12% (b) 74% (c) 3% (d) 60% (e) 154% (f) 270% (4) (a) 0.23 (b) 0.54 (c) 0.05 (d) 0.2 (e) 1.5 (f) 1.04 (5) £1.50 (6) £120 (7) £1,000 (8) £11 (9) £40 (10) 8 litres (11) 12 kg (12) 90 (13) 12 ft (14) £16,100 (15) 12 stone (16) £214,000 (17) £9.60 (18) £1,600 (19) £600 (20) £1.47 (21) £78.89 (22) £31.51 (23) (a) 2.5 m (b) 127.5 m (c) 122.5 m

Exercise 6.4 Page 48
(1) 65% (2) 25% (3) 85% (4) 66 $^2/_3$ % (5) 55% (6) 40% (7) 66 $^2/_3$ % (8) 10% (9) 25% (10) 10% (11) 20% (12) 4%

Exercise 6.6 Page 50
(1) £266.2 (2) £1,157.63 (3) £524.32 (4) £3,070.63 (5) 21,900,000 (6) (a) £726 (b) £1,556.25

(7) (a) £472.06 (b) £2,306.99 (8) (a) £1,425.72 (b) £3,375.20 (9) 22.3 million (10 £3,596 (11) (a) 2.43 kg (b) 1.046 kg (12) £300 (13) £1,500 (14) £1,600 (15) 20 million (16) £11,000 (17) £500 (18) £25 (19) £2.40

Chapter 7

Exercise 7.1.2 Page 54
(1) (a) 300 (b) 360 (c) 3.6 (d) 6,500 (e) 12.7 (f) 103.632 (2) (a) 0.4 (b) 0.213 (c) 1,390 (d) 0.6858 (e) 4.572 (3) (a) 0.3 (b) 15,340 (c) 5,584.2 (d) 104 (e) 482 (4) (a) 0.867 (b) 0.065 (c) 1.362 (d) 1.94312 (e) 0.96475 (5) (a) 4 (b) 0.277 (c) 13.62 (d) 2.8375 (6) 72,846 m^2=7.3 hectares (7) (a) 22 km (b) 16.5 m (8) (a) 55 (b) £7.50 (9) (a) 1.95, 2.1 (b) 25 (10) (a) 60 (b) 576 (c) 3.94 (d) 9.65 (e) 13.4 (11) (a) 42 (b) 2 (c) 10.1 (d) 50.7 (12) (a) 10.872 (b) 1,235.5 (13) 60, £2.7 (14) 9.1 kg, 11 lb (15) (a) 74$ (b) £37, £13

Exercise 7.2.2 Page 57
(1) (a) £2 (b) £1.05 (c) £7.125 (d) £3.60 (2) (a) 16 galls (b) 10 lb (c) 8 m^2 (d) 40 m (3) (a) 86 p (b) £6 (c) 3 p (d) 19 p (4) (a) 5 hours (b) 8 cm (c) 2 /2 hours (5) (a) 20 (b) ^{1}/3 (c) 200 c.c. (6) 22 min (7) 40 m.p.h. (8) 13 km, 4 ^{1}/3 km/hr (9) 13 p (10) 520 c.c. (11) 135 miles, 4 hours (12) (a) 54 m.p.h. (b) 1.30 (13) 13 1 (14) 32 m.p.g., 256 miles (15) £18.75 (16) £57 (17) 2^1/2 hours (18) £40

Exercise 7.3.2 Page 59
(1) (a) £3,380 (b) £9,620 (c) £10,800 (d) £4,368 (e) £3,276 (2) (a) £112 (b) £230 (3) Yes, by 10p per hour (4) £125.55 (5) £89.7, 4 hours (6) £215.8 (7) (a) £1,155 (b) £1,785 (c) £815.5 (d) £967.75 (8) (a) £920 (b) £426 (c) £225 (d) £524 (9) 155 pence in the £ (10) 125 p in the £ (11) £25,300,000 (12) (a) £1,950 (b) £1,950 (13) £10,500 (14) (a) £5.25 (b) £3 (c) £63 (15) (a) £388 (b) £2,710 (16) £5

Exercise 7.4.2 Page 62
(1) (a) 1100 (b) 0730 (c) 1600 (d) 1200 (e) 1945 (f) 2305 (2) (a) 8 a.m. (b) 10.15 a.m. (c) 12.15 p.m. (d) 4.45 p.m. (e) 8.30 p.m. (3) (a) 0300 (b) 1700 (c) 0730 (d) 23.45 (4) (a) 4 /2 hours (b) 3^1/4 hours (c) 11 hrs 38 minutes (5) (a) 7 (b) 31 July, 28 Aug, 4 Sept (6) 13 (7) (a) 5217 (b) 2936 (9) (a) 24 miles (b) 120 miles (c) 138 miles, 159 miles (10) (a) 1 hr 44 min, 5 min (b) 18.41 (c) 2042 (11) (a) £1,577.6 (b) £1,427 (12) £14 (13) (a) 1 hr 40 min (b) 1 hr 5 min

Chapter 8

Exercise 8.1.2 Page 67
(1) 16 (2) p=4, q=28 (3) 7+x (4) £(x+2) (5) 10-x m (6) 180-d (7) 8x (8) 10b (9) 20-x (10) 100h (11) 10p (12) $^{£p}$/8 (13) x+y=10 (14) N=M+5 (15) (a) 4 (b) 14 (c) 3 (d) 6

Exercise 8.1.4 Page 68
(1) LB, 2(L+B) (2) x=68t (3) £(n-m) (4) 180n-360 (5) x+120 (6) 301-s (7) £(30+20m) (8) 2N+5 (9) sr-t^2 (10) £(2.50x+4y) (11) £(10x+8y), £(10x+8y)/18 (12) £54x (13) x=180-y-z (14) 6x+4y+z

Exercise 8.1.6 Page 69
(1) $^{£x}$/10, £(x+30)/11 (2) ^{1}/2x^2, 3x (3) £1.2h (4) T-B, T+C-B (5) Gm/(m+n) cm, Gn/(m+n) cm (6) ^{L}/v cm^2 (7) £(xP+yQ)/(x+y) (8) $\sqrt{s^2/\pi}$

Exercise 8.2.2 Page 70
(1) 17 (2) 5 (3) 42 (4) 18 (5) 14 (6) 2 (7) 2 (8) 154 (9) 32 (10) 6 (11) 28 (12) 26

Exercise 8.2.4 Page 71

(1) 46.5 (2) (a) 60,000 (b) 50 (3) 18 (4) 120 (5) 132 (6) 75, 7 (7) 3 (8) 135 (9) 30 (10) $12^1/4$
(11) 35, -2 (12) 4 (13) 4 (14) 132

Exercise 8.3.2 Page 72

(1) $5x+35$ (2) $4a-4b$ (3) $14x+7y$ (4) $6x-10y$ (5) $8x+18$ (6) $5y-27$ (7) $-2z+22$ (8) $37w-21$ (9) $7x-y$
(10) $8a+11b$ (11) $2x+2y$ (12) $c-14d$ (13) $11r$ (14) $2p+14q$ (15) x^2+6x+5 (16) $y^2+10y+21$ (17)
p^2-p-12 (18) $z^2+2z-24$ (19) $w^2-18w+77$ (20) $q^2-7q+10$

Exercise 8.3.4 Page 74

(1) x^2-y^2 (2) $4x^2-y^2$ (3) $9p^2-4q^2$ (4) $18r^2-8s^2$ (5) $3x^2+19x-40$ (6) $6y^2+13y-28$ (7) $9a^2-4b^2$ (8)
$6w^2+11wz-10z^2$ (9) x^2+6x+9 (10) $4y^2-12y+9$ (11) $x^2+4xy+4y^2$ (12) $9z^2-12zw+4w^2$ (13)
$ac+ad+bc+bd$ (14) $xz+xw-yz-yw$ (15) $6tr+2tp+3sr+sp$ (16) $fj-5fk-3gj+15gk$ (17)
$10yz-14yw+15xz-21xw$ (18) $15sr-10st+3r-2t$

Exercise 8.4.2 Page 75

(1) $y=(x-2)/3$ (2) $d=a+c-b$ (3) $C=5(F-32)/9$ (4) $s=^{rt}/3p$ (5) $a=^y/3x^2$ (6) $m=(59-j)/n$ (7) $a=2b-c$ (8)
$b=2a(d+3)$ (9) $\pi=^{3V}/4r^3$ (10) $X=20Y-2Z$ (11) $x=2z-y$ (12) $p=7/5(q-r)$

Exercise 8.4.4 Page 75

(1) $x=\sqrt{y/4a}$ (2) $r=\sqrt[3]{3V}/4\pi$ (3) $S=\sqrt{3+T+R}$ (4) $h=\sqrt{l^2-4r^2}$ (5) $n=((h+3)/3m)^2$ (6) $l=g(^t/2\pi)^2$ (7)
$b=(c+1)/a+3$ (8) $u=1/(1/f+^1/v)$ (9) $m=T/(g-a)$ (10) $x=(-5a-2b)/2$ (11) $q=p(c-a)/(b-d)$ (12)
$x=(7b-3a)/(a+b)$ (13) $R=rm/(1-m)$ (14) $x=(3y+1)/(y-1)$

Exercise 8.5.2 Page 76

(1) $x(y+z)$ (2) $q(p-3)$ (3) $s(3+r)$ (4) $t(q+r)$ (5) $r(t-5s)$ (6) $x(3y-4z)$ (7) $2(x+2y)$ (8) $3(t-3r)$ (9)
$2r(s+2q)$ (10) $3y(3x-z)$ (11) $3f(2g+3h)$ (12) $7t(1+3q)$ (13) $a(a+b)$ (14) $z(1-z)$ (15) $5x(x-3)$ (16)
$7y(3y-2)$ (17) $4z(2z+c)$ (18) $3d(1+3cd)$

Exercise 8.5.4 Page 78

(1) $(p-q)(p+q)$ (2) $(t-1)(t+1)$ (3) $(2-n)(2+n)$ (4) $(q-3)(q+3)$ (5) $(3-s)(3+s)$ (6) $7(1-s)(1+s)$ (7)
$2(1-2x)(1+2x)$ (8) $9(1-3a)(1+3a)$ (9) $(7t-3s)(7t+3s)$ (10) $(4m-5n)(4m+5n)$ (11) $(y-^1/2)(y+^1/2)$ (12)
$(1/3-z)(1/3+z)$ (13) $(x+2)(x+1)$ (14) $(x+2)^2$ (15) $(x+3)(x+4)$ (16) $(x-3)(x-8)$ (17) $(x+3)(x-2)$ (18)
$(y-4)(y+3)$ (19) $(z+6)(z-1)$ (20) $(w-6)(w+2)$ (21) $(x-8)(x+3)$ (22) $(p+10)(p-3)$ (23) $(2x-1)(x+3)$
(24) $(3x+2)(x-4)$ (25) $(5y+4)(y+1)$ (26) $(6x-1)(x+6)$ (27) $(6a+5b)(a+b)$ (28) $(2p-3q)(3p+2q)$ (29)
$(t-q)(s+r)$ (30) $(x-y)(a-b)$ (31) $(3t-2)(2s+1)$ (32) $(z-3)(2x+y)$

Exercise 8.6.2 Page 79

(1) $^{ac}/bd$ (2) $2x^2/5$ (3) $4x^2/3y^2$ (4) $^{wx}/yz$ (5) $^7/2$ (6) $q^2/2$ (7) $^{11x}/15$ (8) $^{5a}/4$ (9) $(7r-4)/12$ (10)
$(-11x-18)/12$ (11) $(27b-1)/6$ (12) $(7c-25)/12$ (13) $(2y+3)/xy$ (14) $(3bc+4a)abc$ (15) $^1/3x(x-1)$ (16)
$6z^2/(z^2-1)$ (17) $2(x+2)/(3(x-1))$ (18) $2(2-y)/(3(1-y))$ (19) $(7x+3)/x(x+1)$ (20) $(-7z+5)/z(1-z)$ (21)
$(z^2+z+6)/3(z+1)$ (22) $(17y+8)/y(3y+2)$ (23) $(3y+5)/(y-1)(y+3)$ (24) $(-2w+1)/(3w+1)(w+2)$ (25)
$(16p+4)/(2p-1)(2p+3)$ (26) $(-3q+1)/(3-2q)(1-q)$

Chapter 9

Exercise 9.1.2 Page 83

(1) 2 (2) 17 (3) 7 (4) -2 (5) 6 (6) 4 (7) 3 (8) 2 (9) 1 (10) 1 (11) 3 (12) 4 (13) 5 (14) 1 (15)
147 (16) 3

Exercise 9.1.4 Page 84
(1) 14 (2) 36 (3) 2 (4) -70 (5) -6 (6) -6 (7) 8 (8) 24 (9) 11 (10) -60 (11) 29.6 (12) -17 (13) -32 (14) $-1/2$ (15) $3/4$ (16) $2/3$

Exercise 9.2.2 Page 85
 (1) 5, -2 (2) 6, -9 (3) 2, -1 (4) 1, 4 (5) 4, 3 (6) 27, -37 (7) 6, 2 (8) 5, 5 (9) 4, 1 (10) 3, -2 (11) 2, 7 (12) 4,-3 (13) 2, 3 (14) 8, 7 (15) 12, 18 (16) 8, 7

Exercise 9.3.2 Page 87
(1) 1, -6 (2) -3, -4 (3) 9, 1 (4) -4, -15 (5) 30, -2 (6) 1.5, -0.5 (7) 5, -2 (8) 5, 3 (9) 5, -4 (10) 6, 2 (11) 1, -6 (12) 4, 3 (13) -0.349, -2.151 (14) 4.562, 0.438 (15) 1.193, -4.193 (16) 2.303, -1.303 (17) 2.618, 0.382 (18) 3.815, -6.815

Exercise 9.4.2 Page 87
(1) £(20+x/10), 170 miles (2) 16 (3) £(2x+1,500), £13,100, £14,600 (4) 50° (5) 2x-7, 9 (6) 40 (7) (6y-3)=11(y-3), 36 (8) n+1, n+2, n+n+1+n+2=441, n=146 (9) 8 (10) 3x, 6x, £180, £540, £1,080 (11) £(9x+50), 63 (12) 110, 11 years

Exercise 9.4.4 Page 89
(1) £500 (2) 320 (3) 62p (4) x=96, y=24 (5) 21 (6) 260 (7) 90 (8) 22, 5 (9) 58 (10) 120 miles (11) 13 (12) 3 (13) 8 (14) 8, 9 (15) 8, 12 (16) 98 (17) 1.633 m.p.h.

Chapter 10

Exercise 10.1.2 Page 92
(1) $6^3/4$ (2) $4 1/3$ (3) $1 1/4$, $9/7$, 1.3, 1.42 (4) $1/100$, 0.0011, thousandth, millionth, 0 (5) 1 kg jar (6) £3.20 per roll (7) method 2, £400 (8) second (9) yes, by 60p (10) second, by £4 (11) £17 rise (12) England (13) $1/8$ $1/4$ $3/8$ $1/2$ $5/8$ $3/4$ $7/8$ 1

Exercise 10.2.1 Page 93
 (1) a,d,e (2) $4.45 < 4.6 < 14/3 < 19/4$ (3) $0.2 > 1/8 > 0 > -1/8 > -3/17 > -1/4 > -1/2$ (4) 4 or 5 (5) -4 or -3 (6) (a) 1, 2,3,4,5 (b) 2,3,4,5,6,7 (c) 14,15,16,17 (d) 7,8,9 (7) (a) -1,0,1 (b) -7,-6,-5,-4,-3 (c) -1,-2,-3,-4 (d) 3,2,1,0,-1,-2

Exercise 10.2.3 Page 95
(1) (a) $\{x:x<7\}$ (b) $\{x:x>1\}$ (c) $\{x:x \geq 3\}$ (d) $\{x:x<-4\}$ (e) $\{x:x>7\}$ (f) $\{x:x \leq 3\}$ (2) (a) x>12 (b) x<3 (c) x<6 (d) x≤13 (e) x $\geq$ $-1/5$ (f) x>33 (3) (a) $\{x:2<x<6\}$ (b) $\{x:4<x<6\}$ (c) $\{x:-2 \leq x \leq 6\}$ (d) $\{x:2 \leq x \leq 11\}$ (e) $\{x:-5<x \leq -2\}$ (f) $\{x:-2<x \leq 4\}$ (4) (a) $\{x:x \geq 1\}$ (b) $\{x:x \leq 1\}$ (c) $\{x:x>-1\}$ (d) $\{x:x<-6\}$ (5) $x/10 \leq 1$, x≤10 (6) 25y<150, y<6 (7) 30z≥100, z≥$10/3$ (8) $m/50 \leq 5$, m≤250 (9) 17x+24x<500, x<13 (10) 15+0.05x<50, x<700 (11) 4x+4<60, x<14

Exercise 10.3.2 Page 98
(3) (a) y≥1 (b) x+y≥1 (c) x+2y≤2 (d) x≥2y (4) 10x+15y≤100 (5) x+y≤8 (6) 3x+4y≥15 (9) 4X+6Y≥16, 5X+3Y≥11 (10) x>y, 80x+150y≥900, 5y+4x≤44 (11) 4x+3y≤600, 50x+20y≤5,000, y≥50 (12) $x/5 + 5y/6 \leq 240$, $4x/5 + y/6 \leq 250$

Chapter 11

Exercise 11.1.2 Page 102 (1) (a) 110° (b) 120° (c) 10° (d) 130° (2) (a) 60° (b) 80° (c) 30° (3) (a) 180° (b) 90° (c) 1,080° (4) (a) 2 (b) 10 (c) $1 1/2$ (d) $1/3$ (5) 270° (6) (a) 90° (b) 180° (c) 150° (7) 60° (8) 4 hours (9) y=x=145°, z=35° (10) 20°, 160°

Exercise 11.1.4 Page 103 (1) a=c=e=130° b=d=f=50° (2) a=140° b=70° c=60° d=50° (3) 90° (4) a+b=180° (5) 50° (6) c&d

Exercise 11.2.2 Page 106 (1) (a) 80° (b) 75° (c) 70° (d) 65° (2) (a) 30° (b) 110° (c) 85° (d) 70° (3) (a) 140° (b) 60° (4) (a) 7 (b) 4 (5) x=40°, y=100° (6) z=100° (7) 30° (8) Both 61° (9) Equilateral (12) Square (13) Rhombus

Exercise 11.2.4 Page 108 (1) (a) 540° (b) 720° (c) 1080° (d) 3240° (2) (a) 108° (b) 144° (c) 150° (d) 168° (3) (a) 45° (b) 30° (c) 20° (d) 12° (4) 7 (5) 9 (6) 5 (7) 12 (8) 88° (9) 60° (10) 110° (11) (a) 36 (b) 18 (d) 12 (12) 9 (13) 10 (14) 108°, 36°, 72° (15) 120°, 90°, 60°, 30°, rectangle (16) 100° rhombus (17) 108° (18) 3, 4, 6 (19) 150°, 12

Exercise 11.3.2 Page 112 (1) a&c (2) ΔBAC & ΔBED, <ABC=<EBD, <BAC=<BED (3) ΔBAC & ΔBDE, <BAC=<BDE, <BCA=<BED, parallel (4) ΔXPQ & ΔXRS, XP=XR, XS=XQ (5) ΔACD & ΔCAB, <D=<B, <DAC=<ACB, <ACD=<CAB, parallel, parallelogram (7) AC, ΔADC≡ΔABC, ΔADX≡ΔABX, ΔCDX≡ΔCBX, <D=<B (8) a (9) ΔALM & ΔABC (12) (a) 3 (b) 4 (c) 6 (d) 1 (13) 30 cm (14) 80 cm, 160 cm (15) 1.5 m, 1 m

Exercise 11.3.4 Page 115 (10) 1:4, 2, 30 (11) ΔAEF & ΔCDF, 1:9 (12) ΔAXY & ΔABC, 1:3, 64 (13) ΔABC & ΔAQP, <APQ=<C, <AQP=<B, 36, 45 (14) ΔABC & ΔDAB (a) ²⁰/₃ (b) 4

Chapter 12

Exercise 12.1.2 Page 119 (2) 100 m, 314 m (3) 25 cm (4) 7.85 cm (5) 9.4 cm (6) 9.4 cm (7) 127 m, 63.5 m (8) 3.18 cm (9) 7.85 cm (10) 10 cm (11) 1,244 in, 20.7 in (12) 52.5 cm, 26.5 times (13) (a) 50° (b) 140° (c) 150°

Exercise 12.2.2 Page 121 (2) (a) x=90°, y=20° (b) z=45° (c) p=30°, q=120° (4) 70°, 20°, 90° (5) 2.62 (6) 82° (7) 5:4:3 (8) <ACO=90°, congruent, 40° (9) OA, 40° (10) AC=AB, AD=DB, AM=MB, ΔCMA≡ΔCMB, kite (11) <C=<D=100°, <B=80° (12) congruent, equal

Exercise 12.3.2 Page 124 (1) (a) 50° (b) 294° (c) 61° (2) (a) 43° (b) <CDB (c) ΔAXD & ΔBXC, ΔDXC & ΔAXB (3) (a) 124° (b) 106°, 74° (4) (a) 100° (b) 85° (c) 100° (d) 55° (e) 70° (5) 184° (7) 70°, 70°, 35° (8) both 44° (9) 70°, 110°, 70°, parallel (11) rectangle

Exercise 12.4.2 Page 126 (1) (a) 67° (b) 135° (c) 10° (2) 35° (4) 108° (6) square

Exercise 12.4.4 Page 127 (2) similar (5) 10 cm (6) (a) 43° (b) 80° (c) 45° (7) 60°, 57°, 63° (8) 70°, 54°, 56° (9) 20° (10) 82.5°, 97.5°, 97.5°, 82.5° (11) 140°, 70°, 45° (12) (a) 40° (b) 100° (c) 67°

Exercise 12.5.2 Page 130 (1) (a) 4 (b) 1 (c) 5 (d) 10 (2) 7¹/₂ (3) 22 (4) 12 (5) 8 (6) 6¹/₂ m (7) 15 (8) 5 (9) 22.6 km.

Chapter 13

Exercise 13.2 Page 133 (1) 5, 8, 5 (2) 5, 9, 6, ABC & DEF, AD & CF etc (4) 12,4 (5) 27, 3, 12, 12, 4 (6) (a) 4 (b) 12 (c) 8 (7) cube, F, H (9) b&d (10) cuboid (11) (a) pyramid (b) cuboid (c) prism (d) pyramid

Exercise 13.4 Page 135 (2) (a) tetrahedron (b) ΔBED (c) 4, 6, 4

Exercise 13.5 Page 137 (1) (a) 3 (b) 5 (c) 9 (d) ∞ (e) 4 (f) ∞ (g) ∞ (h) 4 (2) (a) 3 (b) 5 (c) 11 (d) ∞
(e) 1 (f) ∞ (g) 1 (h) 4 (4) (a) pyramid, 5, 8, 5 (b) tetrahedron, 4, 6, 4 (5) (a) cuboid, 6, 12, 8 (b)
prism, 5, 9, 6 (c) tetrahedron, 4, 6, 4 (6) BH, DF, EC (7) (a) 8, 12, 6

Chapter 14

Exercise 14.1.2 Page 140 (1) (a) A3 (b) C3, C4, C5, B5, A5 (2) (1,3), (2,2), (3,1), (4,2) (4) (3,2)
(5) (4,3), $(2^1/2, 2^1/2)$ (7) 70 mins, 90 mins, 110 mins
(8) Units | 100 | 400 | 800 | 1200 | 1500 |
    A in £| 25 | 40 | 60 | 80 | 95 |
    B in £| 13 | 37 | 69 | 101 | 125 | By B. Over 500 units.

Exercise 14.1.4 Page 143 (1) (a) 1 (b) 2 (c) -1 (d) $-^1/2$ (2) 3 (3) $^1/2$, x=1, y=4 (4) (a) 4 (b) 1 (c) $^1/2$
(d) $-^3/4$ (5) (a) (3,4) (b) (3,2) (c) (2,3) (d) (4,1) (6) parallel (7) (a) 1.4 (b) 0.75 (c) -5 (8) (a) £0 (b)
£1,350 (c) £4,100 (d) £8,600, £17,250

Exercise 14.2.2 Page 145 (1) (a) 1 mile, 20 min (b) 11 miles, 30 min (c) 10 min (2) (a) $^1/2$ mile (b)
35 min (3) (a) 1 hr, $^1/2$ hr, $1^1/2$ hr (b) 30 m, 40 m, 30 m (c) 30 m.p.h., 80 m.p.h., 20 m.p.h. (d)
$^{100}/3$ m.p.h. (4) (a) 10 min (b) 4 miles (c) 24 m.p.h. (5) (a) 4 min (b) 30 m.p.h. (c) $2^1/2$ min (6) (a)
4 years (b) 1983 (c) 1988 (7) (a) 21°, 2° (b) 3-4 a.m. (c) 2 hours (8) (a) 1980-81 (b) 1982-83 (c) Late
1982 (9) (a) 111 m, 4.8 secs (b) 3.5 secs, 6.1 secs (c) 9.3 secs

Exercise 14.2.4 Page 148 (1) (a) 3 m.p.h (b) 6 m.p.h. (c) $1^1/2$ hours (2) 3 m.p.h., 22 m.p.h. (3) 2
m.p.h., 6 m.p.h., 6 m.p.h. (4) 20 m/sec, 0 m/sec (5) 0.75 ins/month, 0.45ins/month (6) (a) 2,000
galls (b) 10-11, 1 hour (c) 9% (d) 160 galls/hr

Exercise 14.2.6 Page 151 (1) (a) $^1/3$ $^m/s^2$, $^2/3$ $^m/s^2$ (b) 1,125 m (2) (a) $^{2m}/s^2$ (b) 45 m (3) (a) 1 $^m/s^2$
(b) 9 m (4) after $^3/4$ hour, $2^5/8$ miles (5) 2 hours, 100 miles (6) 1.10 & 1.00, at 1.30 (9) c

Chapter 15

Exercise 15.1.2 Page 155
(1)
  x=| -1 | 0 | $^1/2$ | 1 | $1^1/2$ | 2 | 3 |
  y=| 4 | 1 | $^1/4$ | 0 | $^1/4$ | 1 | 4 | gradient=-4

(2)
  x=| $^1/4$ | $^1/2$ | $^3/4$ | 1 | $1^1/2$ | 2 | 3 |
  y=| 9 | 5 | 3.7| 3 | 2.3 | 2 | 1.7| x=0.7

(3)
  x=| -3 | -2 | $-1^1/2$| -1 | $-^1/2$ | 0 | 1 |
  y=| 2 | -1 |-1.$^3/4$| -2 |-$1^3/4$ | -1 | 2 | x=0.4 or -2.4

(4)
  x=| $^1/8$ | $^1/4$ | $^3/8$ | $^1/2$ | $^3/4$ | 1 | $1^1/2$| 2 | 3 |
  y=| $8^1/8$ | $4^1/4$ | 3 | $2^1/2$| 2.1 | 2 | 2.2| $2^1/2$| 3.3 | at x=$^1/2$ grad=-3,
at x=2 grad=$^3/4$

(5)
  x=| 1 | 2 | 3 | 4 |
  y=| 1 | 4 | 7 | 10 |

(6)

x= |0 |1 |1¹/₂| 2 |3 |

y= |1 |-1 |-1¹/₄ | -1 | 1 | 2.6 and 0.4

(7)

x= |-2 |-1 |-¹/₂ |0 | ¹/₂ | 1 | 2 |

y= |-1 | 2 | 2³/₄ | 3 | 2³/₄ | 2 |-1 | 1.4 and -1.4, grad=2

(8)

x=0.7, grad=¹/₄

Exercise 15.1.4 Page 157

(1)

x= |-2 |-1 |-¹/₂ |0 | ¹/₂ | 1 | 1¹/₂ | 2 |

f= |-6 |0 | ³/₈ |0 |-³/₈ | 0 | 1⁷/₈ | 6 | x=-1.3

(2)

x= |-1 |-¹/₂ |0 | ¹/₂ | 1 | 1¹/₂ | 2 |

f= |-3 |-⁵/₈ |0 |-³/₈ |-1 |-1¹/₈ | 0 | x=-1 or 1 or 2

(3)

x= | ¹/₄ | ¹/₂ |³/₄ | 1 | 1¹/₂ | 2 |

y= | -4 |-1³/₄ |-.77| 0 |1.58 | 3¹/₂ | 1.2

(4)

x= | ¹/₄ | ¹/₂ |³/₄ | 1 | 1¹/₂ | 2 | 3 |

f= |16¹/₄ |4¹/₂ |2.5 | 2 | 1.9 | 2¹/₄ |3.1 | 0.6

(6) -¹/₂ (7) (a) 60°, 300° (b) 120°, 240° (c) 50°, 310° (d) 140°,220° (10) (a) 2 (b) -1 (c) -10 (d) -13
fg:x → -6x+8, gf:x → -6x+5

Exercise **15.2.2 Page 160** (1) (a) y=3x (b) y=x+1 (c) y=7-3x (d) y=¹/₂x+¹/₂ (e) y=2x (f) y=-1-2x (2)
(a) y=2x (b) y=¹/₂x+1 (c) y=2-2x (3) P=0.55x+.1, P=4.5 (4) y=¹/₄x+0.4 (5) (a) y=1.4x-0.2 (b)
y=0.75x+0.7 (c) y=-5x+32 (6) (a) 22m (b) 3.5 sec and 0.5 sec (c) 4.1 sec (7) (a) 205 ft (b) 54
m.p.h. (8) x=5 cm (9) 3.3 cm (10) A=12 sq. in.

Exercise **15.3.2 Page 162** (1) 18, 24, 45 (2) £18.50, £22.50 (3) 10, 26, 186

Exercise **15.3.4 Page 163** (1) 1.73 (2) 3.162 (3) 24, 720 (5) primes 5, 17 in A, rest in B

Chapter 16

Exercise 16.1.2 Page 166
(1) 1.5 (2) 8.5 cm, 1.7 cm (3) 54°, 48°, 78°, sum is 180° (5) 5.4, 45° (6) (a) 2.2 cm (b) 63° (c) 4
(7) 81°, 55°, 44° (8) 70° (9) 90° (10) 5.9 cm, 5 cm (11) 15.8 cm, 15.2 cm (12) 3.0. 2.3 (13) 5.1
cm, 83° (14) 7.1 cm, 52° (15) 9.3 cm

Exercise 16.1.4 Page 169
(5) 3.6 cm (6) 2.1 cm (7) 12.7 cm, 44° (9) 7.3 cm (10) 8.6 cm (11) hexagon, right-angled
triangle, 8.7 cm (14) GC=GC'

Exercise 16.2.2 Page 173
(1) (a) 090° (b) 180° (c) 000° (d) 045° (e) 225° (2) (a) E (b) W (c) SE (d) NW (3) (a) 20 km (b) 060°
(4) 56 km, 116° (5) (a) 70 km, 060° (b) Bristol (c) Norwich (6) 22 miles, 027° (7) 025°, 1.9 miles,
72° (8) 126 m (9) 35 m (10) 10.5 km (11) 60.5 miles, 100°

Chapter 17

Exercise 17.1.2 Page 178
(1) (a) 12 (b) 9 (c) $15^3/4$ (d) 8 (2) (a) 14 (b) 12 (c) 16 (d) 18 (3) (a) 1,890 sq ft 186 ft (b) £9.45 (c)
£65.10 (4) £16.10 (5) 6m² £96 (6) 30 cm (7) 625 m² (8) (a) 6 (b) $3^1/2$(c) 4 (d) 6 (9) (a) 2 (b) 3 (c)
2 (d) 2 (10) 12 ft (11) 10 cm (12) 6, 10 (13) 6 (14) 6 (15) 314 m (16) 191 m (17) 40,000 km
(18) 1,634 cm (19) 6 m² (20) 576 sq ft (21) 759 cm² (22) (a) 41 cm (b) 25 cm (23) (a) 16 cm² (b)
2 cm² (c) 8 cm² (24) (a) 6 cm²

Exercise 17.1.4 Page 181
(1) (a) 24 (b) 10 (c) 12 (2) (a) 15 (b) 22.5 (c) 12 (3) (a) 4 cm² (b) 3 cm² (c) 6 cm² (d) 4 cm² (4) 6
(5) $3^1/2$ (6) 14 cm² (7) (a) 28 cm² (b) 12.6 sq ft (c) 452 m² (d) 0.20 sq in (8) (a) 3 cm (b) 1.9 in (c)
1.7 m (d) 1.3 ft (9) 100 sq in (10) 0.215 (11) 3.43 cm²

Exercise 17.2.2 Page 183
(1) (a) 112 (b) 72 (c) 108 (2) 120 (3) 150,000 kg (4) 3.12 m³ (5) 4 cm (6) 0.4 m (7) 18 (8) 512,
5.0625 g (9) (a) 6 cm², 1 cm³ (b) 10 cm², 2 cm³

Exercise 17.2.4 Page 185
(1) (a) 141 (b) 154 (c) 9.42 (2) (a) 7 (b) 162 (3) 77 c.c. (4) 1.96 c.c. (5) 51.5 m · 5,150 cm³ (6)
54 cm² (7) 3.28 m² (8) (a) 7.07 cm² (b) 28,300 cm³ (9) (a) 17,700 cm³ (b) 31,400 cm³ (c) 13,700
cm³ (10) 1.99 cm (11) 2,550 cm (12) 50 cm² (13) 6.37 (14) (a) 2.8125 m² (b) 8.4375 m³ (15)
1,000 cm, 39.8 times

Exercise 17.2.6 Page 188
(1) (a) 524 cm³ (b) 33.5 m³ (c) 1,767 in³ (2) (a) 314 cm² (b) 50.3 m² (c) 707 sq in (3) (a) 151 (b)
352 (c) 39.3 (4) 2.88 cm (5) 2.02 cm (6) (a) 2.22 cm (b) 1.42 in (c) 1.10 mm (7) 11 cm² (8) 50.4
mm³ (9) 2,610,000 m³ (10) 264 cm³ (11) (a) 50.27 (b) 50.29 (12) (a) 0.427 cm (b) 0.043 cm (c)
0.54, 1.9 cm (13) 9.15 cm (14) 35,200 cm (15) (a) 61.3 cm³ (b) 3,063 cm (16) 283 cm/sec (17)
0.75 cm (18) (a) 402 cm³ (b) 2 cm (c) 50.3 cm³ (d) 352 cm³ (19) 1:63 (20) (a) 20 cm (b) 293 cm³
(21) (a) cone (b) rectangle

Exercise 17.3.2 Page 190
(1) (a) 14 (b) 14 (c) 14 (d) 10 (2) (a) 4 cm (b) 5 cm² (c) 4 cm² (d) 7 cm² (3) 163 m, 1,314 m² (4)
16 cm² 9,600 cm³ (5) 6 m², 15 m³ (16) 3,030 cm² (7) 1,636 cm³

Exercise 17.3.4 Page 192
(1) (a) 2.09 (b) 12.4 (c) 23.6 (2) (a) 2.09 (b) 4.97 (c) 15.7 (3) 50 cm², 28.5 cm² (4) 32.6 cm (5)
3.36, 6.71 (6) 60.6 cm², 30.3 cm (7) 64.9 cm³, 132 cm² (8) 11.8 cm³, 1.05 cm³ (9) 6.28 cu. in.
(10) 4 cm (11) 96 litres, 11,300 cm² (12) 2:3 (13) 364 cm³ (14) 47.1 cm² (15) 216° (16) 302 sq
ft, 188 sq ft (17) (a) 8.38 cm (b) 1 $^1/3$ cm (c) 7.89 cm (d) 14.7 cm³ (18) 4:1, 8:1 (19) $^3/4$ (20) 8
(21) 101:100, 102:100, 2%, 3% (22) (a) 8:1 (b) $^7/8$ (23) (a) 1,230 cm³, 550 cm² (b) 1:6 (c) 1:36 (d)
535 cm²

Chapter 18

Exercise 18.1.2 Page 198
(1) (a) 0.391 (b) 0.682 (c) 3.732 (d) 0.901 (e) 1.000 (2) (a) 53.1° (b) 76.1° (c) 32.2° (d) 78.5° (3) (a) 7.71 (b) 8.66 (c) 7.46 (4) (a) 5.14 (b) 4.77 (c) 5.31 (5) (a) 26.6° (b) 36.9° (c) 55.2° (6) (a) 73.4° (b) 53.1° (c) 22.0° (7) 7.13 cm, 3.63 cm (8) 66.2°, 23.8° (9) 35.2°, 54.8° (10) 60.7 m, 56.6 m

Exercise 18.1.4 Page 199
(1) (a) 7 (b) 7.15 (c) 5.85 (d) 7.71 (2) (a) 77.6° (b) 47.2° (3) 7.66, 10 (4) 2.75 (5) 33.2° (6) 53.1°, 126.9° (7) 23.371 in (8) 2.66 m (9) 28° (10) 11.3° (11) 63.4°, 36.9° (12) 33.9° (13) (a) 72° (b) 4.85 cm (c) 7.05 cm (d) 85.6 cm^2

Exercise 18.2.2 Page 202
(1) 89 m (2) 1.4° (3) 1.64 m (4) 31° (5) 26 ft (6) 26.6° (7) 7.66° (8) 62° (9) 153 m (10) 163 m (11) 21.2 m (12) (a) 5.7° (b) 9.95 m (13) 195 m/sec (14) 21.8° (15) 218 km, 335 km (16) 149° (17) 33.3 km

Exercise 18.2.3 Page 203
(1) 186.6 m., 53.8° (2) 81.4 km, 97° (3) 570 m (4) 113 m (5) 955 m (6) 30°, 10.4 ft

Exercise 18.3.2 Page 205
(1) (a) sin 42° (b) -cos 81° (c) -tan 4° (d) sin 62° (e) -cos 77° (2) (a) 150° (b) 139° (c) 156° (d) 168° (e) 133° (f) 110° (3) 30°<x<150° (4) y≥96° (5) 84°<z<90°

Exercise 18.4.2 Page 207
(1) (a) 8.38 (b) 4.64 (c) 5.52 (2) (a) 46° (b) 34° (c) 92° (3) 5.19, 2.19 (4) 43°, 75° (5) 18.6 m, 7.27 m (6) 1,010 m (7) 76° 2.01m (8) 148 miles (9) (a) 9.67 (b) 22.7 (10) (a) 41.4° (b) 134.4° (11) (12) 31.3° (13) 121 miles (14) 40.7 cm (15) 101°, 79°, 12.7 cm (16) 215°, 331° (18) 6.6 m (19) (a) 9.99 (b) 7.42 (c) 10.3 (20) (a) 36° (b) 102° (c) 135° (d) 62° (21) 11.3, 48° (22) 83°

Chapter 19

Exercise 19.1.2 Page 212
(1) (a) 12.04 (b) 13 (c) 15 (d) 7.14 (2) 5.83 miles (3) 10.63 cm (4) 93 ft (5) 8.2 cm (6) 5.66 cm (7) 19.84 (8) 3 in, 5.29 in (9) 50 m, 50.06 m (10) 4 ft (11) 199.75 cm, 0.25 cm (14) 6.32 (15) 3.46 cm, 6.93 cm^2 (16) (a) 1.414 cm (b) 2.236 cm (c) 3.606 cm (17) (a) 2.236 cm (b) 5.831 cm (c) 7.211 cm (d) 5.385 cm

Exercise 19.2.2 Page 215
(1) 10.49, 28.5°, 55.1°, 48.1° (2) 8.60, 5.10, 8.66, 78.7°, 53.9° (3) 5.66 cm, 5.29 cm, 61.9°, 70.5° (4) 10.44 cm, 73.3° (5) 3.54 ft, 34.4°, 45°, 25.1° (6) 141.4 m, 86.6 m, 70.7 m, 35.3°, 45° (7) (a) both 6.93 cm (b) 70.5° (c) 2.31, 4.62 (d) 54.7°

Exercise 19.3.2 Page 217
(1) (a) 4.60 (b) 3.86 (c) 8.45 (2) 8.70 (3) 166 cm^2 (4) 10.2 cm, 248 cm^2 (5) 8.45 cm, 549 cm^2 (6) 9 cm, 5 cm^2 (7) 8.19 cm^2, 98.3 cm^3 (8) 23 cm^2, 230 cm^3

Chapter 20

Exercise 20.1.2 Page 222
(1) (a) 1 (b) 5 (c) 29 (2) (a) 4 (b) 12 (c) 33 (3) (a) 23 (b) 129 (c) 421 (d) 194 (4) (a) 15 (b) 150° (5) (a) 6 (b) 15° (c) 16 (6) (a) £20 (b) £7.50 (9) (a) 24 (b) vanilla (c) 56 (11) (a) 65,000, 40,000

Exercise 20.2.2 Page 226

(1)

Number of letters	1	2	3	4	5	6	7	8	9+
Frequency	12	13	11	13	14	10	7	3	8

(2)

Score	1	2	3	4	5	6
Frequency	13	4	10	12	6	15

Exercise 20.3.2 Page 228

(1) (b) 30-35 (c) 0.382 (2) (b) 20-30 (c) 0.6

Chapter 21

Exercise 21.1.2 Page 232

(1) 6, 5, 7 (2) 27, $26^1/_2$, 19 (3) 118 $^1/_3$ m, 110 m (4) 6 cm, 5.9 cm (5) 103, 103, 21 (6) 2,0 (7) 48 (8) 47.2 g (9) £9,050, £7,450 (10) 2.1, 1 (11) 82.5 p, 82 p (12) $249^1/_4$ g, 249 g (13) 7 hrs, 7.3 hrs, 6.9 hrs (14) 4.44 mill, 4.3 mill, (15) 42 $^7/_9$ hrs, 42 hrs, 34 hrs (16) 3, 3 (17) $25^3/_4$, $25^1/_2$

Exercise 21.2.2 Page 235

(1) 5.184, 5, 5 (2) 1.375 (3) 2.758, 2 (4) 2.076 (5) 1, 1.7475 (6) 0.833, 1 (7) 1, 1.25 (8) 3.035 (9) 20-25, 23.8 (10) 60-70 cm, 67.2 cm (11) £10,600 (12) 2-4 hrs, 4 $^2/_3$ (13) 1 min, 2.7 min (14) 2,510 m (15) 51

Exercise 21.3.2 Page 238

(1) (a) 10, 33, 51,60 (c) 169, 163, 177, 14 (d) $^2/_3$ (2) (a) 7, 22, 50, 69, 80 (c) 77, 69, 89, 20 (d) 0.7
(3) (a) 74, 147, 251, 367, 434, 500 (c) 30, 24, 36, 12 (4) (a) 22, 40, 56, 68, 80 (c) 30, 19, 43, 24
(5) (a) 24, 39, 56, 75, 85, 100 (c) £10,600, £9,100, £12,000, £2,900 (6) (a) 20, 29, 44, 54, 60 (c)
3.5, 2.7, 4.1, 1.4 (7) (a) 4, 17, 26, 37, 41, 50 (c) 15.4, 14.7, 16.1, 1.4 (8) For A, 1, 4, 14, 28,
30, 30: for B, 2, 7, 14, 23, 27, 30 IQ range for A=12, for B=19 (9) For Lucy, 2, 5, 28, 35, 40, 40
for Liz, 5, 13, 27, 32, 36, 40. IQ range for Lucy=4, for Liz=10

Chapter 22

Exercise 22.1.2 Page 242

(1) $^1/_6$ (2) $^1/_2$ (3) $^1/_4$ (4) $^{17}/_{25}$ (5) $^4/_{11}$ (6) $^1/_5$ (7) $^{93}/_{100}$ (8) $^1/_{37}$ (9) $^2/_5$ (10) $^1/_{99}$ (11) $^1/_{13}$ (12)
(a) $^1/_{11}$ (b) $^2/_{11}$ (13) (a) $^1/_8$ (b) $^1/_2$ (14) (a) $^1/_{24}$ (b) $^1/_6$ (c) $^1/_4$ (15) (a) $^1/_{36}$ (b) $^1/_6$

Exercise 22.2.2 Page 244

(1) $^5/_6$ (2) $^{36}/_{37}$ (3) $^3/_4$ (4) (a) $^1/_{25}$ (b) $^{16}/_{25}$ (5) (a) $^1/_4$ (b) $^1/_4$ (c) $^1/_2$ (6) $^2/_3$ (7) $^1/_6$ (8) (a) $^1/_{36}$ (b)
$^1/_6$ (c) $^8/_9$ (d) $^5/_{18}$ (9) (a) $^1/_{36}$ (b) $^1/_6$ (c) $^5/_{12}$ (10) $^1/_{21}$ (11) $^1/_6$ (12) $^{22}/_{57}$

Exercise 22.3.2 Page 246

(1) (a) $^{81}/_{100}$ (b) $^9/_{50}$ (2) (a) $^4/_7$ (b) $^3/_7$ (3) (a) $^{28}/_{171}$ (b) $^{83}/_{171}$ (c) $^{88}/_{171}$ (4) (a) $^9/_{16}$ (b) $^{15}/_{16}$ (5) (a)
$^1/_{36}$ (b) $^{25}/_{36}$ (c) $^{11}/_{36}$ (6) (a) $^{66}/_{325}$ (b) $^{168}/_{325}$ (7) (a) $^1/_{15}$ (b) $^9/_{20}$ (c) $^7/_{60}$ (d) $^3/_{25}$ (8) (a) $^1/_4$ (b) $^1/_8$
(c) $^1/_4$ (9) $^9/_{14}$ (10) (a) $^2/_5$ (b) $^4/_{15}$ (11) (a) $^1/_{17}$ (b) $^4/_{17}$ (c) $^1/_{52}$ (12) (a) $^1/_6$ (b) $^5/_{36}$ (c) $^{25}/_{216}$ (13)
(a) $^1/_2$ (b) $^3/_{20}$ (c) $^1/_{20}$ (d) $^1/_5$ (14) (a) $^3/_{16}$ (b) $^1/_6$ (c) $^1/_{24}$ (d) $^5/_{16}$ (15) (a) $^1/_3$ (b) $^4/_{15}$ (16) (a) $^7/_{15}$ (b)
$^{43}/_{60}$

Chapter 23

Exercise 23.1.2 Page 250
(10) (a) 2 (b) 4 (c) 3 (11) 8008 (12) 6009 (13) translation, 3 right, 1 down (b) trans. 3 right, 2 up (c) D, K (d) reflection in CIOU (e) rotation of 180° (f) 90° clockwise rotation (g) enlargement, factor 2 (15) (a) translation (b) reflection (c) reflection

Exercise 23.1.4 Page 253
(1) (a) (1,-1) (3,-1) (2,-5) (b) (-1,1) (-3,1) (-2,5) (c) (1,1) (1,3) (5,2) (d) (-1,-1) (-1,-3) (-5,-2) (2) (a) (1,2) (1,1) (-3,2) (b) (1,4) (1,3) (-3,4) (c) (-1,-2) (-1,-3) (-5,-2) (d) (5,6) (6,6) (5,2) (e) (0,1) (-1,1) (0,5) (3) reflection in y-axis (4) reflection in y=x (5) rotation of 90° clockwise about (0,0) (6) rot. of 180° about (0,0) (7) (2,-1), (2,1) (5,1) (8) (4,-1) (5,1) (3,2) (9) enlargement, factor 2

Exercise 23.1.6 Page 254
(1) rot. 90° anti cw about B (2) rot 180° about A (3) rot 90° cw about (0,-1) (4) rot. 180° about (-1,-1) (5) reflection in y=1 (6) refl. in x=-2 (7) enlargement factor 4 about B, 16:1 (8) enlargement factor -2 about P, 4 (9) (6,2) (4,3) (4,2) (10) (0,-4) (2,-5) (2,-4) (11) (0,0) (-1.3,1.5) (0.8,0.6) (-0.5,2.1) (12) (0,2) (1.9,0.8) (1,0.3)

Exercise 23.2.2 Page 256
(1) trans of 2 to right (2) trans of 4 down (3) rot of 180° about A (4) rot of 180° about (2,-1) (5) trans 2 to left, 2 down (6) rot of 180° about L (7) trans 2 to left 6 down (8) (a) rot 180° about (0,0) (b) rot 90° cw about (0,0) (c) trans 2 left (d) trans 2 up (e) trans 2 right (f) trans 2 down (g) trans 2 down, $(PQ)^{-1}=Q^{-1}P^{-1}$ (9) (a) refl in x=y (b) nothing (c) rot 90° acw about (0,0) (d) rot 90° cw about (0,0)

Chapter 24

Exercise 24.1.2 Page 260

(1) (a) $\begin{pmatrix} 6 & 2 \\ -1 & 16 \end{pmatrix}$ (b) $\begin{pmatrix} 11 & 0 \\ -4 & 25 \end{pmatrix}$ (c) $\begin{pmatrix} -10 & 26 \\ 19 & 8 \end{pmatrix}$

(2) $AB = \begin{pmatrix} -7 & 34 \\ -11 & 59 \end{pmatrix}$ $BA = \begin{pmatrix} 1 & 6 \\ 15 & 51 \end{pmatrix}$ (3) (a) None (b) DE, EC

(4) $DE = \begin{pmatrix} 38 & 30 & -2 \\ 39 & 27 & 0 \end{pmatrix}$ $EC = \begin{pmatrix} 8 & -5 & 41 \\ 18 & 22 & 33 \end{pmatrix}$ (5) $\begin{pmatrix} 2 & -3 \\ -2^1/2 & 1 \end{pmatrix}$

(6) $\begin{pmatrix} 9 & 32 \\ 16 & 57 \end{pmatrix}$, $\begin{pmatrix} 31 & -28 \\ -42 & 87 \end{pmatrix}$, $\begin{pmatrix} 34 & 60 \\ 30 & 54 \end{pmatrix}$ (8) $A^{-1} = \begin{pmatrix} -7 & 4 \\ 2 & -1 \end{pmatrix}$: x = 42, y = -11

(9) $B^{-1} = \frac{1}{39} \begin{pmatrix} 9 & 2 \\ 3 & 5 \end{pmatrix}$ x = 4, y = 2 (10) x = 1, y = 2 (11) x = y = -1

(12) (a) 1 = (d) -39 $\begin{pmatrix} 59 & -34 \\ 11 & -7 \end{pmatrix}$ (b) 1 = (c) -39 $\begin{pmatrix} 51 & -6 \\ -15 & 1 \end{pmatrix}$

(14) $\begin{pmatrix} -39 & 22 \\ 39 & -21 \end{pmatrix}$ (15) $\begin{pmatrix} -47 & 50 \\ 13 & -13 \end{pmatrix}$

Exercise 24.2.2 Page 263

(1) (a) $\begin{pmatrix} 0 & 1 & 1 \\ 1 & 0 & 1 \\ 1 & 1 & 0 \end{pmatrix}$ (b) $\begin{pmatrix} 0 & 2 & 0 \\ 2 & 0 & 1 \\ 0 & 1 & 0 \end{pmatrix}$ (c) $\begin{pmatrix} 0 & 2 & 1 \\ 2 & 0 & 2 \\ 1 & 2 & 0 \end{pmatrix}$

$\begin{pmatrix} 2 & 1 & 1 \\ 1 & 2 & 1 \\ 1 & 1 & 2 \end{pmatrix}$ $\begin{pmatrix} 4 & 0 & 2 \\ 0 & 5 & 0 \\ 2 & 0 & 1 \end{pmatrix}$ $\begin{pmatrix} 5 & 2 & 4 \\ 2 & 8 & 5 \\ 4 & 2 & 5 \end{pmatrix}$

(2) (a) $\begin{pmatrix} 0 & 0 & 1 \\ 1 & 0 & 1 \\ 0 & 1 & 0 \end{pmatrix}$ (b) $\begin{pmatrix} 0 & 1 & 0 \\ 1 & 0 & 1 \\ 0 & 1 & 0 \end{pmatrix}$ (c) $\begin{pmatrix} 0 & 1 & 1 \\ 1 & 0 & 1 \\ 1 & 1 & 0 \end{pmatrix}$

(4) (a)   64.5, 72, 62   Average in each exam $\begin{pmatrix} 203 \\ 205 \\ 161 \\ 225 \end{pmatrix}$ Total marks

   (b) $w = \begin{pmatrix} 2/5 \\ 2/5 \\ 1/5 \end{pmatrix}$

(5) (a) $T = \begin{pmatrix} 1 \\ 1 \\ 1 \end{pmatrix}$ $NT = \begin{pmatrix} 74 \\ 76 \end{pmatrix}$ (b) $R = \begin{pmatrix} 1, & 1 \end{pmatrix}, RN = \begin{pmatrix} 66, 53, 31 \end{pmatrix}$

   (c) $\begin{pmatrix} 20 \\ 25 \\ 15 \end{pmatrix}$ 3110 P

(6) $P = R = \begin{pmatrix} 1 \\ 1 \\ 1 \\ 1 \\ 1 \end{pmatrix}$ $Q = S = \begin{pmatrix} 1, 1, 1 \end{pmatrix}$

(7) $L = \begin{pmatrix} 2 \\ 5 \\ 6 \end{pmatrix}$ $PL = \begin{pmatrix} 1404 \\ 1503 \\ 1488 \end{pmatrix}$

(8)   Butter $\begin{pmatrix} 1/2 & 2/5 \\ 1/2 & 2/5 \end{pmatrix}$ $\begin{pmatrix} 5 \\ 3 \end{pmatrix}$
     Flour
            A   B

(9) $(1\ 1\ 1\ 1)$ C $C \begin{pmatrix} 0 \\ -1 \\ 50 \end{pmatrix}$

Exercise 24.3.2 Page 267

(1) A (1,-2) (3,-2) (2,-1) B (-1,2) (-3,2) (-2,1) C (2,-1) (2,-3) (1,-2) D (-2,1) (-2,3) (-1,2) E (-2,-1)
(-2,-3) (-1,-2) (2) A (1,-1) (1,-3) (2,-3) (2,-1) B (-1,1) (-1,3) (-2,3) (-2,1) C (1,-1) (3,-1) (3,-2)
(1,-2) D (-1,1) (-3,1) (-3,2) (-1,2) E ((-1,-1) (-3,-1) (-3,-2) (-1,-2) (3) A. reflection in x-axis B. refl.
in y-axis C. 90° cw rotation D.90° acw rotation E. refl. in x=-y (4) F (0,0) (3,0) (3,3) (0,3) G (0,0)
(1/2,0) (1/2,1/2) (0,1/2) H (0,0) (^{1}/2,0) (1/2,1) (0,1) J (0,0) (1,0) (3,1) (2,1) K (0,0) (2,1) (3,3)
(1,2) (5) F. enlargement factor 3 G. enlargement factor ^{1}/2 H. compression along x-axis J shear
parallel to x-axis K. shear parallel to x=y (6) & (7) 9, ^{1}/4, ^{1}/2, 1, 3

(8) $$BA = \begin{pmatrix} 0 & -1 \\ -1 & 0 \end{pmatrix} \quad A^{-1}B^{-1} \begin{pmatrix} 0 & -1 \\ -1 & 0 \end{pmatrix}$$ (9) $$DC = \begin{pmatrix} 1 & 2 \\ 0 & -1 \end{pmatrix} \quad C^{-1} = \begin{pmatrix} 1 & -2 \\ 0 & 1 \end{pmatrix}$$

(10) $A^{17} = A$ (12) $\begin{pmatrix} 4 & 2 \\ -2 & -1 \end{pmatrix}$ (13) $\begin{pmatrix} 2 & 3 \\ 3 & 2 \end{pmatrix}$ (14) $\begin{pmatrix} 2/3 & 4/3 \\ 5/3 & 2/3 \end{pmatrix}$

(15) (14,-5) (16) (-4,1) (17) (0,0) (^{3}/4,-^{1}/8) (^{1}/4,^{1}/8) (-^{1}/2,^{1}/4) (18) b if det. is +ve, d if det. is -ve.

Chapter 25
Exercise 25.1.2 Page 271

(1) (a) $\begin{pmatrix} 7 \\ -6 \end{pmatrix}$ (b) $\begin{pmatrix} 29 \\ -9 \end{pmatrix}$ (c) $\begin{pmatrix} -11 \\ -17 \end{pmatrix}$ (d) $\begin{pmatrix} -3 \\ -8 \end{pmatrix}$

(2) (a) 7.28 (b) 5.10 (c) 21.95 (d) 8.54 (3) x=4 y=7 (4) x=1 y=2

(5) (a) $\begin{pmatrix} 3 \\ 8 \end{pmatrix}$ (b) $\begin{pmatrix} 9 \\ -13 \end{pmatrix}$ (c) x = 1, y = 4 (d) $\lambda = 2,\ \mu = 1$

(8) (a) $\begin{pmatrix} 1 \\ 1 \end{pmatrix}$ (b) $\begin{pmatrix} 1 \\ -2 \end{pmatrix}$ (c) $\begin{pmatrix} 2 \\ 0 \end{pmatrix}$

(9) (a) $\begin{pmatrix} 3 \\ -6 \end{pmatrix}$ (b) $\begin{pmatrix} 6 \\ 2 \end{pmatrix}$ (c) $\begin{pmatrix} 8 \\ -5 \end{pmatrix}$ (10) $\begin{pmatrix} 1 \\ -2 \end{pmatrix}$, $\begin{pmatrix} -1 \\ 2 \end{pmatrix}$

Exercise 25.2.2 Page 273

(1) $\underline{AB} = \begin{pmatrix} 1 \\ 2 \end{pmatrix}$, $\underline{AC} = \begin{pmatrix} 1 \\ -6 \end{pmatrix}$, $\underline{CD} = \begin{pmatrix} 2 \\ 4 \end{pmatrix}$, $\underline{BD} = \begin{pmatrix} 2 \\ -4 \end{pmatrix}$, $\underline{DA} = \begin{pmatrix} -3 \\ 2 \end{pmatrix}$, AB parallel to CD

(2) $\underline{JL} = \begin{pmatrix} 2 \\ 5 \end{pmatrix}$, $\underline{JK} = \begin{pmatrix} 1 \\ -3 \end{pmatrix}$, $\underline{ML} = \begin{pmatrix} -3 \\ 9 \end{pmatrix}$, $\underline{MK} = \begin{pmatrix} -4 \\ 1 \end{pmatrix}$, $\underline{MJ} = \begin{pmatrix} -5 \\ 4 \end{pmatrix}$, JK parallel to ML

(3) 3:1, no (4) no (5) PQ=SR= $\sqrt{13}$, QR=PS=$\sqrt{2}$, no (8) no, no

(9) ABD in straight line $AD = \begin{pmatrix} 9 \\ -3 \end{pmatrix}$ $AB = \begin{pmatrix} 3 \\ -1 \end{pmatrix}$

(10) L, M,
K (11) (0,0) (12) (5,-1) (13) 3 (14) c-b, ^{1}/4 b, ^{1}/4 c, ^{1}/4 (c-b), ^{3}/4 b, c -^{1}/4 b, BC parallel to DE, DB
parallel to AD (15) ^{1}/2 (a+b), ^{1}/2 (b+c), ^{1}/2 (c+d), ^{1}/2 (d+a), ^{1}/2 (c-a), ^{1}/2 (d-b), ^{1}/2 (a-c), ^{1}/2 (b-d),
parallelogram (16) c, b, 2c, c-b (17) e, -a, 2a, 2a+e (20) a, a+b, a-b (21) b+d, ^{1}/2 b, d+ ^{1}/2 b, d XY
parallel to AD (22) b=a+c, c, ^{1}/2 (c-a), ^{1}/2 (a+c), ^{1}/2 (c+a) X=Y (23) ^{1}/2 c, ^{3}/4 b,^{3}/4 b-c, ^{3}/8 b - ^{1}/2 c,
^{1}/2 c+^{3}/8b, ^{3}/8b, ZY parallel to AB (24) ^{1}/2 (a+b), ^{2}/3a, 2b, PQ=^{1}/6a - ^{1}/2 b, PR=1 ^{1}/2 b - ^{1}/2 a, PQR in
a straight line

385

Chapter 26
Exercise 26.1.2 Page 278
(1) 5,3 (2) 4 (3) 5,6 (4) 10 (5) 100, 0.56 (6) 15, 25, 4 (7) 23, 0 (8) 84 (9) 17, 19, 21 (10) 26
(11) x=3, 17, 140, 137/300

Exercise 26.2.2 Page 280
(1) (a) {2,4,6,8,10} (b) {7,9} (c) {1,3,5,6,7,8,9,10} (d) {2,4} (e) {1,2,3,4,5,6,8,10}   (2) (a)
{m.n,...,z} (b) {a,e,i} (c) {m,n,p,q,r,s,t,v,w,x,z} (d) {o,u,y} (e)=(c) (3) B={1,2,6,8} (4) (a)
{AS,AH,AD,AC} (b) Ø (c) {AH} (d) {AH,QH} (e) 4 (f) 48 (g) 8 (h) 44 (i) 12   (5)
A={feb,apr,jun,sept,nov} B={jan,feb,mar,apr,sept,oct,nov,dec}, 31 day months with an r, $x \in A \cap$
B' (6) A' ∪ B'=(A ∩ B)' (7) (a) {x:x>3} (b) {x:x≤-2} (c) {x:-2<x≤3} (d) ξ (e) Ø (f) A' (g) B (h)
B' (i) A (9) P⊂T, S=Re ∩ Rh (11) (a) B⊂A (b) B ∩ C=Ø (14) 86 (15) 8 (18) (b) 2 (c) 3 or 11

Solutions to Test Papers

London Group Paper 1. Total 70 marks

(1) £6.72 [1] (2) $^3/6 = {}^1/2$ [1] (3) 6.08 [1] (4) 56,217 [1]
(5) 35 minutes [1] (6) $^{18}/100 = {}^9/50$ [2] (7) $^7/5$, $1\,{}^2/7$, 1.3, $1\,{}^1/3$ [2]
(8) 50x3x3.14 = 471 m [2] (9) 16 boxes, 4 left over [2] (10) (a) 17, 20 [1] (b) 5, 11 [1] (11)
23-5.32 = 17.68 oz [2] (12) (a) $^4/10 = {}^2/5$ [1] (b) $^{46}/100 = 0.46$ [1].
(13)(a)   [2]   (b) 1 [1]   (14)   [2]

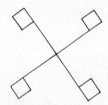

(15) (a) Cube [1] (b) $1\ cm^3$ [2] (16) A is 550÷25 = 22p. [1] B is 420÷20 = 21 p [1] B is cheaper
[1] (17) 3x3x2 = 18 [1] 3x3 = 9 [3]
(18) (a) 90° anticlockwise rotation. [2] (b)   [2]

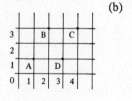

(19) (a) 35 km [2] (b) 020° [2] (c) 180° + 45° = 225° [2]
(20) (a) (1,1) [1]   (b)   [1]

(c) Parallelogram [2]
(d) 2x2 = $4\ cm^2$ [2]

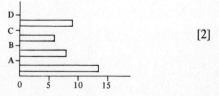

(21)

Type	A	B	C	D
Number	13	8	6	9	[2]

A is most popular [1]
Probability = $^{13}/36$ [2]

(22)

Time	12	1	2	3	4
Number	800	1600	3200	6400	12800	[3]

At 11 o'clock 800÷2 = 400 [1] At 10 o'clock 400÷2 = 200 [1]
(23) 2.6 cm, 1.6 cm, 2.5 cm. [3] 2.6+1.6+2.5 = 6.7 cm [1] (24) (a) $2x^{40}/5 + 3x^{70}/5 = 58$ [2](b)
$2x^{60}/5 + 3x^{48}/5 = 52.8$ [2]

387

London Group Paper 2. Total 84 marks

(1) 42x£4.75 = £199.5 [1] (2) 7-4 = 3° [1] (3) 12,000x(0.35-0.3) = £600 [2]
(4) (a) 7284 [1]

[1]

(5) $^{44}/_{66}$ hrs = 40 mins [2](6) B1, B3, A2, C2 4 squares [2] (7) (a) 8.15 [1] (b) 8.36 [1](8) 56x³/4 =
42 miles ≈ 40 miles [2] (9) (a) 3x£3.10 = £9.30 [1](b) 0.9x£9.30 = £8.37 [1] (10) 1 yolk, 1 tbs
lemon, 2 oz butter [2] (11) 6,340+642 = £6982 [1] 6982+4x402 = £8590 [2] (12) (a) 4x180 = 720
students [3] (b) ^{1}/6 of 360 = 60° [2] (c) 420÷2 = 210 [2] (d) (960 - 720)÷960 = ^{1}/4 [2]
(13) (a) 4 times [2] (b) 15 [2] (14) (a) £1,121 [2] (b) £1.295 [2] (15) (a) 1^1/2 miles [1] (b) 10
minutes [1] (c) 4^1/2 m.p.h. [2]
(16) [3]

BT = 13.5 cm, which represents 135 m. [2]

(17) (a) Monday [2] (b) Tuesday [2] (c) Monday [3] (d) 28 July [3]
(18)

Paper	News	Graphic	Bugle
Number	11	21	9	[2]

Total = 11+21+9 = 41 [2]Most popular, Graphic [1]

(19) (a) £17 (b) £19 (c) £21 [3] [3]

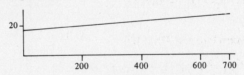

For £27, 600 unit calls [2]
(20) (a) ^{1}/2x3x2 = 3 cm² [2] (b) <ABC = 45° [2][3]

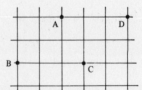

(21) (a) 20x2x3.14 = 125.6 m. [2](b) 125.6÷2 = 62.8 secs [2] (c) 40x2x3.14÷3 = 83.7 secs [3]

London Group Paper 3a. Total 30

(1) d	(2) e	(3) b	(4) d	(5) c	(6) e
(7) d	(8) b	(9) c	(10) b	(11) a	(12) a
(13) b	(14) b	(15) e	(16) e	(17) e	(18) a
(19) c	(20) e	(21) b	(22) c	(23) d	(24) c
(25) b	(26) e	(27) a	(28) d	(29) d	(30) c

london Group Paper 3b. Total 70

(1) (a) 25.62-(5.80+5.45+6.50) = £7.87   [3]
(b) 43.12-26.62 = £17.50   [3]
(c) 0.15x43.12 = £6.47   [3]
(d) 6.47+43.12 = £49.59   [3]

(2) (a)   [4]   [4]

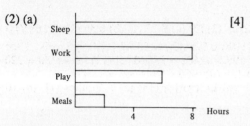 

(c) 360÷3 = 120° [1]
(3) (a) $\sqrt{40^2+50^2}$ = 64.0 m² [3](b) $\tan^{-1}(40/50)$= 39°   [3]
(c) $\tan^{-1}(64.0/30)$= 65° [3](d) ¹/2x40x50 + ¹/2x30x 4100 = 1960.47 m²   [4]

(4) (a) translation 2 to right. [2](b) 180° rotation   [2]
(c) enlargement factor 2 [2](d) reflection about vertical line   [2]

(5) (a) 5x30 = 150 sq ft [2](b) 30x45 - 26x41 = 284 sq ft   [4]
(c) πx3x3 = 28.3 sq ft [3](d) 26x41 - 28.3 = 1,038 sq ft   [2]

(6)   (a)   [4]

(b) ¹/2xπx50² = 3,927 sq miles [5]
(c) After 7.5 hours. For 10 hours [8]

London Group Paper 4. Total 100

(1) (a) F=^{k}/d², 4= ^{k}/25, k=100. Hence F = ^{100}/d²   [2]
     (b) 400=^{100}/d². d² = ¹/4. d=¹/2 or - ¹/2   [2]

(2) (a) ⁴/9x³/8 = ¹/6 [2] (b) ¹/6 + ⁵/9x⁴/8 = ⁴/9 [2] (c) 1-⁴/9 = ⁵/9   [2]

(3)   [4]

Area = ¹/2x3x(10+24) + ¹/2x3x(24+30) = 132 metres. [4]

(4) (a) $a=\sqrt{7^2+24^2}=25$ [1] (b) $b=\sqrt{8^2-2^2}=7.746$ [2] (c) $2c\leqslant11$. So $c = 1, 2, 3, 4, 5$ [5]

(5) <BCD=x. <BAD=$^x/2$. $x+^x/2=117°$. Hence $x = 78°$ [3]

(6) (a) $10x^2-35xy+6xy-21y^2 = 10x^2 - 29xy - 21y^2$ [2]

(b) $a^2 = 2 + x$. $x = a^2 - 2$ [3]

(c) $x = (-3 +/-\sqrt{9-4x1x(-3)})/2 = 0.791$ or -3.79 [4]

(7) (a)   [4]   (b)

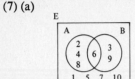

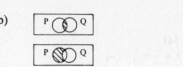

[1]
[2]
[3]

(8) (a) New base area $= 8\pi$. Height $= 8x9\pi/(8\pi)$. It rises 1 cm. [4]

(b) $hx\pi x3^2 = {}^4/3x(1^1/2)^2$. $h = 0.5$ cm [4]

(9) (a)
$$A^{-1} = \begin{pmatrix} 1 & -1 \\ -1^1/2 & 2 \end{pmatrix}$$
[2]

(b) $\begin{pmatrix} x \\ y \end{pmatrix} = A^{-1}\begin{pmatrix} 8 \\ 7 \end{pmatrix} = \begin{pmatrix} 1 \\ 2 \end{pmatrix}$ x=1, y=2

Hence x = 2, y = 4   [3]

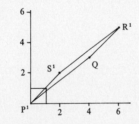

A parallelogram.   [4]

(10) (a) $PQ = \sqrt{200^2+100^2} = 224$ m.   [2]

(b) From P $\tan^{-1}(^{30}/_{100}) = 16.7°$ From Q $\tan^{-1}(^{30}/_{200}) = 8.5°$ [4]

(c) Volume is $^1/3(100x200x^1/2)x30 = 100,000$ m^3 [4]

(11) (a) $^1/2$**b, c-b**, $^1/2$**(c-b)**, $^1/2$**(c+b)** [4]

(b) Similar [2] (c) ΔAZY, ΔZBX, ΔYXC [3] (d) 24÷4 = 6 [2]

(12)

x | -3 | -2 | -1 | 0 | 1 | 2 | 3 |
y | -3$\frac{5}{8}$ | 0 | 1$\frac{5}{8}$ | 2 | 1$\frac{7}{8}$ | 2 | 3$\frac{1}{8}$ | [2]

[4] x = -1.1 [4]
Gradient at x = 2 is $\frac{1}{2}$ [4]

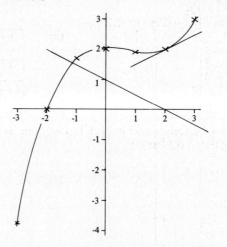

[4]

Midland Group Paper 1. Total 100

(1) $^4/9$ [1] (2) 123 + 27 = £150 [1] (3) 0.17x850 = £144.50 [2] (4) 66x2.54 = 167.64 cm [3]
(5) 0.9x(23+2x9) = £36.90 [2] (6) (a) 5x0.35=£1.75 [1](b) 5÷50=0.1 lb. [1]
(7) 1.1x(6.85+5.55)=£13.64 [2] (8)   [2]

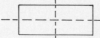

(9) (a) 3x4-7=5 [1] (b) $^{11}/3$x3-7=4 [1] (10) (a) 88 mins. [1] (b) 8 26 [1](11) $^2/6 = ^1/3$ [2]
(12)

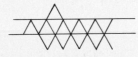

[3]

(13) (a) 17-3=14 [1](b) 8 biros. 2p left. [2] (14) (a) 4 cm^2 [1](b) 2 cm^2 [2] (15) (a) 12x15=£1.80
[1](b) £15.60÷3=£5.20 [2] (16) (a) 5 (b) 8 (c) 5 [3](17) (a) $^3/4$ (b) 0.75 (c) 75% [3] (18) (a)
5x40=200 miles [1](b) 120÷40=3 gallons, 3x1.70=£5.10 [2] (19) 5x7+8+2-27=18 hours [4] (20)
(5+0+40+0+61+14)÷6=20 [2](120+13)÷7=19 [2] (21) 5 cm, [1]100x100 = 10,000km^2 [3] (22) (a)
5÷10=50p [1](b) $^{60}/400=^3/20$ [3] (23) (a) 66° [1] (b) 180-66-66=48° [1] (c) 180-66=114° [2] (24) (a)
30÷6=5 [2](b) 360÷10=36° [2] (25) (a) 30 miles [1] (b) 49 miles [2] (c) 12 minutes [2] (26) (a)
120x10.5=1,260 FF [1](b) 232÷1.45=£160 [2] (c) 840x1.45÷10.5=$116 [3] (27) (a) 10 ft [2]
(b) 2 mins [2](c) 30 secs [2] (28) (a) 8 cm [2]   (b) 1 cm^2 [2](c) (5,3) [3] (29) (a) 1896, 1904 [4]
(b) 14 [2](c) 20 [4]

Midland Group Paper 2. Total 100

(1) 4x73 = 292 [3] (2) 12:1,200 = 1:80 [3](3) 22 times, 0.3 litre left [4] (4) 16x5.50-2x41=£6 [2] (5)
(a) 99,999 [1] (b) -999,999[1] (c) -1x10^{11} [3] (6) £200, £180, £140 [3](7) (44+72)÷2=58% [3] (8)
(a) x=5 [1] (b) x=-1 [2](9) GCD is 70: every 70 days [3] (10) (a) 134+213=347 [2](b) 158+365=523
[2] (11) 2,142 [2](12) (0+45+55+23+0+43+6+34+49+10)÷10=26.5 [2] (13)$\sqrt{20^2+30^2}$=36 m. [2](b)
tan$^{-1}(20/30)$=33.7° [2] (14) (a) $^9/10$ [1]   (b) $^9/10$x$^7/8$=$^{63}/80$ [2](c) $^1/8$x$^9/10$ = $^9/80$ [2] (15) (a) enlargement
factor 2 about (0,0) [2]

[2]

(16) (a) (2+2$^1/2$)÷2x2=4$^1/2$ m^2 [2](b) 4$^1/2$ x3=13$^1/2$ m^3 [1]
(17)   [4]   (18)   [4]

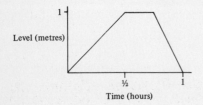

(19) 32, 50, -13 [2] C=5(F-32)/9 [3](20) 5x-2=7(x-2) [3] x=6 [2]
(21)   [2]   (22)   [2]

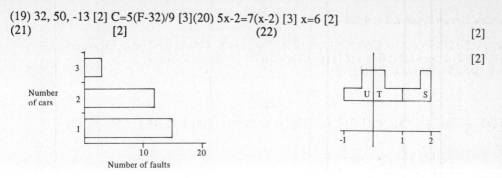

                                                 [2]

Mean=(15+2x12+3x 3)/30=1.6   [2]   translation 2 to left   [2]

(23) (0,-2) and (1,1) [2]gradient=3 [2]
(24)

                                                 [4]

(25) x=90-80=10° [2]y=90-(90-40)=40° [2]
(26)

AB=8.1 [2]   [3]

B to C is 8x50,000 cm = 4 km [2]

Midland Group Paper 3. Total 100

(1) 73 [2]   (2) 17,248+1.12 = £15,400 [2](3) 18x⁵/₆ = 15 kg [2] (4) 8,000x0.9⁵ = £4,724 [3] (5)
$(3x+4)^2=(3x+3)^2+x^2$. [3] x=7. 7, 24, 25 [2] (6) 5x+7y=304, 4x+9y=335, x=23 p, y=27 p. [5] (7)
(a) 8x8+¹/₂x8x2=72 m² [2](b) 72x10=720 m³ [1] (8) (x+2)=2(x-1) → x=4 [3](9) (a) a=$\sqrt{9+16}$ =5 [2]
(b) x=y+2z [2] (10) greatest length 5.35 cm [2]greatest area 5.35x4.25=22.7 cm² [2] (11) ΔOAB
similar to ΔOXY. XY:AB=1:4 [2]ΔOXY area ³⁶/₁₆. ABXY area 33³/₄ [2] (12) (a) y=¹/₄x+³/₄ [3](b)
(0,3) and (2,0). gradient -1 ¹/₂ [3]
(13)

                                                 [3]

(14) (a) (2p+3)/p² [2](b) 5q³/10r³=q³/2r³ [2] (15) (a)¹/₂x2x2=2cm² [1] (b) ¹/₄xπx2²=3.14 cm² [2]
(c) 2(3.14-2)=2.28 cm² [2] (16) (a) Angle=2sin⁻¹(¹/₂/2)=29° [3](b)(ˣ/₂)/2=sin22¹/₂°. x=1.53 ft [3] (17)
d, b+d, d-b, b+¹/₂**d, -b** [4]parallelogram [2] (18) 20+5-2=23% have either. 77% have neither. [3]
(19) (a) $\sqrt{8^2 -7^2}$=3.87 cm [2](b) ¹/₂x3.87x14=27.1 cm² [2] (c) 27.1x20=542 cm³ [1]

(20)
$$A^{-1}=\begin{pmatrix} 1 & -1 \\ -2 & 3 \end{pmatrix}$$

(21)

Frequency

Price in £000's

$(^{1}/4\times16+14+10)/100=^{28}/100$ [2]

[3]

[3]

[3]

(22)

$x=37°$ or $143°$ [2]

(23)

[2]

$$\begin{pmatrix} 3 & 0 \\ 0 & 3 \end{pmatrix}$$

[2]

(24) (a)

[4]

(b) greatest velocity = 30 m/sec [1] (c) time=85 secs [1]distance=$^{1}/2(85+60)\times30=2,175$ m.   [2]

Midland Group Paper 4. Total 100

(1) (a) 27 [2]   (b) 1 [2]   (c) 8 [2]   (d) 12 [2] (e) 6 [2]

(2)   (a)   [4]

(b)
Length | 1 | 2 | 3 | 4 | 5|
Number | 1 | 4 | 9 | 16 | 25| [2]
(c) Squares [2]
(d) 36, 49 [2]
 (3) (a) 6.5 miles [3]   (b) 60° [3](c) Roberttown [2] (4) (a) 13.50+93x0.02=£15.36 [2] (b)

(25-13.5)÷0.02=575 calls [3] (c) 13.5x1.1+20x1.2=£38.85 [3] (5) (a) 11x23=253 [2] (b) 21x14-253=41 [4] (6) (a) 2,4,7,7,4 [2] (b) A must win [2] (c) E must win by at least 2 [3] (7) (a)   [8]   (8)

Goals | 0 | 1 | 2 | 3 | 4 | 5 |
Frequ | 10 | 15 | 7 | 6 | 0 | 2 | [4]

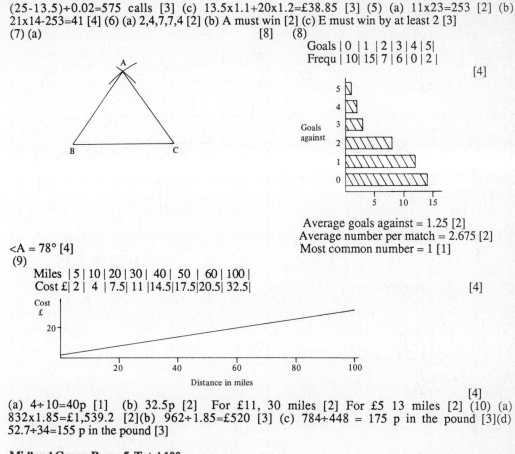

<A = 78° [4]

Average goals against = 1.25 [2]
Average number per match = 2.675 [2]
Most common number = 1 [1]

(9)

Miles	5	10	20	30	40	50	60	100
Cost £	2	4	7.5	11	14.5	17.5	20.5	32.5	[4]

[4]

(a) 4÷10=40p [1] (b) 32.5p [2] For £11, 30 miles [2] For £5 13 miles [2] (10) (a) 832x1.85=£1,539.2 [2](b) 962÷1.85=£520 [3] (c) 784÷448 = 175 p in the pound [3](d) 52.7÷34=155 p in the pound [3]

Midland Group Paper 5. Total 100

(1) (a) 3.6x40+4x3.6x1$\frac{1}{2}$=£165.6 [3](b) (176.4-144)÷(3.6x1$\frac{1}{2}$)=6 hours [4] (2) 300xtan14=75 m [3] tan^{-1}75/200=20.5° [2]$\sqrt{200^2+75^2}$=214 m [2] (3) (a) 2x90=180° [2] (b) 12xcos 40=9.19 cm [3](c) sin^{-1} AD/10 = 66.8° [3] (4) (a) £13.86 (b) £241.56 (c) £1.8 (d) 13.8(e) £255.36 [9] (5) PA=25+0.04u : PB=10+0.06u

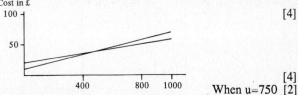

[4]

When u=750 [2]

(6) A has 26, B has 25. [1] B must win [2]: A wins or comes second, A is 3rd and B is not 1st, A is unplaced and B is 3rd or unplaced[4]

(7)

x | -1 | 0 | $^1/_2$ | 1 | $1^1/_2$ | 2 | 3 |
y | 5 | 2 | $1^1/_4$ | 1 | $1^1/_4$ | 2 | 5 | [3]

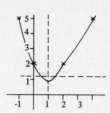

[6]

x=0 or 2 [2]

x=1.7 or 0.3 [2]

(8) (a) 60° cw rotation about O [2](b) reflection in BE [2] (c) 180° rotation about O [2] AFEB is a trapezium [2] FAOE is a rhombus, FACD is a rectangle (many other possibilities) [5] (9) (a) 1 5 10 10 5 1, 1 6 15 20 15 6 1 [3](b) 1, 2, 4, 8, 16, 32, 64 [2] (c) 512 [2]   (d) n [2]   (e) 1 n-1 [2]   (f) 2^{n-1} [2]

(10)   (a)

Score | 1 | 2 | 3 | 4 | 5 | 6 |
Frequ | 10| 6 | 15| 13| 6 | 10| [2]

(b)

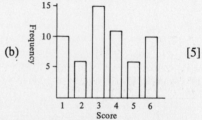

[5]

(c) Average=3.48 [3]

(d) Reasonable frequencies. Die is probably fair [3]

(11) (a) $\sqrt[3]{20}$= 2.71 cm [3](b) 2x2x2=8 [3] (c) 20÷(π16)=0.4 cm [3](d) $\sqrt{10/(2\pi)}$=1.26 cm [4]

Midland Group Paper 6. Total 100

(1) 30x $\geqslant$ 90, 1.7x $\leqslant$ 6 [6]x=3 [3] (2) (a) external angle=30°. 360÷30=12 sides [4](b) $\sqrt{10^2-3^2}$=9.54 cm [4] (3) (a) $^2/_3x^1/_4$=$^1/_6$ [2](b) $^1/_6$+$^1/_3x^7/_8$=$^{11}/_{24}$ [2] (c) $^1/_3$+$^2/_3x^3/_4$=$^5/_6$ [2]

(4) (a) 1 o'clock [1]

(b) at 2.15, flow=3 cm.hr [4]

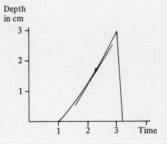

(c) [3]

(5) (a) -4 or -6 [4](b) 0.562 or -3.562 [4]

(6)   [5]

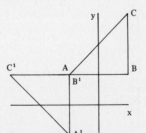

rotation 90° acw about origin [2]

(1,1) [2]

(7) E beat B, D beat B [3]

Team |A|B|C|D|E|
Points |7| 2| 1| 4|6| [3]

Total=20 points [1]   6 schools, 15 games [3]7 schools, 42 points [3]

(8) (a) 1:3 [2] (b) 2 cm [3](c) 16.76 cm^3, 436.63 cm^3 [4] (d) 1:26 [2](e) 1:8 [2]

(9)

x	$^1/_4$	$^1/_2$	$^3/_4$	1	$1^1/_2$	2	3
y	-7.9	$-3^3/_4$	-2.1	-1	0.9	3	8.33	[3]

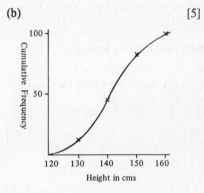

[5]

x=1.5 [2]

 Draw y=1-x, x=1.2 [3]

(10) (a) cos rule gives 7.24 miles [5](b) cos rule gives 77.5° [4] (c) area=$^1/2$x10x11xsin40°=35.4 sq miles [4] (11) (a) **c-b**, $^1/2$**(c-b)**, $^1/2$**(c+b)** [4](b) **b+c, c, b** [4] (c) parallelogram [2](d) 2x16=32 [3] (12) (a) 12, 45, 83, 100 [2]

(b)

[5]

(c) median=141 cm, IQR=(147-134)=13 cm [4]

(d) 35% [2]

Northern Group Paper 1. Total 78

(1) 30-26=4 mins [1](2)50+10=60 feet [1] (3) 360-140-130=90° [1](4) $^5/9$ [1] (5) $^{60}/40$=1$^1/2$ hours [1](6) (a) 2,401 [1] (b) 3,499 [1] (7) 9+4=13 [2](8) $^{12}/16$=$^3/4$ litres [2] (9) 3 mugs, 150 cc left over [2](10) 23 miles [1] 46 m.p.h. [1] (11) $^2/3$x150=100 women [2](12) 15+2x7+20+2x9= £67 [2]
(13)   [2]

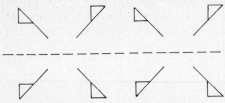

(14) (a) X=1 [1] (b) X=16 [1](15) 4x70x2xπ=1,760 metres [2] (16) (a) 82 min [1]   (b) 12 08[1](c) 10 Oct [1] (17) (a) x=8 [1](b) x=-5 [2] (18) 20x43.2=£8.64 [1] 8.64÷4=£2.16 [1](19) 8+7=15 cm [1] (20) (a) 5 rooms [2](b) 20 rooms [2]
(21) <A=43   (22)    [3]
     <B=38°
     <C=99°   [3]

Sum=180° [1]Base area = 1 cm^2 [1] (23) (a) 325° [2](b) 150 km [2] (24) (a) 5x3.98= £19.90 [2](b) 19.9x0.9= £17.91 [2] (25) (a) 4x3000=12,000 [2]   (b) $^{360}/3$=120° [2](c) 12,000÷6=2,000 [2]
(26)

```
first | 1 | 2 | 3 | 4 | 5 | 6 |
  1   | 2 | 3 | 4 | 5 | 6 | 7 |
  2   | 3 | 4 | 5 | 6 | 7 | 8 | second
  3   | 4 | 5 | 6 | 7 | 8 | 9 |
  4   | 5 | 6 | 7 | 8 | 9 | 10|
  5   | 6 | 7 | 8 | 9 | 10| 11|
  6   | 7 | 8 | 9 | 10| 11| 12| [4]
```

Prob. of 12 = $^1/36$ [2]Most likely is 7, with prob $^1/6$ [3]

(27)
Wickets | 0 | 1 | 2 | 3 | 4 |
 Frequ. | 4 | 6 | 7 | 2 | 1 | [3]

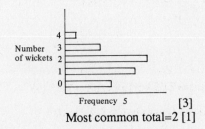

Average=(6+14+6+4)÷20=1$^1/2$ [2]   Most common total=2 [1]

398

Northern Group Paper 2 Total 102

(1) $7 \div (1/5) = 35$ [2] (2) $11 \times 360 \div 30 = 132$ mins [2] (3) $14-77=63$ BC [2] (4) $320 \times 1.1 = £352$ [2] (5) $2 \times 17 + 4 \times 12$ [2] (6) $2 \times 2 + 1.5 + 3.2 = £8.70$ [2] (7) $120 \times 1.08 = £129.60$ [2] (8) 0.1 km in $1/240$ hr = 24 km/hr [3] (9) Prices are £165 and $200 \times 0.8 = £160$. Difference = £5 [3] (10) $2 \times 2.7 + 3.5 = £8.90$ [2] $10 - 8.9 = £1.10$ change [1] (11) (a) 8 [2] (b) 3 [2] Any side square, not a corner. [2] (12) (a) $15 \div (3+2) = 10$ [2] (b) $4 \times (2+1) = 12$ [2] (13) (a) $2 \times 0.95 + 2.2 + 1.9 + 0.5 + 2 \times 0.3 = £7.10$ [2] (b) $7.1 \times 1.1 = £7.81$ [2] (14) Pyramid [1] 8 edges [2] EHD joined to A [2] (15) (a) 24 litres [2] (b) 2.2 gallons [2] (16) (a) $2 \times 2 \times 3 + 2 \times 3/ \times 2 + 3 \times 3^1/2 - 1 \times 1^1/2 = 35$ m^2 [3] (b) $35 \times 2 \div 20 = 3.5$ litres [2] (c) $3.5 \times 3.5 = £12.25$ [1] (d) $12.25 \div 35 = 35$ p per m^2 [2]

(17) [4]

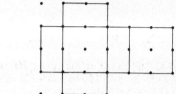

volume $= 2 \times 2 \times 1 = 4$ cm^3 [1]
surface area $= 2 \times 2 \times 2 + 4 \times 2 \times 1 = 16$ cm^2 [2]
(18) (a) $x+20$ [2] (b) $2(x+20)$ [3] $2(15+20) = £70$ [2] (19) £0, £1,350, £3,600, £7,900 [4]

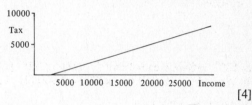

[4]

(20)

Amount in £	10	30	50	70
Mr Jones	5	25	45	65
Mr McKay	8.5	25.5	42.5	59.5	[3]

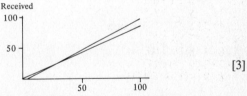

[3]

A cheque for £33 [2]
(21) (a) $144 \div (3 \times 8) = 6$ days [1] (b) $144 \times 3.2 = £460.80$ [2] (c) $144 - (3 \times 8 \times 5) = 24$ hours [2] (d) $460.8 + 12 \times 3.2 = £499.20$ [3]
(22) [4]

Distance $= 2.05$ km [3] bearing $= 103°$ [3]

Northern Group Paper 3. Total 108

(1) $15 \times 23 = £3.45$ [1] (2) $360 - 90 - 90 - 50 = 130°$ [2] (3) $x=9$ [2] (4) £600:£900:£1,200 [3] (5) (a) $\sqrt{1.69} = 1.3$ m [2] (b) $1.69/\pi = 0.733$ m [3] (6) $85 + (1.7 \times 10^{-23} = 5 \times 10^{24}$ [3] (7) pyramid. [2] one for each face. 6 pyramids [2] (8) (a) $\tan^{-1}(2000/5000) = 21.8°$ [2] (b) $\sqrt{2000^2 + 5000^2} = 5,385$ m [2] (9) (a) 9 [1] (b) 1024 [1] (c) 4 [1] (10) (a) 39.5 hrs [3] (b) £94.30 [2] (11) $30 + 25 + 17 - 50 = 22$ had both [3] (12) (a) 15 mill. $\times 1.015 = 15,225,000$ [1] (b) $15 \times 1.015^2 = 15,453,375$ [2] (13) enlargement factor 4 [2] $(1/2, 0)$ [2] (14) (a) **AB=DC, BC=AD** [2] trapezium [2]

399

(15) 2 m.p.h. [1]

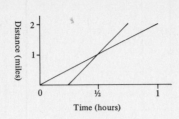

[3]

^{1}/$_2$ hour after John set off [3]

(16) (a)

Romance	Crime	Sci-Fi	Historical
7	3	10	2	[2]

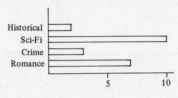

[3]

(17)

[2]+[2]

reflection in y=2 [2]

(18) $\sqrt{4^2+5^2}$=6.4 [2] $\tan^{-1}(5/4)$=51.3° [2]

(19)

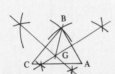

BG=4.6 cm [3] (20) HHH,HHT,HTH,HTT,THH,THT,TTH,TTT [3](a) ^{1}/$_8$ [1] (b) ^{3}/$_8$ [2] (c) ^{7}/$_8$ [2]

(21)

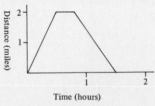

[3]

(22) (a) [2]

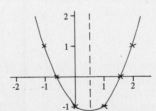

(b) gradient=3 [4]
(c) x=1.6 or -0.6 [4]

(23) $^3/_8{}^2$xπx$^1/_8$=0.055 cu in. [2]36 coins, 0.012 cu in left over [3] Vol. of cube = 5.5 cu in. side = $^3\sqrt{5.5}$=1.77 in [2]50x0.055x11.6=32 oz.[2] Cost = 300x32=£9,600 Value of coins 50x240=£12,000.Profit £2,400 [4]

Northern Group Paper 4. Total 121

(1) (18.05-12)÷275=2.2 p [2](2) $^1/_2$(180-90-60)=15° [2] (3) (a) (x-10)(x+7) [2](b) x^2-3x+2=72 → x=10 or -7 [3] (4) Total time=1 $^1/_3$ hr. BC time=$^7/_{12}$Speed=50/($^7/_{12}$)=85 $^5/_7$ m.p.h. [2] (5) (a) 50x70x11=38,500 cm^3 [1](b) 50x70+0.9=3,889 cm^3 [2]

(6) (a) $A^{-1}=\begin{pmatrix} 1 & -1 \\ -^1/_2 & 1 \end{pmatrix}$ [2] $\qquad C = A^{-1}B=\begin{pmatrix} 1 & -6 \\ 0 & 6^1/_2 \end{pmatrix}$ [2]

(7) (a) (iii)   (b) (i)(c) (iv)   (d) (ii) [3] (8) (a) 200-4x^2 cm^2 [2](b) (20-2x)(10-2x) cm^2 [2] (c) x(20-2x)(10-2x) cm^3 [2] (9) x⩾1, y⩾2, 2x+3y⩽12 [3] (10) (a) $^a/_3$, a-b, $^a/_3$-b, [3] (b) $^1/_2$a-$^1/_2$b, $^1/_2$a+$^1/_2$b [3](c) parallel [2] (11) (a) $^2/_3$ [1]   (b) $^1/_3$x$^1/_3$=$^1/_9$ [2](c) $^1/_3$x^2/3x$^2/_3$=$^4/_{27}$ [3] (12) (a) 7, -5, -8, -13 [4](b) fg:x →4 - 6x [2] (c) (5-1)/3=$^4/_3$ [2]x=(1-y)/2 [3] (13) AX= $\sqrt{5^2+12^2}$=13 cm [2]2π5=10π cm [2] Angle=360x$^{10\pi}$/$_{26\pi}$=138.5° [2]Area=πx13^{2}x^5/13=204.2 cm^2 [2] (14) (a) 3^3=27 [2] (b) 3x+1=2(x+2) → x=3 [2](c) x^{2+3-6}=x^{-1} [2] (15) (a) S ∩ T = ø [2](b) T⊂ S [2]

(16) 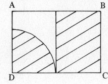 [4]

(17) (a) [2]
(b) [3] (c) cw rotation of 55° about (0,0) [2]

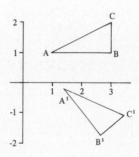

(d) acw rotation of 55°. [2] M takes (-0.2,1.4) to A [2]
(18) (a) [2]
(b) S=7.8-0.78t [2]
(c) S=0 when t=10 mins [2]
(d) approx 1,800 m. [2]

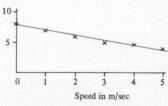

Speed in m/sec

(19) (a) [4]

$(2, 3)\begin{pmatrix} 89 & 95 & 85 \\ 43 & 52 & 41 \end{pmatrix}$ = (307 346 293)

(b) [3]

P $\begin{pmatrix} ^1/_3 \\ ^1/_3 \\ ^1/_3 \end{pmatrix}$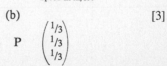

(20) (a) [5] **(d)** [4]

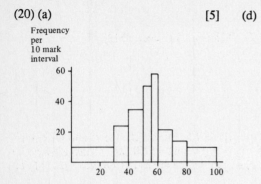

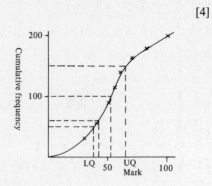

(b) mean=(30x14^1/$_2$+24x34^1/$_2$+..+22x90)/200=50.755 [2]

 (c) cumulative frequencies 30, 54, 89, 114, 143, 164, 178, 200 [2]

(e) median=53 LQ=37 UQ=63 [3] (f) 42% [2]

Southern Group Paper 1. Total 135

(1) 8 teams, 3 left over [2](2) $^1/2 \div 3 = ^1/6$ [2] (3) $^4/7$ [2] $^3/7$ [3](4) 7.80÷6=£1.30 [2]

(5) [3]

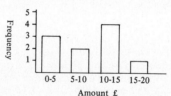

(6) (a) 13, 16 [2](b) 16, 32 [2] (7) (a) 8 km [2](b) 7.5 miles [2] (8) diam.=3 cm [3] circ.=3.14x3=9.42 cm [2] (c) area=9x$^{3.14}/4$=7.065 cm^2 [2] (9) cost=£6.50 [2]   change=7-6.5=50p [2]fraction=$^{50}/650$=$^1/13$ [2] (10) (a) 4 [2] (b) 2 [2] (c) 48 cm^3 [1](d) 4x12+2x16=80cm^2 [2] (11) 11, 23, 8. [3]change "add 1" to "subtract 3" [2]

(12) (a) $^1/4$ [2] (b) [2]

(13) 1.7 cm, 2.1 cm, 1.6 cm, [4] perimeter=5.4 cm [2]<A=48° [2] (14) (a) 6500x0.3=£1,950 [2](b) 3000+1500÷0.3=£8,000 [3] (c) 10500-2100÷0.3=£3,500 [3] (15) (a) 75 p [1](b) to B [1] (c) 110+35=£1.45 [2](d) (110+35+75)x0.75=£1.65 [4] (16) (a) 023° [3]   (b) 3.5 cm [3](c) 1 cm to 5 km [2] (17) 4 & 12. Sum is 16 cm^2 [5]36 & 20. Difference is 16 cm^2[5] Same result. [1]16x0.04=0.64 grams [2]

(18)

Amount £	0-5	5-10	10-15	15-20
Frequ.	3	2	4	1

[3]

[3]
Greatest=3x5+2x10+4x15+1x20=£115 [4]
Average=(4+16+12+6+13)/5=£10.20 [3]

(19) (a) 2x24=48 [2] (b) 48x4x4=768 [2](c) 768x$^1/2$x$^1/4$x$^1/4$=24 cu ft [2] (d) 8x24 stretchers+8x48 headers = 576 [3]8 headers+16 stretchers=24 [3]

(i) [2] (ii) [3]

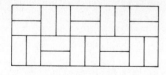

8x12x18=1728 tiles [3]$^1/2$x2x4x1728=6,912 cu in [2]
6,912x0.07=483.84lb[2]6912x0.15=£1,036.80 [2]

Southern Group Paper 2. Total 135

(1) 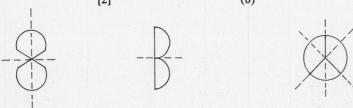 [2]   (2) 90x2÷3=60 times [1]

7x3÷2=10^1/2 times [2] (3) ^{35}/50=^{7}/10 [3](4) 320÷160=2 m [2] 0.1x160=16 m [2] (5) first ^{18}/2^1/2=7.2, second ^{25}/3=8 ^{1}/3 second is denser [4] (6) ^{640}/20=32 ft [2]32x0.32=£205 [2] (7) -11, -16 [2](a) 4 (b) -1 [2] (8) (a) £23.50 [2]   (b) £12.70 [2]36.20x1.15=£41.63 [2] (9) (a) 20.0778 [2] (b) 42.6 [2](c) 480 [2]   (d) 86 [2] (10) (a) 160÷4=40 hours [1] (b) 40x3.5=£140 [2](c) 4x140x1.15=£644 [3] (11) (a) hexagon [1]   (b) rectangle [2](c) rhombus [2] (d) trapezium [2](e) right-angled triangle [2] (12) (a) 10x61=610 [2]   (b) 12x72=864 [2](c) (610+864)÷22=67% [3] (13) (a) 11 lb [2]   (b) 3.2 kg [2](c) 3x11=33 lb [3]

(14) (a)   [2]   (b)   [2]

(a) has rotational symmetry, of order 2. [3]   [3]

[2]

(15) 6-2=4 hours [2]9+4-2=11 00 hours [2] NY is 12+8-15=5 hours behind [3]Flight time=4 hrs. Moscow 3 hrs ahead [3] (16) 4x2=8 prisms [2]   area=^{1}/2x2x3=3 [2] volume=2x3=6 [2] vol of solid=8x6=48 [1] add 8 prisms [3]vol of cuboid=2x48=96 [2] (17) (a) 1 mile [2] (b) 30 min [2](c) 2 miles [2] (d) 40 min [2] (e) 2 m.p.h. [2]   (f) 6 m.p.h. [2](g)3 m.p.h [2]
(18)   [8]

Distance=23 m [5]bearing, T from M=59° [3] bearing, M from T=239° [2] Time taken=23÷0.8=29 secs [2]

Southern Group Paper 3. Total 135

(1) ^{3}/17<^{1}/5<0.22<^{2}/9 [2](2) ^{450}/800=0.5625=56.25% [2] (3) 7.5x10^{-13} [3](4) (a) R=7 [1] (b) S=5 [2] (5) 32x^3/8=12 [2](6) (a) 9π=28.3 sq ft [2] (b) 20x45-9π=871.73 sq ft [3] (7) 180-2(180-108)=36° [4](8) sin^{-1}^{30}/60=30° [3] (9) x+16, 4x+16 [3] 4x+16=2(x+16) [3]x=8 [3] (10) (a) ^{2}/5x^1/6=^{1}/15 [3](b) 300x^3/5x^1/3=60 [3] (11) (a) ^{1000}/4π=79.58 cm [3](b) $\sqrt{1000}$/4π=8.92 cm [4]
 (12) reflection in y=x [2]

404

(13) (a) 2 hrs, 3 hrs [2]

(b) [3]

Distance from Bristol (miles) — Jones, Smith, Time (hrs)

(b) [5]

(c) 1.2 hrs. 48 miles [4]

(14) 5 faces, 8 edges, 5 vertices [3] 3 edges meet at the 4 base vertices [3] surface area=4+4x2=12 cm^2 [4]

each triangle is 2 cm^2 [2]

[6]

(15) gradient=4 [3]
population in 1988=48,000 [2]
population in 1978=8,000 [2]
founded when population zero, in 1976 [3]

[5]

(16) $a^2+b^2=c^2$ [2]
$c=\sqrt{7^2+24^2}=25$ [2]
$b=\sqrt{7^2-4^2}=5.74$ [3]

3 | 4 | 5 |
5 | 12 | 13 |
7 | 24 | 25 |
9 | 40 | 41 | [4]

a=39, b=52, c=65. (many others) [3]

sinP|cosP|tanP|sinQ|cosQ|tanQ|
a/c | b/c | a/b | b/c | a/c | b/a | [3]

sinP=cosQ, cosP=sinQ, tanP=1/tanQ [3]

(17) (a) he asked 24 people [2]
  (b)

number | 0 | 1 | 2 | 3 |
frequ. | 10 | 6 | 5 | 3 | [3]

(c)   [5]

(d)   [6]

(e) Mode=0 [1] (f) mean=(6+5x2+3x3)/24=$^{25}/24$ [3]

Southern Group Paper 4. Total 135

(1) $^{85}/_{47.5}$=1.79 hrs [4](2) (20-10)÷0.25x100=£4,000 [3] (3) (a) $\sqrt{4\pi}$=3.54 cm [3](b) $\sqrt[3]{27x^3/4\pi}$=1.86 cm[3] (4) $^{4\pi8}/_{\pi9}$=$^{32}/_{37}$ cm [3] (5) (a) 6, 120 [2] (b) 4 $^2/_3$ [4] finds mean [3] (6) (a) 3x=2(x+1) →x=2 [3](b) 5^{3+4-6}=5^1 [3] (7) (a) $^7/_{17}$x$^6/_{16}$=$^{21}/_{136}$ [2](b) 2x$^{10}/_{17}$x$^7/_{16}$=$^{35}/_{68}$ [3] (8) (a) square [2]   (b) T=$37^1/_2$ [2](c) S=$^1/_3$ [2] (9) (a) 100-44=56° [2]   (b) $^1/_2$x44=22° [2](c) 180-50=130° [2]
(10)   [4]
 (a) 7 [1]

(b) 39 [3]
(c) 29 [2]
 [4] mean=(64+78+84)/140=1.614 [4]

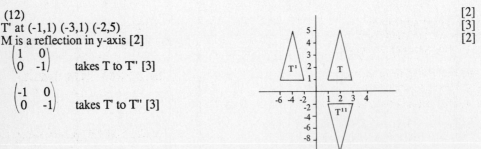

                                                                                   [4]

(11)

x | $^1/_2$ | $^3/_4$ | 1 | $1^1/_4$ | $1^1/_2$ | $1^3/_4$ | 2 |
y |-$5^3/_4$ |-3.4 |-2 | -.84 | $^1/_4$ | 1.35 | 2.5|

(a) x=1.4 [1]
 (b) x=1.7 [3]
 (c) x=0.8 [3]
                                                                                   [4]

(12)   [2]
T' at (-1,1) (-3,1) (-2,5)   [3]
M is a reflection in y-axis [2]   [2]
$\begin{pmatrix} 1 & 0 \\ 0 & -1 \end{pmatrix}$ takes T to T'' [3]

$\begin{pmatrix} -1 & 0 \\ 0 & -1 \end{pmatrix}$ takes T' to T'' [3]

(13) (a) $^{360}/_8$=45° [2](b) 4cos$22^1/_2$=3.70 cm [4] (c) 2x4xsin$22^1/_2$=3.06 cm [4](d)8x3.70x3.06x$^1/_2$=45.25 cm^2 [4] (14) (a) 2 secs, by 400 m. [4](b)$^{1000}/_{500}$cos θ=$^2/$cos θ [3] (c) 200x^2/cosθ =$^{400}/$cosθ [3](d) 500sinθ x^2/cosθ =1000sinθ /cosθ [3] (e) $^{400}/$cosθ =1000sinθ /cosθ → sinθ =0.4 → θ =23.6° [4] (15) (a) 26000x0.1=£2,600 [2](b) 2600÷0.7=£3,714 [3] (c) amount earned=26000x$^{0.12}/_{0.75}$=£4,160.% increase=12% [3]

406

Syllabus Requirements

The topics treated have been ordered into the three levels of the GCSE. The following table gives the syllabus requirements of the four English examining groups.

If a topic, for example percentages, is classified as 1&2&3, then new material has been introduced at each level.

Do not forget that the syllabus for level 2 includes that for level 1, and the syllabus for level 3 includes that for level 2. So for example questions on a level 1 topic can still be set in papers for level 2.

L = London Group. M = Midland Group. N = Northern Group syllabus A. S = Southern Group.

	L	M	N	S
Numbers, the four operations, number patterns directed numbers, fractions and decimals.	1	1	1	1
HCF, LCM, Rationals & irrationals.	2	2	2	2
Calculator, simple checks, addition of fractions, conversion to decimals.	1	1	1	1
Arithmetic of mixed Numbers	2	2	2	2
Rounding	1	1	1	1
Sig figs, dec. places, appropriate limits of accuracy	2	2	2	2
Squares, cubes, square roots	1	1	1	1
Integer powers, standard form	2	2	2	2
Fractional powers, laws of indices	3	3	3	3
Ratio and scale, idea of proportion	1&2&3	1&2	1&2&3	1&2&3
Direct and inverse proportion	2	2	2	2
Use of α, proportionality to powers	3		3	3
Percentages	1&2&3	1&2&3	1&2&3	1&2&3
Measures of weight, area, length, area, volume, money, 24 hour clock, change of units rates, tax, interest, wages, salaries, VAT, use of tables and charts	1	1	1	1
Use of letters for numbers, substitution	1	1	1	1
Expansion of brackets, simple factors	2	2	2	2
Change of subject, variable occuring once	2	2		2
Change of subject, variable occuring twice	3	3		3
Harder factorizations	3	3	3	3
Algebraic fractions	3	3		3
Equations in 1 unknown	2	2	2	2
Simultaneous and quadratic equations	3	3	3	3
Quadratic equation formula	3	3		3

Order	1	1&2	1&2	1&2
Inequalities in 1 variable		3	3	3
Inequalities in 2 variables		3	3	3
Points, lines, angles, polygons, tessellation	1	1	1	1
Alternate, corresponding angles	2	2	2	2
Angles of polygons, similarity, congruence	2	2		2
Vocabulary of circles	1	1	1	1
Angle in semicircle, tangents, chords	2	2	2	2
Angle theorems	3		3	3
Alternate segment, intersecting chords	3			
Vocabulary of solids, nets	1&2&3	1&2&3	1&2&3	1&2&3
Co-ordinates, interpretation of graphs	1	1	1	1
Gradient, solutions of equations	2	2	2	2
Area under straight line graph	3	3	3	3
Area by trapezium rule	3		3	
Acceleration	3		2	
Algebraic graphs, intersecting graphs	2&3	2&3	2&3	2&3
Flow charts			1	1
Composition of functions			3	
Use of instruments, scale drawings	1	1	1	1
Locus problems	2	2	2	2
Perimeter & area of rectangle, triangle. Circumference of circle. Volume of cuboid	1	1	1	1
Area of parallelogram, circle. Volume of cylinder	2	2	2	2
Area of trapezium, volume of prism	2	2	3	3
Arc length and sector area	3	3	2	3
Volume and surface area of sphere. Volume of cone and pyramid	3	3	3	3
Sin, cos, tan for acute angles. Elevation, depression, bearings	2&3	2&3	2&3	2&3
Trig for angles above 90°		3	3	
Sine and cosine rules		3		
Pythagoras in 2 dimensions	2	2	2	2
Pythagoras in 3 dimensions	3	3	3	3
Statistical data. Interpretation and constr- uction of tables, bar-charts, pictograms	1	1	1	1
Construction of pie-charts, frequency tables and histograms with equal intervals	2	2	2	2
Histograms with unequal intervals			3	
Averages, mean, mode, median	1&2	1&2	1&2	1&2
Cumulative frequency, quartiles		3	3	

Probability of a single event	1	1	1	1
Probabilities of combined events	2&3	2&3	2&3	2&3
Single transformations	1	1	1	1
Combined transformations	2	3	3	3
Matrix operations, information and transformation matrices	3	3	3	3
Arithmetic of vectors	3	3	2	3
Co-ordinate vectors, vector geometry	3	3	3	3
Set problems and Venn diagrams	3		2	3
Set notation	3		3	3

Index